Countercurrent Chromatography

CHROMATOGRAPHIC SCIENCE SERIES

A Series of Monographs

Editor: JACK CAZES
Cherry Hill, New Jersey

1. Dynamics of Chromatography, *J. Calvin Giddings*
2. Gas Chromatographic Analysis of Drugs and Pesticides, *Benjamin J. Gudzinowicz*
3. Principles of Adsorption Chromatography: The Separation of Nonionic Organic Compounds, *Lloyd R. Snyder*
4. Multicomponent Chromatography: Theory of Interference, *Friedrich Helfferich and Gerhard Klein*
5. Quantitative Analysis by Gas Chromatography, *Josef Novák*
6. High-Speed Liquid Chromatography, *Peter M. Rajcsanyi and Elisabeth Rajcsanyi*
7. Fundamentals of Integrated GC-MS (in three parts), *Benjamin J. Gudzinowicz, Michael J. Gudzinowicz, and Horace F. Martin*
8. Liquid Chromatography of Polymers and Related Materials, *Jack Cazes*
9. GLC and HPLC Determination of Therapeutic Agents (in three parts), *Part 1 edited by Kiyoshi Tsuji and Walter Morozowich, Parts 2 and 3 edited by Kiyoshi Tsuji*
10. Biological/Biomedical Applications of Liquid Chromatography, *edited by Gerald L. Hawk*
11. Chromatography in Petroleum Analysis, *edited by Klaus H. Altgelt and T. H. Gouw*
12. Biological/Biomedical Applications of Liquid Chromatography II, *edited by Gerald L. Hawk*
13. Liquid Chromatography of Polymers and Related Materials II, *edited by Jack Cazes and Xavier Delamare*
14. Introduction to Analytical Gas Chromatography: History, Principles, and Practice, *John A. Perry*
15. Applications of Glass Capillary Gas Chromatography, *edited by Walter G. Jennings*
16. Steroid Analysis by HPLC: Recent Applications, *edited by Marie P. Kautsky*
17. Thin-Layer Chromatography: Techniques and Applications, *Bernard Fried and Joseph Sherma*
18. Biological/Biomedical Applications of Liquid Chromatography III, *edited by Gerald L. Hawk*
19. Liquid Chromatography of Polymers and Related Materials III, *edited by Jack Cazes*
20. Biological/Biomedical Applications of Liquid Chromatography, *edited by Gerald L. Hawk*
21. Chromatographic Separation and Extraction with Foamed Plastics and Rubbers, *G. J. Moody and J. D. R. Thomas*
22. Analytical Pyrolysis: A Comprehensive Guide, *William J. Irwin*
23. Liquid Chromatography Detectors, *edited by Thomas M. Vickrey*
24. High-Performance Liquid Chromatography in Forensic Chemistry, *edited by Ira S. Lurie and John D. Wittwer, Jr.*
25. Steric Exclusion Liquid Chromatography of Polymers, *edited by Josef Janca*

26. HPLC Analysis of Biological Compounds: A Laboratory Guide, *William S. Hancock and James T. Sparrow*
27. Affinity Chromatography: Template Chromatography of Nucleic Acids and Proteins, *Herbert Schott*
28. HPLC in Nucleic Acid Research: Methods and Applications, *edited by Phyllis R. Brown*
29. Pyrolysis and GC in Polymer Analysis, *edited by S. A. Liebman and E. J. Levy*
30. Modern Chromatographic Analysis of the Vitamins, *edited by André P. De Leenheer, Willy E. Lambert, and Marcel G. M. De Ruyter*
31. Ion-Pair Chromatography, *edited by Milton T. W. Hearn*
32. Therapeutic Drug Monitoring and Toxicology by Liquid Chromatography, *edited by Steven H. Y. Wong*
33. Affinity Chromatography: Practical and Theoretical Aspects, *Peter Mohr and Klaus Pommerening*
34. Reaction Detection in Liquid Chromatography, *edited by Ira S. Krull*
35. Thin-Layer Chromatography: Techniques and Applications. Second Edition, Revised and Expanded, *Bernard Fried and Joseph Sherma*
36. Quantitative Thin-Layer Chromatography and Its Industrial Applications, *edited by Laszlo R. Treiber*
37. Ion Chromatography, *edited by James G. Tarter*
38. Chromatographic Theory and Basic Principles, *edited by Jan Åke Jönsson*
39. Field-Flow Fractionation: Analysis of Macromolecules and Particles, *Josef Janca*
40. Chromatographic Chiral Separations, *edited by Morris Zief and Laura J. Crane*
41. Quantitative Analysis by Gas Chromatography: Second Edition, Revised and Expanded, *Josef Novák*
42. Flow Perturbation Gas Chromatography, *N. A. Katsanos*
43. Ion-Exchange Chromatography of Proteins, *Shuichi Yamamoto, Kazuhiro Nakanishi, and Ryuichi Matsuno*
44. Countercurrent Chromatography: Theory and Practice, *edited by N. Bhushan Mandava and Yoichiro Ito*
45. Microbore Column Chromatography: A Unified Approach to Chromatography, *edited by Frank J. Yang*
46. Preparative-Scale Chromatography, *edited by Eli Grushka*
47. Packings and Stationary Phases in Chromatographic Techniques, *edited by Klaus K. Unger*
48. Detection-Oriented Derivatization Techniques in Liquid Chromatography, *edited by Henk Lingeman and Willy J. M. Underberg*
49. Chromatographic Analysis of Pharmaceuticals, *edited by John A. Adamovics*
50. Multidimensional Chromatography: Techniques and Applications, *edited by Hernan Cortes*
51. HPLC of Biological Macromolecules: Methods and Applications, *edited by Karen M. Gooding and Fred E. Regnier*
52. Modern Thin-Layer Chromatography, *edited by Nelu Grinberg*
53. Chromatographic Analysis of Alkaloids, *Milan Popl, Jan Fähnrich, and Vlastimil Tatar*
54. HPLC in Clinical Chemistry, *I. N. Papadoyannis*
55. Handbook of Thin-Layer Chromatography, *edited by Joseph Sherma and Bernard Fried*
56. Gas–Liquid–Solid Chromatography, *V. G. Berezkin*
57. Complexation Chromatography, *edited by D. Cagniant*

58. Liquid Chromatography–Mass Spectrometry, *W. M. A. Niessen and Jan van der Greef*
59. Trace Analysis with Microcolumn Liquid Chromatography, *Milos Krejcl*
60. Modern Chromatographic Analysis of Vitamins: Second Edition, *edited by André P. De Leenheer, Willy E. Lambert, and Hans J. Nelis*
61. Preparative and Production Scale Chromatography, *edited by G. Ganetsos and P. E. Barker*
62. Diode Array Detection in HPLC, *edited by Ludwig Huber and Stephan A. George*
63. Handbook of Affinity Chromatography, *edited by Toni Kline*
64. Capillary Electrophoresis Technology, *edited by Norberto A. Guzman*
65. Lipid Chromatographic Analysis, *edited by Takayuki Shibamoto*
66. Thin-Layer Chromatography: Techniques and Applications, Third Edition, Revised and Expanded, *Bernard Fried and Joseph Sherma*
67. Liquid Chromatography for the Analyst, *Raymond P. W. Scott*
68. Centrifugal Partition Chromatography, *edited by Alain P. Foucault*
69. Handbook of Size Exclusion Chromatography, *edited by Chi-san Wu*
70. Techniques and Practice of Chromatography, *Raymond P. W. Scott*
71. Handbook of Thin-Layer Chromatography: Second Edition, Revised and Expanded, *edited by Joseph Sherma and Bernard Fried*
72. Liquid Chromatography of Oligomers, *Constantin V. Uglea*
73. Chromatographic Detectors: Design, Function, and Operation, *Raymond P. W. Scott*
74. Chromatographic Analysis of Pharmaceuticals: Second Edition, Revised and Expanded, *edited by John A. Adamovics*
75. Supercritical Fluid Chromatography with Packed Columns: Techniques and Applications, *edited by Klaus Anton and Claire Berger*
76. Introduction to Analytical Gas Chromatography: Second Edition, Revised and Expanded, *Raymond P. W. Scott*
77. Chromatographic Analysis of Environmental and Food Toxicants, *edited by Takayuki Shibamoto*
78. Handbook of HPLC, *edited by Elena Katz, Roy Eksteen, Peter Schoenmakers, and Neil Miller*
79. Liquid Chromatography–Mass Spectrometry: Second Edition, Revised and Expanded, *W. M. A. Niessen*
80. Capillary Electrophoresis of Proteins, *Tim Wehr, Roberto Rodríguez-Díaz, and Mingde Zhu*
81. Thin-Layer Chromatography: Fourth Edition, Revised and Expanded, *Bernard Fried and Joseph Sherma*
82. Countercurrent Chromatography, *edited by Jean-Michel Menet and Didier Thiébaut*

ADDITIONAL VOLUMES IN PREPARATION

Modern Chromatographic Analysis of Vitamins, Third Edition, Revised and Expanded, *edited by Andre P. De Leenheer, Willy E. Lambert, and Jan Van Bocxlaer*

Micellar Liquid Chromatography, *Alain Berthod and Celia Garcia-Alvarez-Coque*

Countercurrent Chromatography

edited by

Jean-Michel Menet

Rhône-Poulenc Rorer, Inc.
Vitry-sur-Seine, France

Didier Thiébaut

Ecole Supérieure de Physique
et de Chimie Industrielles
de la Ville de Paris
Paris, France

MARCEL DEKKER, INC. NEW YORK • BASEL

Library of Congress Cataloging-in-Publication Data

Countercurrent chromatography / edited by Jean-Michel Menet, Didier Thiébaut.
p. cm. — (Chromatographic science ; v. 82)
Includes index.
ISBN: 0-8247-9992-5 (alk. paper)
1. Countercurrent chromatography. I. Menet, Jean-Michel. II. Thiébaut, Didier. III. Series.
QP519.9.C68C 1999
543′.0894—dc21 99-32805
CIP

This book is printed on acid-free paper.

Headquarters
Marcel Dekker, Inc.
270 Madison Avenue, New York, NY 10016
tel: 212-696-9000; fax: 212-685-4540

Eastern Hemisphere Distribution
Marcel Dekker AG
Hutgasse 4, Postfach 812, CH-4001 Basel, Switzerland
tel: 41-61-261-8482; fax: 41-61-261-8896

World Wide Web
http://www.dekker.com

The publisher offers discounts on this book when ordered in bulk quantities. For more information, write to Special Sales/Professional Marketing at the headquarters address above.

Current printing (last digit):
10 9 8 7 6 5 4 3 2 1

PRINTED IN THE UNITED STATES OF AMERICA

Foreword

Countercurrent chromatography (CCC) which originated in our laboratory at the National Institutes of Health in the late 1960s has recently started to bloom. Development of CCC was preceded by the invention of the coil planet centrifuge, which dates back to the early 1960s. Like other techniques, both were developed with the timely help and contributions of many individuals without whom my dream would have been aborted long ago. In the following story, I attempt to thank several of them, especially those who have participated without receiving credit.

In the 1950s, when I was a student at Osaka City University Medical School in Japan, I had an ambition to research lymphocytes because at that time their function was not well documented. They were morphologically classified simply into large, medium, and small lymphocytes. After graduating from medical school, I served one year in a rotating internship at the Yokosuka U.S. Naval Hospital, where I was sponsored by Dr. John Featherstone, the Commander and a radiologist. In spite of his holding the second highest rank in the hospital, he assisted me in all large and small details. When I asked him to help me select a hospital where I could do further research on lymphocytes, he kindly wrote many letters of application and delivered them to my intern room to ask for my signature! I chose a 4-year residency in pathology at Cuyahoga County Metropolitan General Hospital in Ohio because it promised to provide a research opportunity in my second year of residency. One year later, I started research on a lymphocyte-stimulating factor from irradiated rats with my peer resident, Dr. Fay B. Weinstein, and together we published a preliminary report on the subject in the *Journal of the National Cancer Institute*. In order to continue research on the lymphocyte function, it was first necessary to isolate lymphocytes. Since the conventional density gradient method with a short centrifuge tube was found to be inefficient, I conceived the idea of a centrifuge embodying a long coiled tube that underwent planetary motion, subjecting cells to an Archimedean screw force. The motion of particles through a coiled tube in this device, called a *coil planet centrifuge* was then mathematically analyzed by Fay's husband, Dr. Marvin A. Weinstein, a brilliant physicist.

Two years later, I moved to Michael Reese Hospital in Chicago, where I received help from Dr. Lloyd Arnold, a professor of Biology at Loyola University. In spite of his busy schedule, he tried to construct a coil planet centrifuge at his own expense. Unfortunately, the project was prematurely terminated due to the expiration of my Fulbright Scholarship travel fund.

In 1963, I returned to Osaka City University Medical School as a lecturer in the Department of Physiology. Professor Eiichi Kimura, the head of the department, introduced me to Mr. Yonezo Kubota, the founder and president of Kubota Centrifuge Corporation in Tokyo. Mr. Kubota enthusiastically designed and constructed a prototype of the coil planet centrifuge with his own hands and delivered it to our research laboratory in the spring of 1964. I still remember my excitement on that day when Professor Kimura, Dr. Aoki (an associate professor), and I were chatting together in front of this first prototype. I said, "Tomorrow I will start an experiment with the coil planet centrifuge and continue it so long that you may be surprised." After Mr. Kubota retired, I handed the project to a small company called Sanki Engineering Ltd., Kyoto. I owe many thanks to Mr. Reizo Matsumoto, who helped me to find this excellent company. Its president, Mr. Nunogaki, and his younger brother Yoshiaki Nunogaki, were very interested in the device and quickly built a prototype that surpassed the original in both design and function. The separation of latex particles of cell size was successful and the results were published in *Nature* in 1966. At that time Dr. Aoki had been working on a countercurrent distribution method and we started to explore the capabilities of the coil planet centrifuge in this direction.

In 1967, Dr. K. Nunogaki and I presented a paper on this matter at the 7th International Conference on Medical and Biological Engineering held in Stockholm, Sweden. During the conference, I met Dr. Robert Bowman from the National Institutes of Health and accepted his invitation to visit his laboratory in Bethesda, Maryland. After I delivered a talk on the coil planet centrifuge, Dr. Bowman invited me to join his research group. In 1968, I became a Visiting Scientist, working in his laboratory to develop a new chromatographic method, and the first paper on CCC was published in *Science* in 1970. Since then, the development of CCC has continued steadily.

Over these three decades, research on lymphocytes has advanced remarkably, revealing their important immunological functions as B cells and T cells. Unfortunately, the coil planet centrifuge originally built for purification of these cells has missed an opportunity to be part of this research; instead it has evolved into a chromatographic technique called *high-speed CCC* and is used for separation and purification of a wide variety of natural and synthetic products. I hope that its utility will be extended even further in the coming years.

Yoichiro Ito

Preface

Separative methods are based on the differences between the physical properties of the products that are contained in a mixture. Some of them rely on the differences in partitioning of the products between two phases, such as liquid-liquid extraction and partition chromatography. Countercurrent chromatography (CCC) combines the intrinsic advantages of both methods. On one hand, it does not use any solid stationary phase, consequently preventing potential denaturation, and on the other it relies on a series of equilibria between immiscible liquid mobile and stationary phases to lead to a selective and efficient separation of the mixture of products.

All CCC devices that are currently available derive from the so-called *Craig apparatus*, which consisted of separator funnels connected in a series. The sequence of mixing then decanting between funnels of two immiscible liquid phases led to a progressive impoverishment in each solute and to a separation in different funnels following a countercurrent distribution.

The first improvement of this type of apparatus was to retain the stationary liquid phase in a column using a specific force field (Earth gravity for instance) and to percolate an immiscible liquid mobile phase. Later, the improvement was related to the use of a centrifugal force field to retain the stationary liquid phase inside the column. It led to an improved mixing of the two phases increasing the efficiency and allowing the use of higher flow-rates of mobile phase, thereby decreasing the duration of the separations.

Three types of CCC devices were developed and identified by the mode of retention of the liquid stationary phase and the design of the column. The first type is the centrifugal partition chromatography (CPC) which originates from droplet counter-current chromatography (DCCC) and rotation locular countercurrent chromatography (RLCCC). Modern devices are called Sanki apparatus from the name of the Japanese manufacturer. Sanki apparatus use a so-called hydrostatic mode to retain the stationary phase in channels thanks to a centrifugal force that is constant in intensity and direction. This mode is characterized by a good retention of the stationary phase of many two-phase solvent systems, including viscous systems such as aqueous two-phase systems.

The second type of CCC device is type J High Speed Counter-Current Chromatography (HSCCC) and the retention mode is called hydrodynamic. The planetary motion generates a centrifugal force varying in intensity and direction that retains the stationary phase into the coiled tubings but also leads to a series of mixing and settling inside the column. These characteristics increase the efficiency of the separations as compared to CPC, but at the expense of the retention of the stationary phase for viscous systems.

A third type of CCC device has recently been developed: *cross-axis* HSCCC apparatus. It combines the advantages of the two previous types. On one hand, its hydrodynamic mode of retention of the stationary phase guarantees a good efficiency of the separations, and on the other it retains all the solvent systems developed for CCC. This apparatus is particularly suited for biological separations in aqueous two-phase polymer (ATPS) systems that are difficult to handle with the two previous types.

The common name for these three modern devices is counter-current chromatographs, but we think that centrifugal liquid-liquid chromatography (CLLC) could also be a good term for this technique as it describes its key features.

This book is divided into three main parts, on theoretical approaches, description of main devices, and practical description of recent applications for the third one. The first theoretical approach discusses the properties of two-phase solvent systems with regard to their polarities and to such key physical parameters as interfacial tension, density, and viscosity. Refined parameters are then defined, such as the capillary wavelength and the settling velocities, and guidance is given on predicting the behaviors of solvent systems inside a column of a type J HSCCC device. The influence of temperature is also discussed. The second theoretical approach thoroughly describes the implementation of experimental design methodology in the retention of the stationary phase inside type J HSCCC and *cross-axis* devices. It has been applied to various solvent systems and experimental conditions related to the mobile and stationary phases and to apparatus characteristics and has helped in optimizing the running conditions for a maximum stable retention of the stationary phase. The second part (Chapters 3 and 4) gives a global view of the three main CCC devices—type J and *cross-axis* devices for the hydrodynamic retention mode and the Sanki type for the hydrostatic retention mode. Finally, the third part gathers recent and original applications, such as purification of macromolecules with aqueous two-phase solvent systems, inorganic separations, chiral separations, and direct separation of natural products from crude extracts.

As the editors of this book we have made every effort to demonstrate the position of CCC among separative techniques. The advantages of CCC are amplified when it is used as a preparative chromatographic method: the sample to be injected rarely requires prepurification (crude samples such as fermentation broths can be directly injected), there is no irreversible adsorption in the stationary phase (such a case becomes an extraction step on a CCC apparatus), and no denaturation may occur related to interactions with solid surfaces. It should be kept in mind that even so the choice of a solvent system is related solely to the partition coefficients of the solutes to be separated, it also determines the behavior inside the CCC apparatus and the possible retention of its chosen stationary phase. The *cross-axis* device consequently increases the range of applications for CCC, as it retains all the solvent systems so that the chromatographer need concentrate only on the design of the solvent system with respect to the separation quality and not on its potential behavior inside a CCC column.

All the chapters herein were written by experts in their field. They were asked to give key characteristics of main available CCC devices, to describe which theoretical approaches are useful, and to give selected examples to demonstrate the potential of CCC. We know they have succeeded and that this book will be useful for CCC users at all levels, including newcomers in the field seeking to acquire start-up knowledge.

We would like to thank all our contributors, who devoted their time to writing unique and original comprehensive chapters. We have enjoyed working with them and sincerely hope our common work will make this book a reference in the field of countercurrent chromatography. We would also like to give special thanks to our families, who have supported and encouraged us even when we worked on the book during weekends and late evenings for the ‘‘beauty of science.’’

Dr. Menet: I would like to add a special dedication, for our last book of the Twentieth century on countercurrent chromatography, to the genuine inventor of such a unique technique, Dr. Yoichiro Ito. I worked for more than a year in Dr. Ito's laboratory at the National Institutes of Health and was amazed by his continuous enthusiasm, his sharp sense of observation, and his ideas to transpose experimental observations into equations. I can definitely say I became a CCC addict during those fruitful months.

Jean-Michel Menet
Didier Thiébaut

Contents

APPLICATIONS

Contributors

Alain Berthod Laboratoire des Sciences Analytiques, Centre National de la Recherche Scientifique, University of Lyon 1, Villeurbanne, France

Piotr S. Fedotov Vernadsky Institute of Geochemistry and Analytical Chemistry, Russian Academy of Sciences, Moscow, Russia

Jacques Goupy ReConFor, Paris, France

Geewananda Gunawardana Pharmaceutical Products Division, Abbott Laboratories, Abbott Park, Illinois

Yoichiro Ito Laboratory of Biophysical Chemistry, National Heart, Lung, and Blood Institute, National Institutes of Health, Bethesda, Maryland

Y. W. Lee Chemistry and Life Sciences, Research Triangle Institute, Research Triangle Park, North Carolina

Ying Ma Laboratory of Biophysical Chemistry, National Heart, Lung, and Blood Institute, National Institutes of Health, Bethesda, Maryland

Tatiyana A. Maryutina Vernadsky Institute of Geochemistry and Analytical Chemistry, Russian Academy of Sciences, Moscow, Russia

James McAlpine Phytera Inc., Worcester, Massachusetts

Jean-Michel Menet Process Chemistry and Biochemistry, Rhône-Poulenc Rorer, Inc., Centre de Recherche de Vitry-Alfortville, Vitry-sur-Seine, France

Marie-Claude Rolet-Menet Laboratory of Analytical Chemistry, Faculty of Pharmacy, University of Paris V, Paris, France

Boris Ya. Spivakov Vernadsky Institute of Geochemistry and Analytical Chemistry, Russian Academy of Sciences, Moscow, Russia

Karine Talabardon Laboratoire des Sciences Analytiques, Centre National de la Recherche Scientifique, University of Lyon 1, Villeurbanne, France

Didier Thiébaut Laboratoire Environnement et Chimie Analytique, Ecole Supérieure de Physique et de Chimie Industrielles de la Ville de Paris, Paris, France

1

Characterization of the Solvent Systems Used in Countercurrent Chromatography

Jean-Michel Menet
Rhône-Poulenc Rorer, Inc.,
Vitry-sur-Seine, France

Marie-Claude Rolet-Menet
Faculty of Pharmacy, University of Paris V,
Paris, France

I. INTRODUCTION

The purpose of this chapter is to introduce the reader to the various ways of choosing a solvent system through its important characteristics. The characterizations based on phase diagrams, physical parameters, and behavior inside a countercurrent chromatography (CCC) column will be described. Then some theoretical parameters will be introduced to predict the behavior of the solvent systems so that the three ways of characterizing a solvent system can be obtained before any experiment inside a CCC device.

II. CHARACTERIZATION BASED ON PHASE DIAGRAMS

A biphasic solvent system is chosen to carry out a separation by CCC according to various criteria. One is the solubility of the sample in the solvent system. Another is the ability of the solvent system to be modified to adjust partition coefficients and selectivity factors to obtain a satisfactory separation of the products. A change in the proportion of the various solvents making the biphasic solvent system allows such an adjustment but the relative volumes of the two phases also vary, so that in some cases it can lead to a single-phase system. Consequently, it is necessary to use a tool that characterizes a solvent system with regard to its single-phase or two-phase composition. In the literature, the numerous biphasic solvent systems used in CCC are frequently made of three, sometimes four or more solvents. Many examples of phase diagrams have already been gathered by Foucault and published [1] and we therefore redirect the reader to that book for a list of available redrawn phase diagrams.

Liquid–liquid equilibria are governed by the phase rule:

$$F = C - P + 2$$

where C is the number of components, P is the number of phases (two in the case of CCC), and F is the number of degrees of freedom, i.e., the number of independent variables permitted by the system. Ternary liquid–liquid systems, made of three different solvents, have 3 degrees of freedom according to the phase rule. At a given temperature and pressure, the composition of one phase determines the composition of the other phase at equilibrium.

A. Use of a Phase Diagram

There are two representations to obtain the composition of a mixture of three solvents A, B, and C, with, in this case, A miscible in all proportions with B, A miscible in all proportions with C, but B and C nonmiscible. The first one is shown in Figure 1 with a typical equilateral triangular diagram. Point A represents 100% of A, B represents 100% of B, and C 100% of C. Any point H inside the triangle stands for a mixture of the three components. Its composition can be read according to two modes:

1. Along axes inclined at an angle of 60° so that the sum of volumes V_A, V_B, and V_C must always be 100 (Figure 1a),
2. or by the perpendiculars to axes HL, HJ, and HK (Figure 1b).

The second representation is given in Figure 2 as an orthogonal representation of the previous triangular diagram. The percentages of only two of the solvents, e.g., A and C, are specified. The composition of the mixture of the three components (represented by point H) can be read by the perpendiculars to the axes HK and HJ. The percentage of the third solvent is directly deducted because $VA + VB + VC = 100$.

In both representations, the region NPFQM (limited by the line called "binodal") is the unstable or two-phases region, in which mixtures such as O automatically split into two layers of composition given by points P and Q. The "tie line," one example of which is PQ, connects the points representing compositions of two phases in equilibrium. The relative volumes of the separate phases can

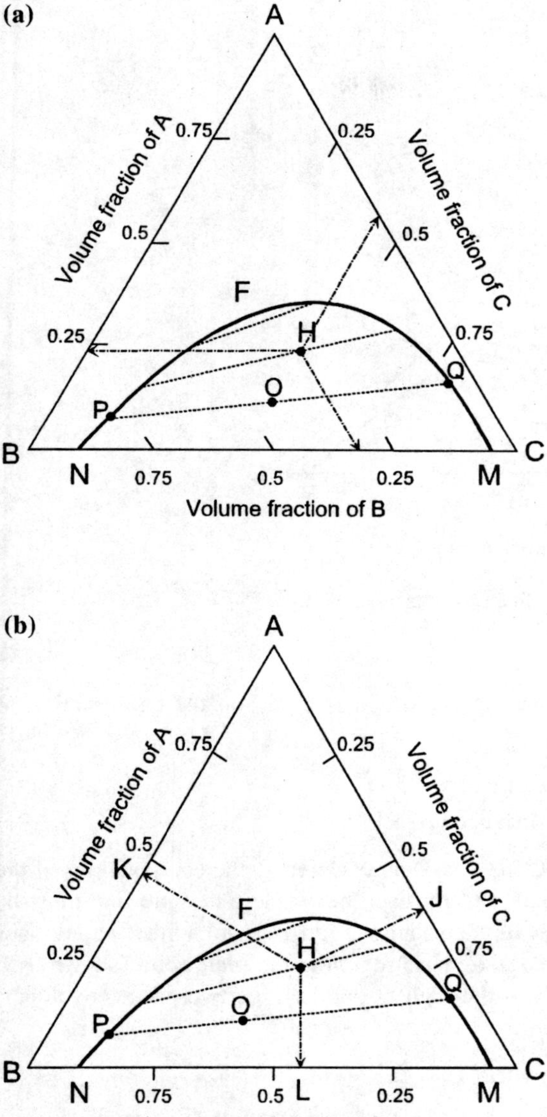

Figure 1 Ternary diagrams: equilateral triangular representation.

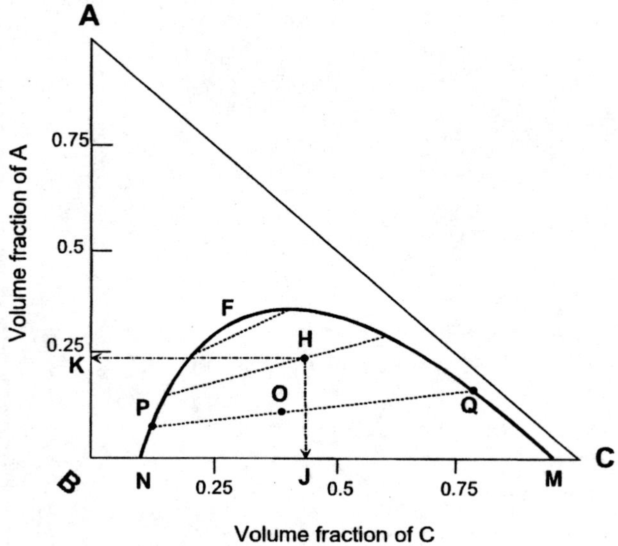

Figure 2 Ternary diagrams: orthogonal representation.

be estimated by the inverse lever rule, which is based on the conservation of mass for the components:

$$\frac{\text{Volume of phase 1}}{\text{Volume of phase 2}} = \frac{\text{distance O} \rightarrow \text{Q}}{\text{distance O} \rightarrow \text{P}}$$

When the volume of solvent C in the system is changed, the compositions of the mixtures Q and P are modified, and for each new case a new tie line must be drawn. In general, the tie lines on the triangular diagram are neither parallel nor horizontal, and the limit of the two-phase region at the plait point (shown as F in Figure 2) is not necessarily at the highest point on the two-phase envelope.

B. Types of Phase Diagrams

Several types of equilibria between three solvents can be described. Each diagram shown in Figures 1 to 5 is obtained for a single temperature; it should be kept in mind that the locations of the binodal curve and the tie lines depend on the temperature. Four types may be defined:

Type 0: A, B, and C are totally miscible with each other in pair, but an area exists in the ternary diagram in which the system is biphasic (Figure 3).

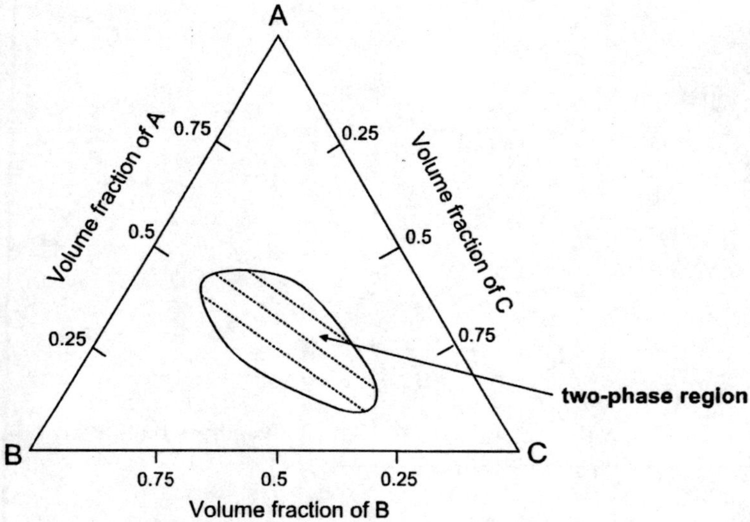

Figure 3 Type 0 ternary diagram.

The unique example recently introduced in centrifugal partition chromatography is the system made of water, dimethylsulfoxide, and tetrahydrofuran [2].

Type 1: This case is shown by Figures 1 and 2. An example is the system made of chloroform, methanol, and water.

Type 2: A is miscible in all proportions with B, but B and C, and A and C are only partially miscible. This case is illustrated by Figure 4, in which regions a and c stand for a single-phase system whereas area b leads to a two-phase system. A mixture represented by point O will automatically split into layers of compositions P and Q, as indicated by the tie line. A variation of this type is one showing two unconnected two-phase regions similar to the single two-phase region of Figure 4. An example is the system made of ethyl acetate, butanol, and water.

Type 3: Each solvent is only partially miscible with each of the two others. This case is illustrated by Figure 5. Area a stands for single-phase systems, whereas area b leads to two-phase systems and c is the three-phase area. Any point in area named c represents a mixture that spontaneously splits into three phases of compositions E, F, and G. Mixtures such as H, represented by points in area b, split into two layers of compositions J and K, as indicated by the appropriate tie line.

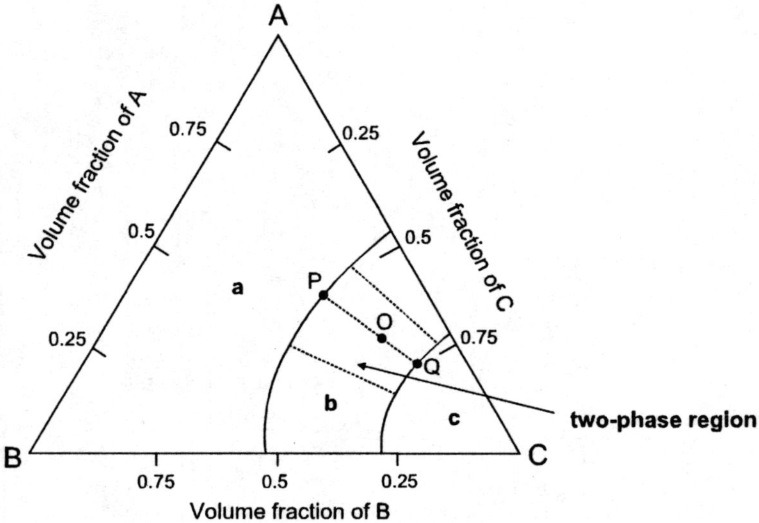

Figure 4 Type 2 ternary diagram.

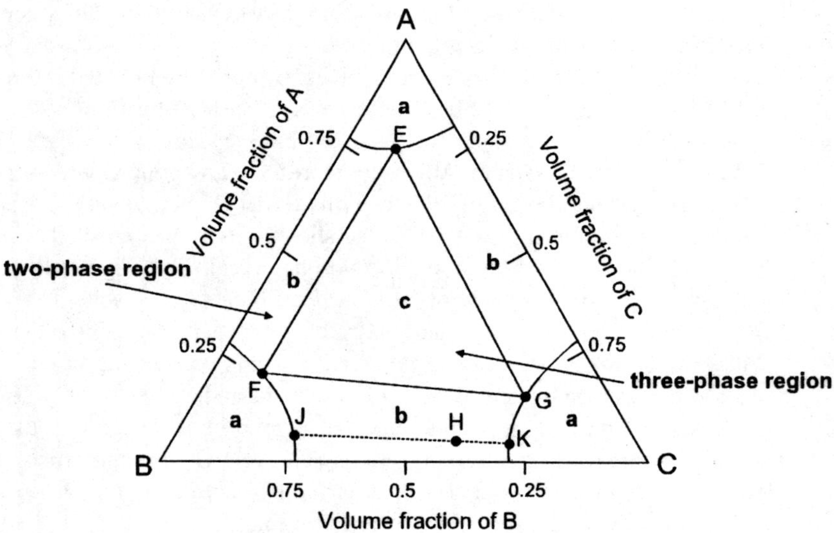

Figure 5 Type 3 ternary diagram.

III. CHARACTERIZATION BASED ON PHYSICAL PARAMETERS

A first approach to characterize a solvent system used for CCC relies on the physical properties of the two liquid phases. Two concepts are useful: (1) polarity, well known among chemists when choosing a solvent system and a stationary phase for high-performance liquid chromatography (HPLC), and (2) the basic physical properties of the two-phase liquid solvent system, i.e., density, interfacial tension, or the viscosity. Tables of values are given as a quick reference to the reader for the most common solvents used in CCC.

A. Polarity

This parameter is commonly used for HPLC in order to choose the stationary phase, i.e., the column, and the mobile phase. However, until now the term ''polarity'' had not been precisely defined. The polarity of a solvent depends on the permanent dipole moment of the solvent molecules, on the dielectric constant, and on the sum of molecular properties responsible for the interactions between the molecules of the solvent. The latter includes coulombic, inductive, hydrogen bonding, and electron pair donor and acceptor interaction forces [3]. Various polarity scales have been established, which are explained in the following paragraphs, and a comparison between these classifications is criticized.

1. Hildebrand Solubility Parameter, δ

Hildebrand and Scott have linked the polarity of a solvent to its solubilization capability by defining the total solubility parameter δ as the work that is necessary to separate two molecules of the solvent [4]:

$$\delta = \sqrt{\frac{\Delta h_v - RT}{V_m}} \tag{1}$$

where Δh_v is the molar enthalpy of vaporization and V_m the molar volume of the solvent, assuming that its vapor is ideal. This parameter calculated from the value of the enthalpy is consequently experimental. In order to better take into account the complexity of molecular interactions, partial solubility parameters have been introduced; they are measured from solubility experiments:

$$\delta_0 = \sqrt{\delta_d^2 + \delta_n^2 + \delta_h^2} \tag{2}$$

where δ_d corresponds to dispersion interactions, δ_n corresponds to dipole–dipole interactions, and δ_h corresponds to hydrogen bonding interactions; δ_0 is called the calculated total solubility parameter. These values have been gathered for many solvents by Barton [5].

A simpler expression may be used for the δ parameter by using the Beerboweer approximation:

$$\delta = 3.74 \sqrt{\frac{\gamma}{3\sqrt{V_m}}} \tag{3}$$

with γ the surface tension of the solvent.

The values extracted from Ref. 3 are gathered in Table 1 for common solvents used to make two-phase solvent systems for CCC. A small value indicates a low-polarity solvent, such as cyclohexane ($\delta = 15.8\ J^{1/2}\ cm^{-3/2}$), whereas a higher value means a higher polarity solvent, such as methanol ($\delta = 29.3\ J^{1/2}\ cm^{-3/2}$) or water ($\delta = 48.6\ J^{1/2}\ cm^{-3/2}$).

2. Snyder Adsorption Solvent Strength Parameter, ε_0

According to Snyder [6], the total interaction of a solvent molecule with a sample molecule is the result of four interactions: dispersion, dipole, hydrogen bonding, and dielectric. A stronger attraction between solvent molecules and sample molecules is the consequence of a larger combination of these interactions. The ability of a sample or a solvent molecule to interact in all four ways is referred to as the "polarity" of the compound. Thus "polar" solvents preferentially attract and dissolve "polar" molecules. Accordingly, the strength of a solvent is directly related to its polarity. For normal phase liquid chromatography and adsorption chromatography, solvent strength increases with polarity.

The eluent strength parameter, ε_0, was consequently defined in the framework of adsorption chromatography, which corresponding simplified mechanism is described by the Snyder-Soczewinski theory [7,8]. The parameter ε_0 is the adsorption free energy of the solvent molecules per unit of area of adsorbed solvent under standard activity conditions (highly activated silica, very low water content of the adsorbent). It is written as

$$\varepsilon_0 = -\frac{\Delta G_0}{2.3\ RTA_M} \tag{4}$$

where A_M is area of adsorbent occupied by 1 mole of mobile phase, R is constant for ideal gazes, T is temperature in Kelvin, and ΔG_0 is the variation of free adsorption energy of 1 mole of mobile phase. ε_0 is therefore an energy per unit of area.

The values given in Table 1 for various solvents are related to the eluent strength on alumina [7]. Low-polarity solvents show a small value, e.g., heptane ($\varepsilon_0 = 0.01$), whereas higher polarity solvents, such as methanol ($\varepsilon_0 = 0.95$) or water ($\varepsilon_0 > 0.95$), show a higher value.

Table 1 Physical Properties of Solvents Commonly Used in CCC

Solvent	Hildebrand solubility parameter (δ)	Snyder adsorption solvent strength parameter (ε_0)	Rohrschneider and Snyder solvent polarity parameter (P')	Reichardt polarity parameter (E_T)
n-*Heptane*	14.7	0.01	0.2	1.2
n-Hexane	15	0.01	0.1	0.9
n-Pentane	14.9	0.00	0	0.9
Cyclohexane	15.8	0.04	−0.2	0.6
Carbon disulfide		0.15	0.3	
Carbon tetrachloride	17.6	0.18	1.6	5.2
Triethylamine		0.54	1.9	
Toluene	18.3	0.29	2.4	9.9
p-Xylene		0.26	2.5	
Chlorobenzene		0.30	2.7	
Bromobenzene		0.32	2.7	
Ethyl ether	15.4	0.38	2.8	11.7
Benzene	18.8	0.32	2.7	11.1
n-Octanol	20.9	0.5	3.4	54.3
Methylene chloride	20	0.42	3.1	30.9
i-Pentanol	22.1	0.61	3.7	56.8
1,2-Dichloroethane	20.4	0.44	3.5	32.7
t-Butanol	25.2	0.7	4.1	50.6
n-Butanol	27.2	0.7	3.9	60.2
n-Propanol	24.4	0.82	4.0	61.7
Tetrahydrofuran	18.2	0.57	4.0	20.7
Ethyl acetate	18.2	0.58	4.4	22.8
i-Propanol	23.7	0.82	3.9	54.6
Chloroform	18.9	0.40	4.1	25.9
Dioxane	20	0.56	4.8	16.4
Pyridine		0.71	5.3	
Acetone	18.6	0.56	5.1	35.5
Ethanol	26	0.88	4.3	65.4
Aniline		0.62	6.3	
Acetic acid	20.6		6.0	64.8
Acetonitrile	24.1	0.65	5.8	46
Dimethylformamide	24.2		6.4	40.4
Dimethylsulfoxide	24	0.75	7.2	44.4
Methanol	29.3	0.95	5.1	76.2
Ethylene glycol		1.11	6.9	
Water	48.6	> 0.95	10.2	100

The value of ε_0 for a solvent mixture, ε_{0ab}, can be related to the values of the solvents A and B (ε_{0a} and ε_{0b}) and to the mole fraction of B in the binary mixture according to:

$$\varepsilon_{0ab} = \varepsilon_{0a} + \frac{\log[N_b\, 10^{\alpha n_b(\varepsilon_{0b}-\varepsilon_{0a})} + 1 - N_b]}{\alpha n_b} \quad (5)$$

where α is a constant that depends on the adsorbent activity (water content) and varies between 0.6 and 1.0, N_b is the molar fraction of solvent B in the mixture, and n_b depends on the molecular size of the molecule of solvent B; n_b varies from 4 to 6 [9].

3. Rohrschneider and Snyder Solvent Polarity Parameter, *P′*

Many experimental determinations have been carried out by Rohrschneider [10] to measure the liquid–gas partition coefficients for four test solutes, i.e., *n*-octane, ethanol, dioxane, and nitromethane. These values have been used by Snyder [6] to classify 75 solvent systems according to the solvent polarity parameter referred to as P'. The latter may be further explained by introducing three parameters—x_e, x_d, and x_n—which evaluate the abilities of a solvent to give hydrogen bonds as an acceptor or as a donor of protons or to give dipole–dipole interactions. The test solutes were ethanol (e), dioxane (d), and nitromethane (n). The solvent polarity parameter may be written as

$$P' = p'_e + p'_d + p'_n \quad (6)$$

where

$p'_e = x_eP'$ = polarity corresponding to the proton acceptor ability
$p'_d = x_dP'$ = polarity corresponding to the proton donor ability
$p'_n = x_nP'$ = polarity corresponding to dipole–dipole interactions, and
$x_e + x_d + x_n = 1.$

The dimensionless values of P' given in Table 1 show that low-polarity solvents, such as heptane ($P' = 0.2$), lead to a small value, whereas higher polarity solvents, such as methanol ($P' = 5.1$) or water ($P' = 10.2$), show a higher value.

The polarity P' of a solvent mixture is the arithmetic average of the P' values of the pure solvents in the mixture, weighted according to the volume fraction of each solvent. For a binary mixture made of two solvents A and B, the resulting solvent polarity parameter P' may be computed according to

$$P' = \Phi_a P'_a + \Phi_b P'_b \quad (7)$$

where Φ_a and Φ_b are the volume fractions of solvents A and B in the mixture, and P'_a and P'_b refer to the solvent polarity parameter for each of the two pure solvents.

4. Reichardt Polarity Index, E_T

Reichardt has introduced a solvent polarity parameter, referred to as E_T, based on the variation of transition energy for the longest wavelength absorption band of a dye [3]. The latter is known as the ''Reichardt dye'' and its structure is shown in Figure 6.

Similarly, an analog that is penta-substituted with *tert*-butyl group is used for nonpolar solvents. The maximum of absorbance is obtained at a λ_{max} wavelength, which gives the transition energy corresponding to that band:

$$E_T(\text{kcal/mol}) = 28{,}590/\lambda_{max} \tag{8}$$

The transition energy, E'_T, which is determined when using the analog, a more ''hydrophobic'' dye, is related to the transition energy E_T of the Reichardt dye according to:

$$E_T = \frac{E'_T - 3.434}{0.9143} \tag{9}$$

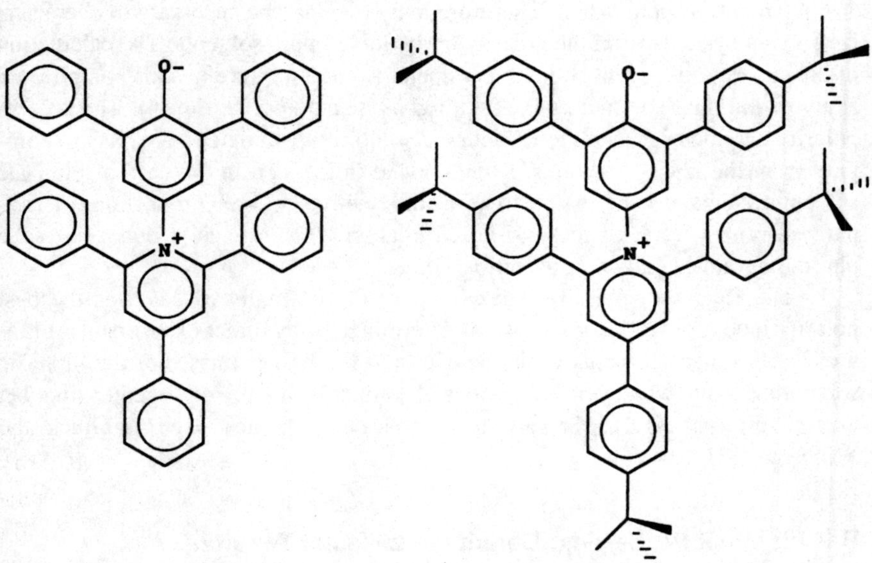

I II

Figure 6 Structure of the Reichardt dye (I) and of its penta-substituted *tert*-butyl analog (II).

This polarity parameter E_TN is normalized using water ($E_T = 63.1$) and tetramethylsilane (TMS, $E_T = 30.7$):

$$E_TN = \frac{E_T(\text{tested solvent}) - E_T(\text{TMS})}{E_T(\text{water}) - E_T(\text{TMS})} \tag{10}$$

Consequently, as indicated in Table 1, low-polarity solvents show an E_TN close to zero, e.g., heptane ($E_TN = 0.012$), whereas higher polarity solvents, such as methanol ($E_TN = 0.762$) or water ($E_TN = 1$), show a higher value.

5. Which Polarity Index for Countercurrent Chromatography?

Snyder and Kirkland have shown good correlations between the polarity scales of Hildebrand et al. [6]. Berthod et al. have also studied the three polarity indexes of Hildebrand (δ), Snyder (P'), and Reichardt ($E_TN * 100$) [11,12]. The three scales are in an average correspondence, as some solvents show some differences between the scales. For instance, butanol-1 leads to $E_T = 60.2$ and $\delta = 17.8$ J$^{1/2}$ cm$^{-3/2}$, whereas furfural leads to a lower E_T value, i.e., 50, but a higher δ value, i.e., 23.6 J$^{1/2}$ cm$^{-3/2}$.

The four scales described in the previous paragraphs may be considered as equivalent to some extent. Therefore, only one may be necessary to characterize solvent systems from the polarity of the mixed pure solvents. The calculation of the adsorption solvent strength parameter ε_0 for a mixture reveals complicated as two additional parameters are required as mentioned for Eq. (5). The solvent polarity parameter P' and the Hildebrand solubility parameter δ are easy to compute from the law of mixtures, as mentioned in Eq. (7). In the case of biphasic solvent systems, it is necessary to know the relative volumic proportions of each pure solvent in each of the two liquid phases, which is easily done when the phase diagram is known (cf. Section II).

The Reichardt polarity index, namely E_T, is finally revealed as the best polarity index for CCC for its ease of determination by direct experimental ultraviolet (UV) measurements in the heavier and the lighter phases of the biphasic solvent systems, whatever the number of pure solvents mixed together may be. One example of its use for solvent systems design is described by Gluck and Wingeier [13].

B. Physical Properties: Density, Interfacial Tension, and Viscosity

Basic physical properties may also be used for solvent systems characterization. As for the Reichardt polarity index, they can be calculated from direct experimental measurements for a biphasic solvent system, whatever the number of mixed pure solvents. Five values may be obtained: the interfacial tension between the two liquid phases, the densities of the heavier and the lighter liquid phases, and

Table 2 Values of Interfacial Tension, γ, Densities of Lighter and Lower Phase, ρ, and Dynamic Viscosities for These Two Liquid Phases, η, for Selected Solvent Systems[a]

	Interfacial tension (dyn/cm)	Density (g/cm³)		Dynamic viscosity (cP)	
Solvent system (v/v)	γ	ρ_1	ρ_2	η_1	η_2
Hexane/water (1:1)	51.1	0.66	1.00	0.41	0.95
Ethyl acetate/water (1:1)	6.8	0.92	0.99	0.47	0.89
Chloroform/water (1:1)	31.6	1.50	1.00	0.57	0.95
Hexane/methanol (1:1)	1.2	0.67	0.74	0.50	0.68
Ethyl acetate/acetic acid/water (4:1:4)	3.5	0.94	1.01	0.76	0.81
Chloroform/acetic acid/water (2:2:1)	12	1.35	1.12	0.77	1.16
n-Butanol/water (1:1)	1.8	0.85	0.99	1.72	1.06
n-Butanol/0.1 M NaCl (1:1)	4	0.85	0.99	1.66	1.04
n-Butanol/1 M NaCl (1:1)	5	0.84	1.04	1.75	1.04
n-Butanol/acetic acid/water (4:1:5)	<1	0.90	0.95	1.63	1.40
n-Butanol/acetic acid/0.1 M NaCl (4:1:5)	<1	0.89	1.01	1.68	1.25
n-Butanol/acetic acid/1 M NaCl (4:1:5)	1	0.88	1.05	1.69	1.26
sec-Butanol/water (1:1)	<1	0.87	0.97	2.7	1.67
sec-Butanol/0.1 M NaCl (1:1)	<1	0.86	0.98	1.96	1.26
sec-Butanol/1 M NaCl (1:1)	3	0.84	1.03	1.91	1.29
DMSO/heptane (1:1)	10.8	1.08	0.65	2.16	0.40
DMF/heptane (1:1)	3.1	0.89	0.66	0.80	0.41
Toluene/water (1:1)	36.0	0.86	1.00	0.53	0.89
o-Xylene/water (1:1)	36.1	0.88	1.00	0.74	0.89
Methanol/acetic acid/heptane (1:1:1)	2.1	0.88	0.69	0.92	0.41
Chloroform/ethyl acetate/water/methanol (2:2:2:3)	1.3	1.09	0.93	0.68	1.66

[a] Subscript 1 refers to the phase rich in the first solvent used for the biphasic solvent system.

their viscosities. The values measured for various solvent systems [14] are shown in Table 2 for biphasic solvent systems and not for pure solvents, which would be of little interest for CCC.

IV. CHARACTERIZATION BASED ON THE BEHAVIOR INSIDE THE CCC COLUMN

Apart from all of the parameters and their values given in the previous section, another approach to solvent system characterization may be based on their behaviors inside the columns of the various countercurrent devices. One should clarify here that "behavior" only refers to the best combinations of experimental parameters (e.g., choice of lighter or heavier phase, choice of the inlet into which the mobile phase is pumped, etc.) in order to retain the maximum amount of stationary phase inside the column. However, the actual behavior inside the column refers to complex hydrodynamic phenomena; the optimization for the highest retention of the stationary phase that is described in the following paragraphs is only experimental and is often based on a limited number of experiments, carried out on some solvent systems. Moreover, the optimization is dependent on the type of CCC apparatus that is under investigation.

A. Sanki-Type Apparatus

The principle behind the Sanki-type apparatus has been precisely described in previous publications [1,15] and in a chapter of this book. The apparatus is a centrifuge in which cartridges or plates were installed. Two rotating seals are required to allow the flow of the liquid phase; one stays at the top of the centrifuge, the other at the bottom.

Whatever the solvent system may be, the optimization for the best retention of the stationary phase is quite simple. Among the four possibilities, only two lead to a good retention of the stationary phase inside the cartridges or the plates. They are based on the combination of lighter mobile phase pumped from the bottom to the top seal, also called the "ascending" mode, and heavier mobile phase pumped from the top to the bottom seal, also called the "descending" mode.

As the two possibilities are common to all solvent systems, no characterization may be deduced from the choice of mobile phase and of pumping mode on a Sanki-type apparatus.

B. Type J Apparatus

The principle of type J apparatus is precisely described in Chapter 2 and we also refer the reader to the book edited by Mandava and Ito in the same series as this book [16]. Two main parameters for using a two-phase solvent system with this apparatus are the choice of the heavier or the lighter mobile phase and the pumping mode, i.e., from the tail to the head or the head to the tail of the column. The designer of this type of device, Dr. Ito, has tried various solvent systems in order to ascertain the best combinations of the two main parameters. He observed that among the four possibilities only two led to the best retention of the stationary phase. However, the two optimal conditions were dependent on the nature of the solvent system and, for some solvent systems, on the geometrical dimension of the apparatus.

Ito has consequently decided to carry out a systematic study on 15 solvents [16]. His first conclusion was that only one condition was optimal among the two pumping modes for a given phase, i.e., lighter or heavier phase. The second conclusion was that the pumping modes to be used are reversed if the liquid phase is chosen as lighter instead of heavier, or vice versa.

Two groups of solvent system have been defined. One, called "a" group, gathers solvent systems for which the two best combinations are the pumping of the lighter phase from the tail to the head of the column or the pumping of the heavier phase from the head to the tail of the column. The other group, named "b" group, gathers solvent systems for which the two best combinations are the pumping of the lighter phase from the head to the tail of the column or of the heavier phase from the tail to the head of the column. The best combinations are reversed between the two groups.

However, there was a need to define a third group c to take into account the behavior of some solvent systems for which optimal combinations depend on the geometrical dimensions of the apparatus. The discriminating parameter was found to be the β ratio of the coil radius on the distance between the two axes of rotation (see Chapter 2). For β values smaller than 0.3, solvent systems belonging to group c behave like the solvent systems of group b. On the contrary, for β values greater than 0.3, they behave as solvent systems of group a.

The first column of Table 3 gathers the 15 solvent systems studied by Ito. Instead of using a, b, or c letters, Ito has observed that the groups were more or less related to the hydrophobicity of the organic phase; hence the names "hydrophobic" for group a, "hydrophilic" for group b, and "intermediate" for group c.

One should keep in mind that this classification was obtained at room temperature, for a given type of apparatus (similar to PC Inc.) and for a Teflon tubing of 1.6 mm i.d. It is called Ito's Classification. The main drawback of this classification comes from the experimental determination of the three groups. For a solvent system not previously studied, either by Ito or in the literature, the experimenter has to carry out four experiments in order to determine the two best combinations and consequently the group to which it belongs.

C. Cross-Axis-Type Apparatus

The general principle of cross-axis apparatus is described in the corresponding part of the second chapter of this book. Contrarily to the two previous CCC devices, four main parameters have to be considered. Two are common to the other types of CCC units, i.e., choice of a lighter or a heavier phase and pumping mode, from tail to head or from head to tail. Two additional parameters intervene: the pumping direction, from the inside to the outside "flask" of the core or reverse, and the rotation direction, clockwise or anticlockwise.

Table 3 Values of Capillary Wavelength and Characteristic Settling Velocities for the Main Two-Phase Solvent Systems Studied by Ito

Solvent systems (v/v)		λ_{cap}	Characteristic settling velocities (m/s)	
Type	System	(cm)	V_{low}	V_{up}
Hydrophobic	Hexane/water (1:1)	0.246	138.56	70.12
	Ethyl acetate/water (1:1)	0.198	16.35	9.77
	Chloroform/water (1:1)	0.159	42.02	63.36
Intermediate	Hexane/methanol (1:1)	0.083	2.79	2.18
	Ethyl acetate/acetic acid/water (4:1:4)	0.142	5.49	5.22
	Chloroform/acetic acid/water (2:2:1)	0.145	12.94	17.97
	n-Butanol/water (1:1)	0.072	1.32	1.94
	n-Butanol/0.1 M NaCl (1:1)	0.107	3.03	4.41
	n-Butanol/1 M NaCl (1:1)	0.100	3.61	5.49
	n-Butanol/acetic acid/water (4:1:5)	<0.090	>0.75	>0.84
	n-Butanol/acetic acid/0.1 M NaCl (4:1:5)	<0.058	>0.74	>0.93
Hydrophilic	*n*-Butanol/acetic acid/1 M NaCl (4:1:5)	0.049	0.47	0.69
	sec-Butanol/water (1:1)	<0.063	>0.47	>0.69
	sec-Butanol/0.1 M NaCl (1:1)	<0.058	>0.64	>0.91
	sec-Butanol/1 M NaCl (1:1)	0.080	1.96	2.69

The same designer of this type of apparatus as for type J device, Dr. Ito, has applied the same procedure in order to classify various solvent systems by varying the main running parameters [17,18]. However, it has proved difficult to withdraw clear and precise conclusions from all of the results because of the number of running parameters.

The methodology of experimental designs has consequently emerged as the rational method to use for this purpose; it is easy to use and it gives the effects of the parameter and their interactions [19]. The first application of experimental design to a cross-axis type of device was exposed by Goupy et al. [20] and a thorough but global interpretation is given in Chapter 2.

The first conclusion drawn from examination of the results of the experimental designs is that the recommended combinations do not depend on the nature of the solvent system. The best combinations are precisely described in Chapter 2. Consequently, no classification among solvent systems may be derived from their behaviors inside a cross-axis device.

D. Conclusion

Among the three types of CCC units studied, two lead to a nonspecific behavior inside their columns. Only the type J unit enables the definition of a classification based on the experimental behavior of a solvent system inside the column. However, in order to employ a characterization method of a given solvent system, one has to understand the influence of the internal diameter of the tubing and of the temperature. Moreover, a prediction of the group to which a solvent system belongs to would be helpful in defining Ito's Classification as a reliable characterization mean of a solvent system.

V. USE OF HYDRODYNAMICS TO CORRELATE CHARACTERIZATIONS BASED ON PHYSICAL PARAMETERS AND ON OBSERVED BEHAVIOR INSIDE A CCC COLUMN

In Sections III and IV of this chapter, two ways of characterizing a solvent systems have been exposed. The first refers to physical properties of the two liquid phases of the system. The second refers to Ito's Classification obtained from the experimental behavior of the solvent system inside a type J column. It therefore seemed necessary to find a way to gather the two ways of characterizing a solvent system. The first approach was too rough to be considered as a reliable mean of characterizing a solvent system, even if its advantage is that no experiment on a CCC device is required. The second approach also cannot be considered as reliable because it depends on geometrical dimension and temperature. Moreover, it requires four experiments on a CCC unit.

Hydrodynamics has consequently emerged as the method embodying the two ways of characterizing a solvent system. Its use is based on the definition of simple parameters, involving a few of the physical properties described in section III.B, which bears a physical mean and allows a simplified description of the phenomena occurring.

The capillary wavelength and settling velocities have been retained as interesting parameters to step forward the description of the behavior of a solvent system inside a CCC column. Moreover, they have also enabled a simple prediction of the effect of the temperature on the behavior of solvent systems and therefore on Ito's Classification.

A. Definition of "Theoretical" Parameters

1. Capillary Wavelength

This "theoretical" parameter λ_{cap}, has been precisely described by Menet et al. [14,21,22]. The capillary wavelength is a mean of describing the microscopic behavior at the interface between two immiscible liquids. It stands for the wavelength of the deformations that may occur at the interface of the two liquids or represents the mean diameter of drops of one liquid in the other. Actually, it compares the relative intensities of interfacial tension forces, which tend to smooth all deformations at the interface, and of gravity forces, which have the reverse effect. Its definition is as follows:

$$\lambda_{cap} = 2\pi \sqrt{\frac{\gamma}{|\Delta\rho| g}} \tag{11}$$

with γ the interfacial tension, $\Delta\rho$ is the difference in densities between the two phases, and g is the gravity force. For common liquids, its average value is 1 cm in Earth's gravitational field. For CCC devices, the g field has to be replaced by the effective g^* field, which gathers the Earth's gravitational field and the centrifugal force field generated by the motion of the apparatus. As the capillary wavelength is used here only for the purpose of comparison, g^* may be replaced by $100g$, commonly obtained on CCC units.

This theoretical value was later used by Foucault for the theory of Centrifugal Partition Chromatography (CPC) (Sanki-type apparatus) as the Bond number [1]. However this parameter is dimensionless, which prevents easy interpretation, so we will use only our former capillary wavelength.

2. Settling Velocities

As the capillary wavelength only enables description of the formation of droplets of one liquid in another, it seemed interesting to introduce other theoretical parameters to better describe the dynamic phenomena occurring inside a CCC column. Two of these are presented here, V_{low} for the fall of a droplet of the heavier liquid phase (lower) in the continuous lighter one (upper) and V_{up} for the rise of a droplet of the lighter liquid phase in the continuous heavier one, and are defined as follows:

$$V_{low} = \frac{\gamma}{\eta_{up} \dfrac{2 + 3\dfrac{\eta_{low}}{\eta_{up}}}{3 + 3\dfrac{\eta_{low}}{\eta_{up}}}} \tag{12a}$$

and

$$V_{up} = \frac{\gamma}{\eta_{low} \dfrac{2 + 3\dfrac{\eta_{up}}{\eta_{low}}}{3 + 3\dfrac{\eta_{up}}{\eta_{low}}}} \tag{12b}$$

where γ is the interfacial tension between the two liquid phases, η_{up} the dynamic viscosity of the lighter phase, and η_{low} the dynamic viscosity of the heavier phase.

It is interesting to note that as neither V_{up} nor V_{low} depends on g, these velocities do not depend on the selected angular velocity of rotation. This is because the field intensity influences the size of the capillary wavelength and the sedimentation velocity of the droplet in the same way.

B. Correlation Between Physical Parameters and Ito's Classification

1. Is a Correlation Possible?

In order to determine if a correlation can exist between Ito's Classification and the values of the previous ''theoretical'' parameters, the 15 solvent systems used for the design of the classification have been studied. The values of interfacial tension, densities, and dynamic viscosities were extracted from Table 2 to calculate the values of the capillary wavelength and the settling velocities. Results are gathered in Table 3.

At first sight, the values of the capillary wavelengths are smaller for hydrophobic solvent systems than those for hydrophobic ones: the first family of solvent systems tends to form small droplets of one phase in the other one, leading to a more stable emulsion. Consequently, their stationary phase is less retained in the CCC column than the hydrophobic solvent systems. This phenomenon is well known in CCC [23].

From the calculated values of the capillary wavelength given in Table 3, the definition of ranges for each group of solvent systems appears possible, even if some overlay might occur:

		λ_{cap}	≤ 0.072	Hydrophilic
0.072	$\leq$	λ_{cap}	≤ 0.090	*Hydrophilic or intermediate*
0.090	$\leq$	λ_{cap}	≤ 0.145	Intermediate
0.159	$\leq$	λ_{cap}		Hydrophobic

where λ_{cap} is in cm.

The computed values of settling velocities demonstrate that ranges may be defined, even if the hydrophilic and intermediate groups share a small part of their ranges:

		V_{low}	≤ 1.32	Hydrophilic
1.32	$\leq$	V_{low}	≤ 1.96	*Hydrophilic or intermediate*
1.96	$\leq$	V_{low}	≤ 12.94	Intermediate
16.35	$\leq$	V_{low}		Hydrophobic

also:

		V_{up}	≤ 1.94	Hydrophilic
1.94	$\leq$	V_{up}	≤ 2.69	*Hydrophilic or intermediate*
2.69	$\leq$	V_{up}	≤ 9.77	Intermediate
9.77	$\leq$	V_{up}	≤ 17.97	*Intermediate or hydrophobic*
17.97	$\leq$	V_{up}		Hydrophobic

where V_{low} and V_{up} are in m/s.

The ranges that are defined previously show that it is possible to correlate, to some extent, the physical properties of solvent systems with their behaviors inside the column of a type J apparatus. As this correlation can be a means of characterizing a solvent system, a set of six solvent systems, not previously studied by Ito, have undergone the same procedure for testing the reliability of the correlation.

2. How Reliable Is the Correlation?

Six original solvent systems have been introduced: dimethylsulfoxide (DMSO)/heptane (1 : 1, v/v), dimethylformamide (DMF)/heptane (1 : 1, v/v), toluene/water (1 : 1, v/v), *o*-xylene/water (1 : 1, v/v), heptane/acetic acid/methanol (1 : 1 : 1, v/v), and chloroform/ethyl acetate/water/methanol (2:2:2:3, v/v). Their physical properties have already been given in Table 2, which allowed the calculation of their capillary wavelengths and settling velocities displayed in Table 4. It was then possible to use the various ranges described previously to classify each new solvent system within the three groups defined by Ito. All of the results are gathered in Table 5. The CCC device used for the determination of the hydrodynamic behavior of each solvent system had β values larger than 0.3 (which is the case for most commercial CCC devices). It is consequently impossible to differentiate intermediate and hydrophobic solvent systems, as indicated in the last column of Table 5. Since all of the six original solvent systems behave as so-called hydrophobic or intermediate systems, the wrong conclusions are drawn from the physical parameters investigated when a hydrophilic behavior (letter *b*) is predicted.

Table 4 Values of Capillary Wavelength and Characteristic Settling Velocities for the Six Original Two-Phase Solvent Systems

Solvent systems (v/v)	λ_{cap} (cm)	Characteristic settling velocities (m/s) V_{low}	V_{up}
DMSO/heptane (1:1)	0.101	28.48	6.96
DMF/heptane (1:1)	0.074	8.52	4.97
Toluene/water (1:1)	0.322	77.58	51.13
o-Xylene/water (1:1)	0.348	57.48	49.59
Heptane/acetic acid/methanol (1:1:1)	0.067	5.71	2.97
Chloroform/ethyl acetate/water/methanol (2:2:2:3)	0.057	1.03	2.12

Among the six new solvent systems, two are made of an organic phase (toluene or *o*-xylene) added to water. All of the classification methods predict for these two systems the right behavior, which is that of a hydrophobic system. These organic phase/water systems are very similar to those investigated by Ito. Two other solvent systems are completely organic: DMSO/heptane and DMF/heptane. The λ_{cap} and V_{low} and V_{up} parameters give the right prediction for both systems. The lower λ_{cap} value for the system containing DMF probably reflects the higher capability of this system to form an emulsion compared to conventional systems. In practice, the DMF/heptane system is less convenient for high-speed

Table 5 Results from Classifications of Six "new" Solvent Systems Using Interfacial Tension, Capillary Wavelength, Characteristic Settling Velocities, and Experimental Settling Times: Comparison with Experimental Behaviors

Solvent system (v/v)	λ_{cap}	V_{low}	V_{up}	Experimental
DMSO/heptane (1:1)	c	a	c	a or c
DMF/heptane (1:1)	c or b	c	c	a or c
Toluene/water (1:1)	a	a	a	a or c
o-Xylene/water (1:1)	a	a	a	a or c
Heptane/acetic acid/methanol (1:1:1)	b	c	c	a or c
Chloroform/ethyl acetate/water/methanol (2:2:2:3)	b	b	c or b	a or c

[a] Hydrophobic solvent system.
[b] Hydrophilic solvent system.
[c] Intermediate solvent system.

CCC than the DMSO/heptane system [24]; its retention of stationary phase is limited to 50%, which is lower than the average value (80%) for hydrophobic solvent systems. A problem occurs with the three-component solvent system heptane/acetic acid/methanol. The characteristic settling velocities give the right behavior whereas the capillary wavelength leads to the wrong prediction. The higher reliability of V_{low} and V_{up} compared to that of λ_{cap} may be explained from the physical definitions of these parameters given in the theoretical introduction. λ_{cap} reflects the dynamics of formation of droplets at the interface of the two liquid phases. V characterizes the dynamics of the migration of droplets of size λ_{cap}. Consequently, the characteristic settling velocity seems to be a better parameter than the capillary wavelength and thus better describes the hydrodynamic phenomena involved inside the CCC columns. The last new solvent system prepared from four solvents, chloroform/ethyl acetate/water/methanol, leads to wrong predictions for the classification methods based on λ_{cap} and V_{low}. One reason could lie in the difficulties met for measuring the interfacial tension, as the aqueous phase of this system is the upper one, contrarily to the five other original solvent systems. An underestimation of γ lowers both λ_{cap} and V and leads to considering the solvent system as more hydrophilic than it actually is.

C. Influence of Temperature

As we saw in the previous sections, the behavior of a solvent system fundamentally originates from its physical properties. Temperature is revealed as a key parameter for CCC. On one hand, it modifies the physical properties of the solvent system and, consequently, its behavior inside the column. On the other hand, it controls the partition coefficient of the solutes between the two liquid phases. The conclusion of these assertions is that the temperature should be regulated in the countercurrent chromatographs to obtain reproducible experiments.

Unfortunately, studies on the influence of temperature on the separations carried out in countercurrent chromatographs are anecdotal in the literature. Ito discussed the potential benefits of an increase of the temperature [25] and Conway presented some results at the Pittsburgh Conference in 1988 [26]. Two examples of separations carried out at temperatures higher than room temperature were described by Knight et al. [27,28] for successful purification of peptides at 50°C and 45°C, respectively.

1. Experimental Study

We carried out investigations on the influence of temperature. The apparatus was a type J countercurrent chromatograph. Its features are listed in Table 6. The chosen rotation speed was 1900 rpm and the flow rate of the mobile phase was 1 mL/min. The walls of the apparatus were covered by polystyrene plates. The

Table 6 Characteristics of CCC3000[a]

Rotation plane	Vertical
Orbital radius	3.8 cm
Planetary radius	
r_{min}	2.1 cm
r_{max}	3.0 cm
Rotation speed	
ω_{max}	3000 rpm
$\omega_{average}$	1700 rpm
Internal diameter of column tubing	0.8 mm
Internal volume of the column	18 mL
β_{min}	0.55
β_{max}	0.77
Maximum intensity of the force field	500 g
Type	Three-column

[a] Pharma-Tech Research Corp., Baltimore, Maryland.
Note: Values were determined in the authors' laboratory.

original cover of the device was replaced by a small radiator mounted with a fan. A circulating bath was used to set the temperature of the radiator. Preliminary tests on the bath temperature and the fan rotation speed were carried out to determine the conditions required on these parameters to obtain a given temperature. Solvent tanks and the tubing between the pump and the inlet of the apparatus also plunged in the tank of the cooling bath.

The β value of the countercurrent chromatograph was higher than 0.5, which meant that the butanol-1/water system should behave like a hydrophobic

Table 7 Values (%) of Retention of Stationary Phase for the Butanol-1/Water System at Five Temperatures[a]

	Butanol-1	Butanol-1	Water	Water
Mobile phase	head-to-tail	tail-to-head	head-to-tail	tail-to-head
12°C	10	0	14	6
22°C	17	1	14	7
37°C	14	11	26	0
50°C	17	25	43	0
60°C	16	47	48	0

[a] Apparatus: CCC3000, Pharma-Tech.

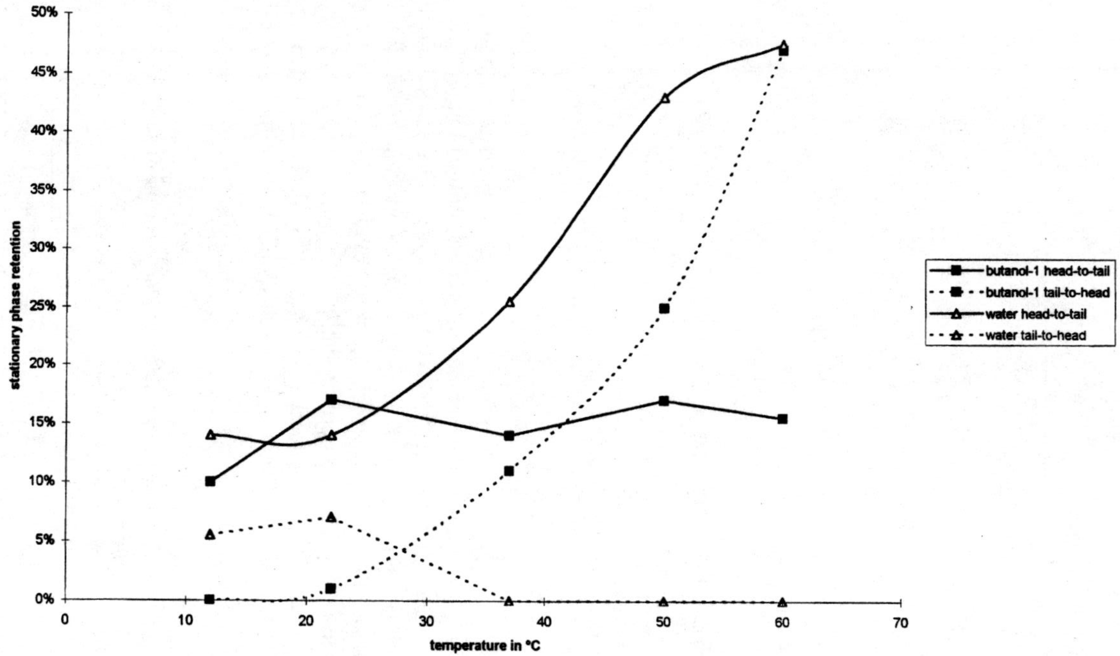

Figure 7 Influence of temperature on the retention of the stationary phase of the butanol-1/water system. The measure is achieved after 25 min of pumping of the mobile phase.

Table 8 Values of Interfacial Tension, γ, Densities of Lighter and Lower Phase, ρ, and Dynamic Viscosities for These Two Liquid Phases, η, for Butanol-1/Water System[a]

Solvent system (v/v)	Temp. (°C)	Interfacial tension (dyn/cm)	Density (g/cm³)		Dynamic viscosity (cP)	
		γ	ρ_1	ρ_2	η_1	η_2
Butanol-1/water (1:1)	20	1.8	0.85	0.99	1.7	1.1
Butanol-1/water (1:1)	60	1.5	0.85	1.00	1.25	0.6

[a] Subscript 1 refers to the phase rich in the first solvent used for the biphasic solvent system. All values obtained by our laboratory.

solvent system. The butanol-1/water solvent system was studied at five temperatures: 12°C, 22°C, 37°C, 50°C, and 60°C. The observed values are given in Table 7. The retention of the stationary phase is plotted vs. the temperature in Figure 7. The two curves, which are butanol-1 as the mobile phase pumped from the tail to the head of the column and water as the mobile phase pumped from the head to the tail of the column, increase with the temperature to reach about 50% at 60°C. They are characteristic of hydrophobic solvent systems. The two complementary curves show different shapes. Butanol-1 as mobile phase pumped from the head to the tail of the column led to a quite constant value of the retention of the stationary phase in a 10% to 15% range, while the curve with water as mobile phase pumped from the tail to the head of the column quickly decreased to 0 as soon as 37°C. For the butanol-1/water system, it was therefore observed that retention of stationary phase in the hydrophobic mode increased with temperature.

Table 9 Values of Capillary Wavelength and Characteristic Settling Velocities for the Butanol-1/Water System at 20°C and 60°C

Solvent systems (v/v)	Temp. (°C)	λ_{cap} (cm)	Characteristic settling velocities (m/s)	
			V_{low}	V_{up}
Butanol-1/water (1:1)	20	0.072	1.32	1.95
Butanol-1/water (1:1)	60	0.063	1.55	2.86

2. Interpretations

According to Ito, intermediate solvent systems behave as hydrophobic ones for β values higher than 0.5 and as hydrophilic ones for β values smaller than 0.5. Butanol-1/water is classified as intermediate and the β values of CCC3000 (a type J countercurrent chromatograph) ranged from 0.55 to 0.77, so that the behavior of this solvent system was that of hydrophobic solvent system. However, as the minimum β value of the apparatus was close to 0.5, the retention of the stationary phase in the hydrophobic pumping modes (1% and 14%) was not really different at room temperature from these in the hydrophilic pumping modes (17% and 7%). Increasing the temperature just widened the gap between the retention of the stationary phase in the hydrophobic mode and that of the hydrophilic mode, so that a 60°C temperature led to about 47% for the hydrophobic pumping mode and 16% and 0% in the hydrophilic pumping mode. In order to better understand such a behavior, the physical parameters previously defined were used to compute the capillary wavelength and both settling velocities; the values are gathered in Table 8. They show the expected decrease in interfacial tension and dynamic viscosity with temperature. The values were then used to calculate the capillary wavelength and both settling velocities, given in Table 9.

The capillary wavelength decreases with the temperature, which prevents its use as an explanatory variable, as small values are characteristic of hydrophilic solvent systems. On the contrary, both settling velocities increase with the temperature. V_{low} increases from 1.32 to 1.55, but the butanol-1/water system remained within the limits of hydrophilic or intermediate range. V_{up} increases from 1.95 to 2.86, which indicates that the butanol-1/water system evolves from a hydrophilic or intermediate family to an intermediate family. This is consistent with the increase of the retention of stationary phase with the temperature for the hydrophobic mode, as the classification based on the settling velocity V_{up} (rise of a droplet of the lighter liquid in the continuous heavier one) indicates that the behavior of the solvent system is more pronounced toward a hydrophobic one. Once again, the classification based on V_{up} was revealed as more reliable than that based on capillary wavelength and V_{low}.

VI. CONCLUSION

This chapter has shown that there are many ways of characterizing biphasic solvent systems. A biphasic system can be first characterized by its phase diagram, which helps in determining the two-phase region, the relative volumes of the two phases and their compositions. Then physical parameters give further information on the solvent system. They include the polarity parameter and other classical physical parameters such as interfacial tension or viscosity. The previous characterization methods are normally sufficient to directly use the chosen solvent system on CCC devices such as Sanki-type chromatographs and cross-axis devices.

However, type J devices require a better knowledge of the characteristics of the solvent system. The experimenter can choose to do preliminary experiments before using the solvent system in a satisfactory way, or he or she can calculate the velocity of a droplet of the lighter phase in the heavier one from the interfacial tension and the viscosities of both phases and determine a priori the behavior of the solvent system inside the CCC column. Consequently, all of the tools detailed in this chapter can be used for choosing a solvent system, optimizing its composition, and using it successfully in a CCC device.

REFERENCES

1. A. Foucault, *Centrifugal Partition Chromatography*, Chromatographic Science Series, Vol. 68, Marcel Dekker, New York, 1995.
2. A. Foucault, P. Durand, E. Camacho Frias, and F. le Goffic, *Anal. Chem. 65*:2154 (1993).
3. C. Reichardt, *Solvents and Solvent Effects in Organic Chemistry*, VCH, Weinheim, 1988.
4. J. H. Hildebrand and R. L. Scott, *Regular Solutions*, Prentice Hall, Englewood Cliffs, NJ, 1962.
5. A. F. M. Barton, *Chem. Rev. 6*:731 (1975).
6. L. R. Snyder and J. J. Kirkland, *Introduction to Modern Liquid Chromatography*, 2nd ed., John Wiley and Sons, New York, 1979, p. 257.
7. L. R. Snyder, *Principles of Adsorption Chromatography*, Marcel Dekker, New York, 1968.
8. E. Soczewinski, *Anal. Chem. 41*:179 (1969).
9. R. Rosset, M. Caude, and A. Jardy, *Chromatographies en Phases Liquide et Supercritique*, Masson, 1991, p. 267 (1991).
10. L. Rohrschneider, *Anal. Chem. 45*:1241 (1973).
11. A. Berthod and N. Schmitt, *Talanta 40*:1667 (1993).
12. A. Berthod, J.-M. Deroux, and M. Bully, *Modern Countercurrent Chromatography*, American Chemical Society Books, Symposium Series 593 (W. D. Conway and R. J. Petroski, eds.), 1995, p. 25.
13. S. J. Gluck and M. P. Wingeier, *J. Chromatogr. 547*:69 (1991).
14. J. M. Menet, D. Thiébaut, R. Rosset, J. E. Wesfreid, and M. Martin, *Anal. Chem. 66*:168 (1994).
15. M. C. Rolet, D. Thiébaut, and R. Rosset, *Analusis 20*:1 (1992).
16. N. B. Mandava and Y. Ito, *Countercurrent Chromatography*; Theory and Practice, Chromatographic Science Series, Marcel Dekker, New York, 1988.
17. Y. Ito and T.-Y. Zhang, *J. Chromatogr. 449*:135–151 (1988).
18. Y. Ito, *J. Chromatogr. 538*:67 (1991).

19. J. Goupy, *Methods for Experimental Designs Principles Applications for Physicists and Chemists*, Elsevier, Amsterdam, 1993.
20. J Goupy, J.-M. Menet, K. Shinomiya, and Y. Ito, in *Modern Countercurrent Chromatography* (W. D. Conway and R. J. Petrosky, eds.), *ACS Symposium Series 593*, 1995 p. 47.
21. J. M. Menet, D. Thiébaut, and R. Rosset, Oral presentation, Prediction of hydrodynamic solvent system behavior in countercurrent chromatography using capillary wavelength, Pittsburgh Conference, March 1992, New Orleans, LA.
22. J. M. Menet, D. Thiébaut, R. Rosset, J. E. Wesfreid, and M. Martin, Poster, The capillary wavelength as the key parameter for describing solvent system behavior in counter-current chromatography, 19th International Symposium on Chromatography. Aix-en-Provence, France, September 1992.
23. J. L. Sandlin and Y. Ito, *J. Liq. Chromatogr. 8*:2153 (1985).
24. M. Bully, DEA de Chimie Analytique, Université Claude Bernard, Lyon I (1990).
25. Same as ref. 16, except p. 377.
26. W. D. Conway, Pittsburgh Conference, New Orleans, LA, 22–26 February 1988; oral presentation.
27. M. Knight, Y. Ito, A. M. Ask, C. A. Tamminga, and T. N. Chase, *J. Liq. Chromatogr. 7*(13):2525 (1984).
28. M. Knight, Y. Ito, P. Peters, and C. diBello, *J. Liq. Chromatogr. 8*(12):2281 (1985).

2

Experimental Designs Applied to Countercurrent Chromatography: Definitions, Concepts, and Applications

Jacques Goupy
ReConFor, Paris, France

Jean-Michel Menet
Rhône-Poulenc Rorer, Inc., Vitry-sur-Seine, France

Didier Thiébaut
Ecole Supérieure de Physique et de Chimie Industrielles de la Ville de Paris, Paris, France

Countercurrent chromatography (CCC) is a separation method based on the partition of solutes between two liquid phases as they interact in a thin tube under a centrifugal force field. One source of interest in this method is that no solid matrix is required to retain the stationary phase. However, optimal settings of the apparatus are difficult to find because many variables or factors are involved. Experimental design methodology (EDM) enables one to solve this problem, and the solution to optimize an 11-factor study is described.

The first section covers the fundamental concepts of EDM, along with definitions and simple examples related to CCC problems. The second section deals with type J chromatography and describes how experimental designs help to clarify the experimental system and the influence of parameters such as temperature on the classification. The third section deals with the cross-axis apparatus;

it shows how experimental designs can provide just four optimized conditions and indicates how parameters influence the retention of the stationary phase.

I. DEFINITIONS AND CONCEPTS

In centrifugal liquid–liquid chromatography (CLLC), the stationary phase occupies up to 95% of the total volume of the column: this ratio, called retention of the stationary phase, plays a key role on the number of theoretical plates and the resolution. Two types of CLLC units are available, i.e., hydrostatic and hydrodynamic [1], depending on the way in which equilibrium between the liquid stationary and mobile phases is reached. Hydrostatic CLLC, often called centrifugal partition chromatography [2,3], is a relatively easy technique and will not be discussed here. Hydrodynamic CLLC is somewhat more complex because the optimized conditions for the best retention of the stationary phase depend on the nature of the solvent system.

Two types of hydrodynamic chromatographs are available. The type J chromatograph [4] has been on the market since 1980. The column is made from a Teflon tube wound on a cylinder in one or several layers. The column rotates on its axis around a parallel axis. Many solvent systems have been studied and a classification based on three groups has been designed [5] according to the conditions leading to the highest retention of stationary phase. However, this classification does not yet have a complete and satisfactory physical and chemical background [6].

The cross-axis type of chromatograph became available in 1992 [7]. The principle is the same as that of type J unit, except that the two axes are perpendicular. Owing to its complex geometry and motion it involves more running parameters than the coil planet centrifuge (CPC) or type J unit, and there is a lack of comprehensive investigation to understand the effect of operating parameters and to obtain the maximum stationary phase retention.

The experimental design method is well suited to this problem for two reasons. The first is its ability to accurately give the optimized experimental conditions in a minimal number of experiments. The second lies in the results of the calculations, which often help in understanding the way in which factors are involved and how they are linked to intrinsic physicochemical parameters.

A. Importance of Experimental Designs for CCC

The experimental design method provides the experimenter with the means to extract the maximum amount of information from a system under investigation. This approach combines practical knowledge of the phenomenon to be studied and theoretical knowledge of EDM. Experimenters generally understand the phenomenon under investigation but are not aware that powerful tools are available to help them to organize trials and interpret experimental results. These tools were discovered by R.A. Fisher [8,9] in 1915 and improved by others, such as Yates [10], Youden [11], and Plackett and Burman [12]. Experimenters who want to know more should read Goupy [13,14] or Box [15].

Experimental design method is very useful when there are many factors to study and when the phenomenon is complex. These two conditions are encountered in CLLC studies. In our studies, we had to manage 11 parameters or factors:

1. Coiling up of the chromatographic column (left-handed or right-handed)
2. Density of the mobile phase (heavy or light)
3. Elution direction (toward the central axis or opposite)
4. Direction of rotation (clockwise or anticlockwise)
5. Coil position (L or X-1.5L)
6. Coil diameter (5 or 10 cm)
7. Elution mode (head-to-tail or tail-to-head)
8. Number of layers (1 or 6)
9. Rotational speed (400 or 800 rpm)
10. Temperature (7°C or room temperature)
11. Nature of solvent system

All these factors are defined in the literature [4]. Experimental design method can make two contributions to CLLC. The first is practical. Analysis of all of the data gives a minimum number of experimental conditions to try to obtain the greatest retention of stationary phase (*Sf*). The experimenter who designs his own solvent system can be confident that the selected testings will furnish the highest *Sf* values. This procedure has been studied for the multiparameter cross-axis CPC and is covered in the third chapter. The second is the calculation of the effects and interactions, which should be closely linked to the physicochemical properties of the solvent systems and the geometrical dimensions of the apparatus. EDM can help provide a better understanding of the phenomena involved in the column. Such an interpretation has been applied to type J CPC to classify the solvent systems and the paramount influence of temperature on cross-axis CPC to give a partial understanding of the influence of the different parameters.

B. Ito's Type J Apparatus

This section uses examples of experiments described by Ito to optimize the stationary phase retention of CPC type J prototype [4]. The general principle of this

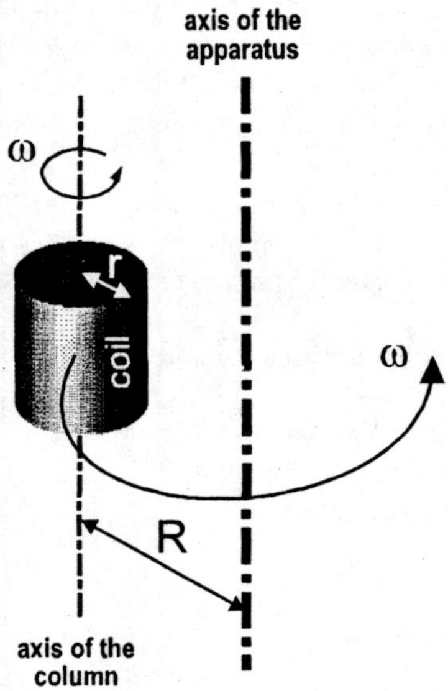

Figure 1 General principle of type J apparatus.

apparatus is shown in Figure 1. The coil is Teflon tubing, wound on a core cylinder of radius *r*. The general motion of the coil is a synchronous planetary motion, as it rotates around its axis, which itself rotates around a parallel fixed axis at a given distance *R*. The β parameter is defined as the ratio of *r* over *R* [16]. The rotational speeds are the same, which allows the use of narrow Teflon tubes (generally 0.8 mm i.d.) for connections without any twisting and consequently no complex and expensive rotary seals.

The coiled tube on the cylinder forms a helix; when the latter rotates around its own axis, the ends of the column may be labeled "head" or "tail," depending on the direction of winding (left-handed or right-handed) and the direction of rotation (clockwise or anticlockwise).

The two fundamental running factors for this apparatus are the choice of a heavy or light mobile phase (factor 2) and the elution mode, head-to-tail or tail-to-head (factor 7). Other factors include the radii *r* and *R*, the rotational speed ω, the internal diameter of the coiled Teflon tubing, the temperature, and the nature of the solvent system.

C. First Example of Factorial Designs

The first example uses two factors to examine several important basic concepts that are used in all experimental designs: response, factors, levels, experimental domain, and experimental matrix.

We assume that an experimenter wants to study the influence of the coil diameter and rotational speed on the retention of a stationary phase in the CLLC chromatograph.

1. Response

Response is the measure that interests the experimenter. In this example, the response is the percentage of stationary phase, *Sf*, retained in the chromatographic column measured at the hydrodynamic equilibrium, as a percentage of the total column volume.

2. Factors

The factors are chosen by the experimenter and can be continuous or discrete. Continuous factors (temperature, rotational speed, and coil diameter) vary continuously from one value to another. Discrete factors vary from step to step. Examples of discrete factors are elution mode, direction of rotation, or coil position. If the investigator assigns only two levels to each independent factor studied, the experiments can be organized as factorial designs whether the factors are continuous or discrete. Continuous and discrete factors can be mixed in this sort of design. The study of Ito's prototype involves discrete factors, which take only two levels. It is thus possible to use factorial design and to mix continuous and discrete factors. Factors with three or more levels must be analyzed using the statistical analysis of variance (ANOVA) method [17]. It then becomes necessary to distinguish between discrete and continuous factors.

Factors must be independent so that all combinations of levels are possible and the experimenter can run a complete factorial design. If factors are not independent, not all combinations of levels are possible and fewer trials are possible. The experimenter then runs a fractional factorial design.

In this first example, the factors are independent, continuous, and have two levels. These factors are as follows:

Coil diameter (factor 6)
Rotational speed of the apparatus (factor 9)

The other eight factors are fixed at a constant level throughout the experiments.

3. Experimental Domain

Each continuous factor has lower and upper limits. The lower limit is the low level and the upper limit is the high level (Table 1).

The experimental design may be shown graphically (Figure 2). The graphical representation gives a good idea of the experimental point locations for two or three factors. Experimental designs may also be shown as tables or matrices

Table 1 Experimental Domain

Parameter	Factor no.	Low level	High level
Coil diameter (cm)	6	5	10
Rotational speed (rpm)	9	400	800

for any number of factors. The matrix representation is useful when the number of factors is four or more.

4. Experimental Design: Graphical Representation

An orthogonal Cartesian system is used to represent factors with the first axis for the first factor (axis x_1 for coil diameter) and the second axis for the second factor, etc. Experimental design theory states that the best experimental points are at the corners of the experimental domain [13]: points A, B, C, and D are the four trials of the experiment.

$$A\begin{cases}5\text{ cm}\\400\text{ rpm}\end{cases}\quad B\begin{cases}10\text{ cm}\\400\text{ rpm}\end{cases}\quad C\begin{cases}5\text{ cm}\\800\text{ rpm}\end{cases}\quad D\begin{cases}10\text{ cm}\\800\text{ rpm}\end{cases}$$

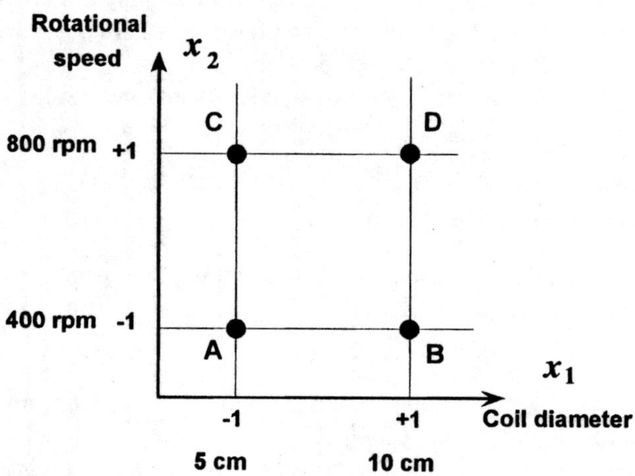

Figure 2 First example: experimental domain.

Table 2 Experimental Matrix

Trial no.	Coil diam.	Rotational speed	Response
1 (A)	−1	−1	$y_1 = 39$
2 (B)	+1	−1	$y_2 = 32$
3 (C)	−1	+1	$y_3 = 51$
4 (D)	+1	+1	$y_4 = 44$
Level (−)	5 cm	400 rpm	
Level (+)	10 cm	800 rpm	

By convention, −1 is used to indicate the low level of each factor and +1 for the high level. Trial A is then trial −1, −1; trial B is trial +1, −1; and so forth.

5. Experimental Design: Matricial Representation

The trials can also be shown as an experimental matrix (Table 2). The result of each trial is indicated in the response column. The factor levels are indicated in a table below the experimental matrix.

D. The Concept of Main Effect

The raw responses y_i are not easy to interpret. Experimenters generally want to know the main effect of each factor independently of the other factors. By definition, the main effect of a factor is half the difference between the mean y_+ of the responses at the high level of this factor and the mean y_- of the responses at its low level (Figure 3). The main effect E_d of the diameter is then

$$E_d = \frac{1}{2}[y_+ - y_-]$$

$$E_d = \frac{1}{2}\left[\frac{1}{2}(y_2 + y_4) - \frac{1}{2}(y_1 + y_3)\right] \qquad (1)$$

$$E_d = \frac{1}{4}[-y_1 + y_2 - y_3 + y_4]$$

This formula gives the main effect of the coil diameter as if it acts alone using the responses in Table 2.

The main effect of the coil diameter is numerically evaluated using the responses in Table 2.

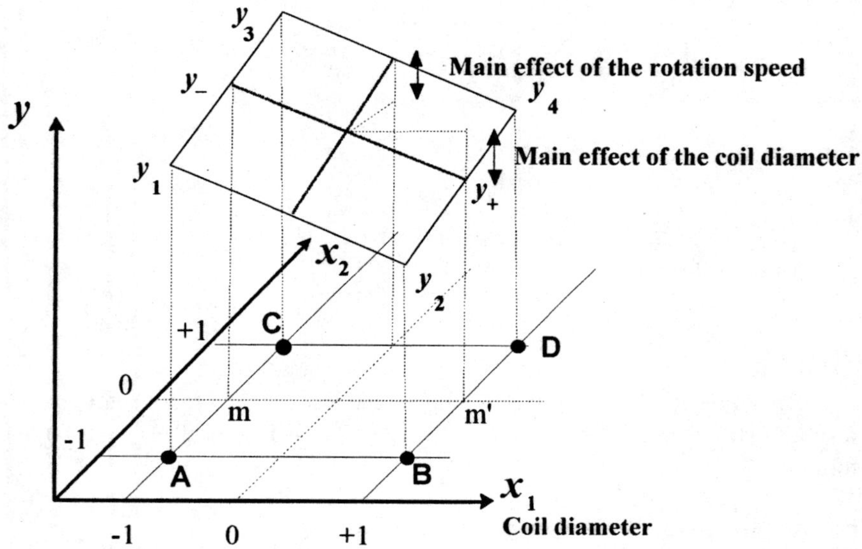

Figure 3 First example: retention of the stationary phase vs. the rotation speed and the coil diameter.

$$E_d = \frac{1}{4}[-39 + 32 - 51 + 44] = \frac{-14}{4} = -3.5$$

The main effect of the diameter is −3.5%, hence the retention drops of two times 3.5% when the diameter of the coil increases from 5 cm to 10 cm. The retention is higher when the diameter is smaller (Figures 3 and 4).

A similar expression gives the main effect E_s of the rotational speed on the retention as if it acts alone.

$$E_s = \frac{1}{4}[-y_1 - y_2 + y_3 + y_4] \tag{2}$$

$$E_s = \frac{1}{4}[-39 - 32 + 51 + 44] = \frac{24}{4} = 6$$

Thus, the retention increases two times 6% when the rotational speed passes from 400 rpm to 800 rpm. The retention is higher when the rotational speed is higher (Figure 3).

Figure 3 shows the experimental domain and the surface response. It explains the variation of the retention as a function of the two factors studied, i.e.,

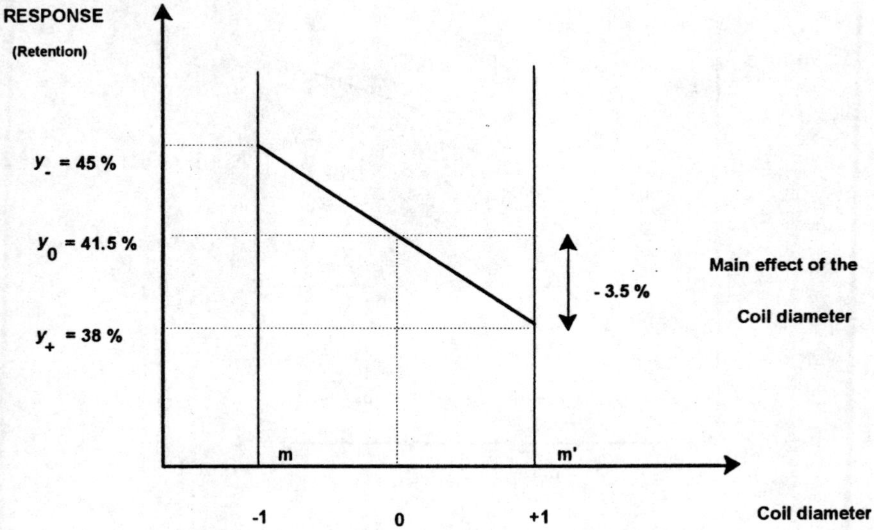

Figure 4 First example: retention of the stationary phase vs. the coil diameter.

coil diameter and rotational speed. Greater retention is obtained when the coil diameter is small and the rotational speed is high.

The plane $mm'y_+y_-$ can be extracted from Figure 3. This plane indicates the change in the retention as a function of the coil diameter and shows the main effect of this factor (Figure 4).

E. The Concept of Interaction

If the effect of a factor is the same at any level of the other factor, then there is no interaction between the two factors, as in the first example. At level -1 of the rotational speed the effect of the coil diameter is -3.5%, and at level $+1$ of the rotational speed the effect of the coil diameter is also -3.5%. But if this is not the case and the effect of a factor is not the same at all levels of the other factor, there is an interaction between the two factors.

By definition, the interaction between two factors (factors 1 and 2) is half the difference between (1) the effect of factor 1 at the high level of factor 2 and (2) The effect of factor 1 at the low level of factor 2.

The interaction between factors 1 and 2 is indicated by E_{12}. At the high level of factor 2, the effect of factor 1, $E_{1(2^+)}$, is (Figures 5 and 6):

$$E_{1(2^+)} = \frac{1}{2}[y_4 - y_3]$$

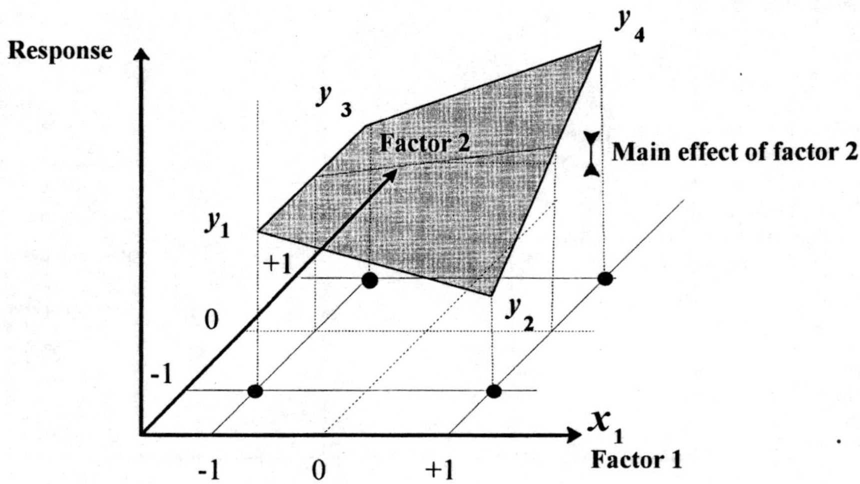

Figure 5 First example: response surface showing an interaction between factors 1 and 2.

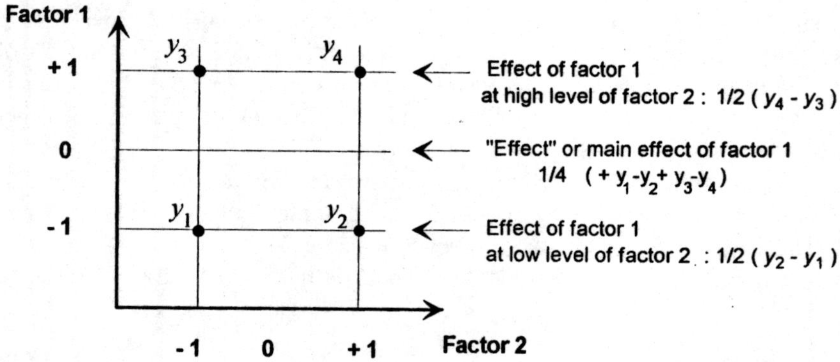

Figure 6 First example: diagram for the definition of an interaction.

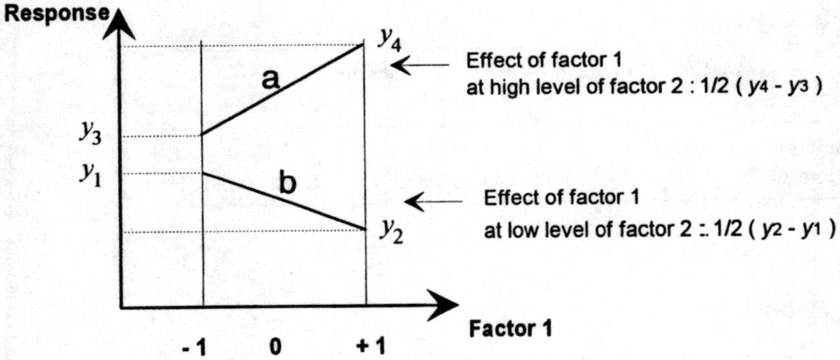

Figure 7 First example: interaction revealed between factors 1 and 2.

while at the low level of factor 2, the effect of factor 1, $E_{1(2^-)}$, is

$$E_{1(2^-)} = \frac{1}{2}[y_2 - y_1]$$

The interaction E_{12} is defined as half the difference of these two effects:

$$E_{12} = \frac{1}{2}\left[\frac{1}{2}(y_4 - y_3) - \frac{1}{2}(y_2 - y_1)\right]$$

This can be simplified to

$$E_{12} = \frac{1}{4}[+y_1 - y_2 - y_3 + y_4] \tag{3}$$

This formula looks very much like the one used to calculate the main effect.

Figure 7 shows the effect of factor 1 at a high level of factor 2 (straight line a) and the effect of factor 1 at a low level of factor 2 (straight line b). If the slopes of these two lines are different there is an interaction. Interactions are usually smaller than the main effects. But this is not so in CLLC, where the interactions may be greater than the main effects.

F. Second Example: Interaction

This example is taken from Ito [4]. The type J CPC apparatus was filled with hexane and water solvents, for a "hydrophobic" system according to the Ito Classification.

Table 3 Experimental Domain

Parameter	Factor no.	Level−	Level+
Mobile phase density	2	Light	Heavy
Elution mode	7	Tail to head (T)	Head to tail (H)

1. Response

The response is the percentage of stationary phase retained in the chromatographic column at hydrodynamic equilibrium.

2. Factors

The factors are as follows:

Density of the mobile phase (factor 2)
Elution mode (factor 7)

The other eight factors were fixed at a constant level during all of the experiments.

3. Experimental Domain

The experimental domain is given in Table 3.

4. Experimental Matrix

The experimental matrix is described in Table 4. Experimental design and results (graphical representation) are shown in Figure 8. Axis x_1 is used for the density of the mobile phase and axis x_2 for the elution mode.

Table 4 Experimental Matrix

Trial no.	Mobile phase density	Elution mode	Response
1	−1	−1	90
2	+1	−1	0
3	−1	+1	0
4	+1	+1	88
Level (−)	Light	T	
Level (+)	Heavy	H	

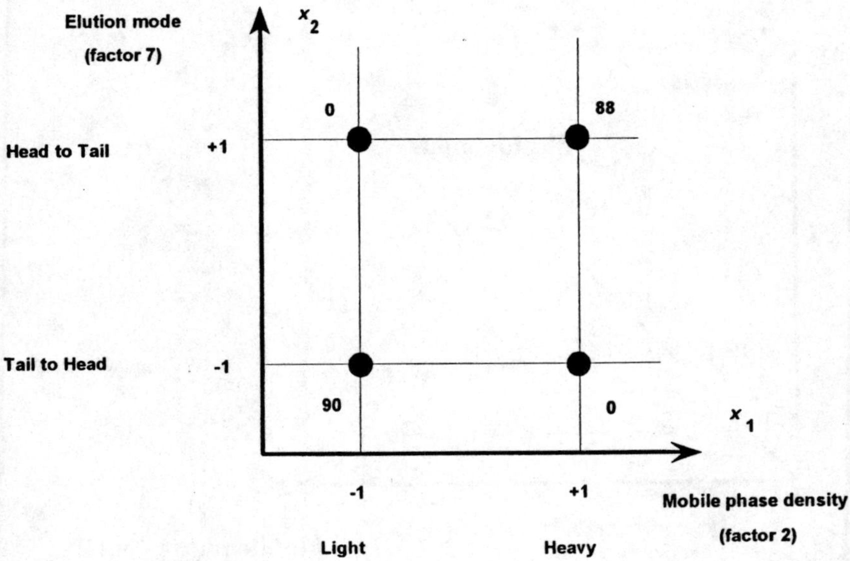

Figure 8 Second example: experimental domain and results of the retention of the stationary phase.

5. Interactions

The factors are indicated by numbers (1, 2, 3, 4, etc.) and second-order interactions are indicated by two numbers (12, 13, 23, etc.). The interaction between the density (factor 2) and the elution mode (factor 7) is indicated by 27. Formulas (1), (2), and (3) can be used to calculate the main effects and the interaction:

$$\text{Main effect of factor 2} = \frac{1}{4}[-90 + 0 - 0 + 88] = \frac{-2}{4} = -0.5$$

$$\text{Main effect of factor 7} = \frac{1}{4}[-90 - 0 + 0 + 88] = \frac{-2}{4} = -0.5$$

$$\text{Interaction 27} = \frac{1}{4}[+90 - 0 - 0 + 88] = \frac{178}{4} = +44.5$$

The interaction is important (Figures 9 and 10) because the effect of factor 2 (density of the mobile phase) at a high level of factor 7 (elution mode) is very different from the effect of factor 2 at a low level of factor 7.

The main effects of factors 2 and 7 are small because the means of the responses at high and low levels are approximately the same. Thus, the combinations of levels are very important for good or bad retention. Good retention is obtained when factors 2 and 7 are together at high levels. Good retention is also

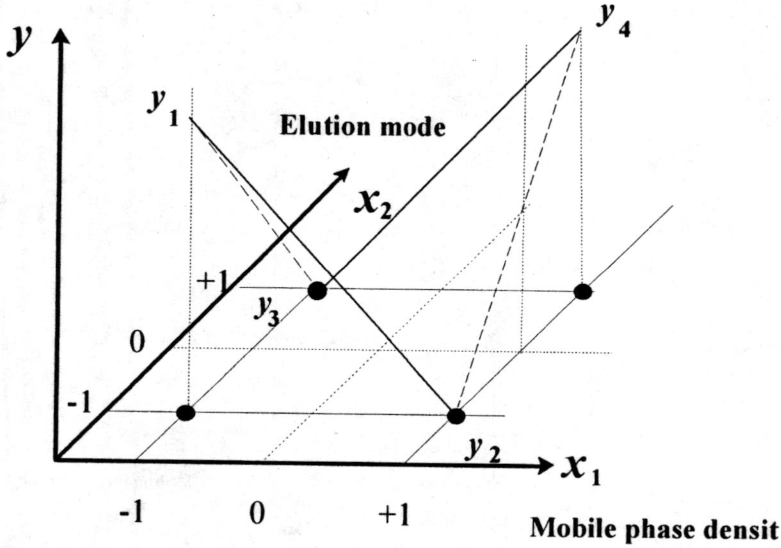

Figure 9 Second example: response surface showing an interaction between factors 1 and 2.

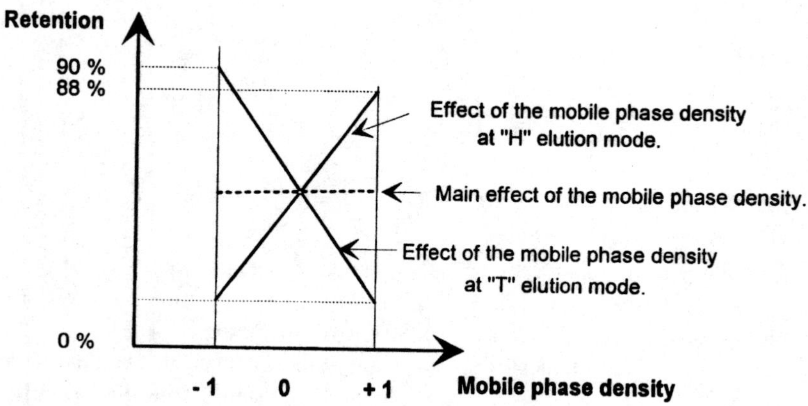

Figure 10 Second example: interaction revealed between factors 1 and 2.

obtained when factors 2 and 7 are together at low levels. We can indicate the high level of factor 2 by 2^+ and the low level by 2^-, and do the same for the other factors. The best combinations of levels are then 2^+7^+ and 2^-7^-. Interaction 27 is positive in both cases.

G. Third Example of Experimental Design 2 [3]

An experimental design with k factors, each at two levels, is written by 2^k, where k indicates the number of factors and 2 the number of levels taken by each factor. An experimental design with two factors, each at two levels, is written as 2^2 and an experimental design with three factors, each at two levels, is written as 2^3. The third example is also from Ito [4]. The graphical representation of this 2^3 design is shown in Figure 11.

1. Response

The response is the percentage of stationary phase retained in the chromatographic column at hydrodynamic equilibrium.

2. Factors and Domain

The three factors were

Density of the mobile phase (factor 2)
Diameter of the coil (factor 6)
Elution mode (factor 7)

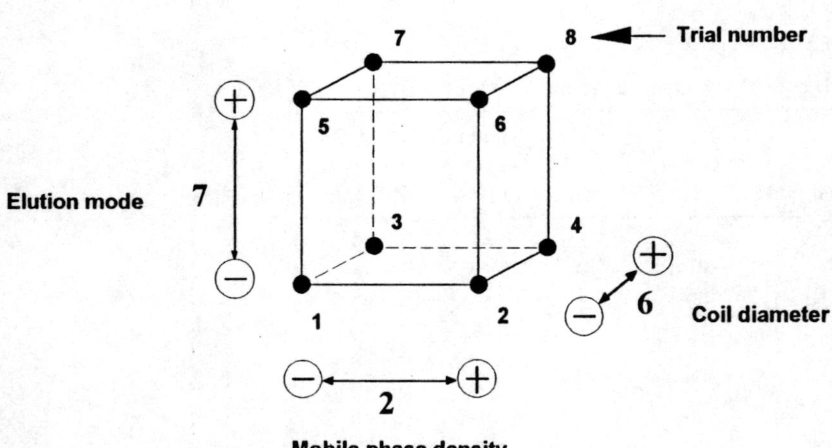

Figure 11 Third example: graphical representation of 2^3 design.

Table 5 Experimental Domain

Parameter	Factor no.	Level−	Level+
Mobile phase density	2	Light	Heavy
Coil diameter	6	5 cm	10 cm
Elution mode	7	Tail to head (T)	Head to tail (H)

and their variations are given in Table 5. The other seven factors were fixed at a constant level during all of the experiments.

3. Experimental Matrix

The experimental matrix is given in Table 6.

4. Experimental Design: Graphical Representation

The experimental design is shown in Figure 11.

5. Experimental Design and Results (Graphical Representation)

The main effects and interactions can be calculated from a relationship similar to (1) or (3), but with eight responses for each main effect and interaction [13]. The main effect of factor 1 is calculated from

Table 6 Experimental Matrix

Trial no.	Density (2)	Coil diam. (6)	Elution mode (7)	Response
1	−	−	−	92
2	+	−	−	10
3	−	+	−	90
4	+	+	−	0
5	−	−	+	0
6	+	−	+	90
7	−	+	+	0
8	+	+	+	88
Level (−)	Light	5	T	
Level (+)	Heavy	10	H	

Table 7 Table of Effect

Effects and interactions	Value
2	0.75
6	−1.75
7	−1.75
26	−1.25
27	43.75
67	1.25
267	0.75

$$E_1 = \frac{1}{8}[-y_1 + y_2 - y_3 + y_4 - y_5 + y_6 - y_7 + y_8]$$

A 2^3 experimental design has three main effects, three interactions of order 2, and one interaction of order 3. The values of the three main effects in our specific example are written **2**, **6**, and **7** (in bold), the three second-order interactions as **26**, **27**, and **67**, and the third-order interaction as **267**. Table 7 shows the results of the calculation.

There is just one significant interaction and the main effects of the three factors are negligible. This result is easy to interpret with the help of Figure 12. There is a large second-order interaction at the high level of factor 6 as in the second example. There is the same significant second-order interaction at the low level of factor 6. The resulting interaction is high.

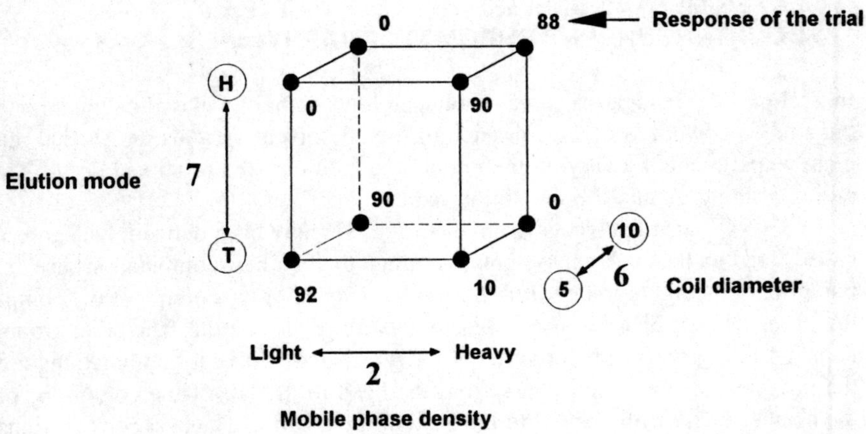

Figure 12 Third example: results of the retention of the stationary phase.

The main effect **2** (**6** or **7**) is small because the means of the responses are the same at the high and low levels of factor 2 (6 or 7). The difference is small.

Interaction **26** is small because the effect of factor 2 (or 6) is the same at high and low levels of factor 6 (or 2).

Interaction **267** is small because the interaction **26** (or 67 or **27**) at the low level of factor 7 (or 2 or 6) is opposite to interaction 26 (or **67** or **27**) at the high level of factor 7 (or 2 or 6).

It is therefore important to examine the combinations of levels. Retention is high when factor 2 is at a high level at the same time as factor 7:

2^+7^+

Retention is also high when factor 2 is at low level at the same time as factor 7:

2^-7^-

Hence, retention is high when the product of the levels of factors 2 and 7 is positive.

H. Conclusion

Principles of experimental design method applied to CCC have been presented using selected examples. This presentation will be useful to understand the second and third sections of this chapter. Real CCC problems involving many factors will be treated using 2^n experimental designs.

II. EXPERIMENTAL DESIGNS APPLIED TO TYPE J COUNTERCURRENT CHROMATOGRAPHY

In Section IV of Chapter 1 it was explained how Ito built a classification among the studied solvent systems. For each of the 15 solvent systems, he carried out many experiments by varying the pumping conditions, the rotational speed, the radius of the coil, and the revolution radius.

Two groups of solvent system have consequently been defined. One group, called ‘‘a,’’ gathers solvent systems for which the two best combinations are the pumping of the lighter phase from the tail to the head of the column or the pumping of the heavier phase from the head to the tail of the column. The other group, named ‘‘b,’’ gathers solvent systems for which the two best combinations are the pumping of the lighter phase from the head to the tail of the column or of the heavier phase from the tail to the head of the column. The best combinations are reversed between the two groups. However, there was a need to define a third group, ‘‘c,’’ to take into account the behavior of some solvent systems in which optimal combinations depend on the geometrical dimensions of the apparatus. The discriminating parameter was found to be the β ratio of the coil radius on the distance between the two axes of rotation (we refer the reader to Chapter 4 by Ito and Menet). For β values smaller than 0.3, solvent systems belonging to group c behave like the solvent systems of group b. On the contrary, for β values greater than 0.3, they behave like solvent systems of group a.

In order to more thoroughly study the effects of the factors and of their possible interactions, EDM appears as particularly suitable.

A. Two-Level Factorial Designs

Our simple approach was based on EDM with full factorial designs related to two-level factors.

1. Choice of Factors

Two factors were chosen because they are the fundamental factors of Ito's Classification. The first one, numbered **2**, is the choice of the mobile phase, i.e., heavier or lighter. The second one is numbered **7** and stands for the elution mode, i.e., from the head to the tail or from the tail to the head of the column. These numbers were chosen in order to remain consistent with the studies carried out on the cross-axis apparatus.

Other factors that were varied by Ito were not kept. They included the rotational speed, the flow rate of the mobile phase, the radius of the coil, and the revolution radius.

2. Choice of Response

Only one response was studied: the retention of the stationary phase, expressed as the percentage of the internal volume of the column, as it is characteristic of the behavior of a solvent system inside the column.

3. Experimental Domain

As each factor could take only two values, their levels are indicated through minus and plus signs. These low and high levels were chosen arbitrarily. The resulting experimental domain is gathered in Table 8.

Table 8 Experimental Domain

	Factor no.	Level−	Level+
Mobile phase density	**2**	Lighter	Heavier
Elution mode	**7**	Tail-to-head (T)	Head-to-tail (H)

4. Solvent Systems

The solvent systems are the 15 systems described in Chapter 1. Their names are given in Table 9.

5. Mathematical Model

The underlying hypothesis behind our study is the linear relationship between the factor and the response in the chosen experimental design. In our case, factors are Boolean and values between −1 and +1 are nonsense. However, the response may still be written as

$$y = a_0 + a_2x_2 + a_7x_7 + a_{27}x_2x_7 \tag{1}$$

where y is retention of the stationary phase
a_0 is the mean of the four measurements of the retention of the stationary phase
a_2 is the effect of factor **2**
a_7 is the effect of factor **7**
a_{27} is the interaction between factors **2** and **7**
x_2 stands for the level of factor **2** and
x_7 stands for the level of factor **7**

Table 9 The 15 Solvent Systems Studied

Solvent system	Volume ratio (v/v)	Number
Hexane/water	1:1	1
Ethyl acetate/water	1:1	2
Chloroform/water	1:1	3
Hexane/methanol	1:1	4
Ethyl acetate/acetic acid/water	4:1:4	5
Chloroform/acetic acid/water	2:2:1	6
n-Butanol/water	1:1	7
n-Butanol/0.1 M NaCl	1:1	7a
n-Butanol/1 M NaCl	1:1	7b
n-Butanol/acetic acid/water	4:1:5	8
n-Butanol/acetic acid/0.1 M NaCl	4:1:5	8a
n-Butanol/acetic acid/1 M NaCl	4:1:5	8b
sec-Butanol/water	1:1	9
sec-Butanol/0.1 M NaCl	1:1	9a
sec-Butanol/1 M NaCl	1:1	9b

These x values can only take -1 or $+1$ values. The concepts of effect and interaction have been previously defined in Chapter I.

In our case, the definition of the effects and their interactions is simple. The effect of the choice of the heavier or lighter phase is half the difference between the mean of the retention of the stationary phase when the mobile phase is the heavier one and the mean of the retention of the stationary phase when the mobile phase is the lighter one. In the same way, the effect of the elution mode is half the difference between the mean of the retention of the stationary phase when the mobile phase is pumped from the head to the tail of the column and the mean of the retention of the stationary phase when the mobile phase is pumped from the tail to the head of the column.

The interaction between both factors is defined as half the difference between the effect of the mobile phase when it is pumped from the head to the tail and the effect of the mobile phase when it is pumped from the tail to the head. It is also half the difference between the effect of the elution mode when the mobile phase is the heavier one and the effect of the elution mode when the mobile phase is the lighter one.

B. Results

The original results were all obtained by Ito and are described in Ref. 5. They were in fact given as plots of the retention of the stationary phase vs. the rotational speed [18]. Two revolution radii, i.e., $R = 10$ cm and $R = 20$ cm, along with three coil radii, i.e., $r = 2.5$ cm, $r = 5$ cm, and $r = 7.5$ cm, were used. Rotational speeds included 200, 400, 600, and 800 rpm for $R = 20$ cm, while they were equal to 400, 600, 800, and 1000 rpm for $R = 10$ cm. In order to compare the influence of the two revolution radii, we considered only the results obtained at 800 rpm.

Experimental design methodology applied to these results gave the effects of the choice of a heavier or lighter mobile phase and of the elution mode along with their interaction for each of the 15 solvent systems for both revolution radii and at 800 rpm. Tables 10 and 11 gather the computed results. The numbers were already given in Table 9.

A positive value for the effect of factor **2** means the use of factor **2** at its high level, i.e., a heavier mobile phase, increases the response, i.e., retention of the stationary phase. Interpretations are the same for the other factors and their interaction.

These tables allowed some observations. For all of the solvent systems, the interaction between the two factors has more influence than the effects themselves. Indeed, the influences of the choice of the mobile phase and of the pump-

Table 10 Values of the Effects and Interaction of Factors **2** and **7** for the 15 Solvent Systems Studied by Ito at $R = 20$ cm and $\omega = 800$ rpm

		Effect of the choice of the mobile phase: factor 2			Effect of the elution mode: factor 7			Interaction 27 between the two factors		
System	Group	$\beta = 0.125$	$\beta = 0.25$	$\beta = 0.375$	$\beta = 0.125$	$\beta = 0.25$	$\beta = 0.375$	$\beta = 0.125$	$\beta = 0.25$	$\beta = 0.375$
1	a	−2.75	−3.25	−1.25	−3.75	−3.25	−1.25	39.75	42.75	45.25
2	c	5.5	−10	0.75	−6.5	8	0.25	−11.5	29.5	40.75
3	a	−2	−2.5	0.25	−4	−1.5	−0.75	34	40.5	40.25
4	c	−0.75	−1	1	2.75	−2.5	−2.5	−6.25	30	35.5
5	a	2.75	2.5	−0.75	9.25	2.5	−0.75	22.75	41	41.25
6	c	−4.5	−5	4.5	7	1	2.5	−27	−14.5	25.5
7	c	−7	−5	2.75	8.5	7	−0.25	−31.5	−26	39.25
7a	c	−8.5	−4.25	1.25	10.5	4.25	2.75	−30.5	−21.75	37.75
7b	c	−7.75	1	2.5	10.25	−4.5	1	−29.25	−10	40.5
8	b	0.25	4.25	2	1.75	−2.75	1	−24.25	−36.25	−28.5
8a	b	−1	2.5	−1.5	3	−2.5	2	−29.5	−39.5	−28.5
8b	b	−1.25	−5.5	−2.25	2.25	5.5	−7.25	−37.25	−32.5	9.75
9	b	0.5	3.5	0.5	−0.5	−2.5	1.5	−18	−34	−29
9a	b	1.25	3.25	1.5	0.75	−3.25	1	−27.25	−38.75	−29.5
9b	b	−2.75	−5	−3.5	4.75	5	7	−34.75	−32	−10

Table 11 Values of the Effects and Interaction of Factors **2** and **7** for the 15 Solvent Systems Studied by Ito at R = 10 cm and ω = 800 rpm

		Effect of the choice of the mobile phase: factor 2			Effect of the elution mode: factor 7			Interaction 27 between the two factors		
System	Group	β = 0.25	β = 0.5	β = 0.75	β = 0.25	β = 0.5	β = 0.75	β = 0.25	β = 0.5	β = 0.75
1	a	2.00	0.50	−1.25	−0.50	−0.50	−1.25	35.50	43.50	43.75
2	c	14.50	−1.25	1.50	−6.00	0.75	−1.00	−9.50	26.25	33.50
3	a	4.25	0	0.75	−2.25	−2.50	0.25	36.75	40.50	36.75
4	c	8.25	3.25	−0.75	7.25	2.75	1.75	1.25	26.25	25.75
5	a	4.25	4.75	1.50	7.75	5.25	0	29.25	37.75	40.5
6	c	2.50	9.5	1.75	2.5	7	1.75	−12	30	34.75
7	c	6.25	3	4	−1.25	2.5	7.5	−3.25	36	26
7a	c	5.25	3	3.75	−0.75	1	5.25	−3.25	39	29.25
7b	c	7.75	3.75	2.75	−2.25	2.25	4.25	−5.75	38.25	31.75
8	b	6	4.75	9	1	−2.75	5	−32.5	−29.25	−9.5
8a	b	5	3.5	8.5	1.5	−0.5	14.5	−32	−20.5	1.5
8b	b	9	3.25	5.25	−3	2.75	8.75	−7.5	16.25	25.25
9	b	6.75	4.75	8.75	1.25	−2.75	2.75	−32.25	−25.75	−12.75
9a	b	5.50	7.5	5.75	−1.5	−4.5	6.25	−30	−26	−7.75
9b	b	5.5	7.75	2.5	−1	7.25	7.5	−7	12.25	25

ing mode are very low, in the order of 2–5%. For $R = 10$ cm, the value of the interaction between the choice of the mobile phase and the elution mode increases with β. The same rule applies in most cases with $R = 20$ cm.

A strong correlation between the sign of the interaction **27** and the classification built by Ito is observed. Indeed for $R = 10$ cm and $\beta = 0.25$, the values for group a are 35.5, 36.8, and 29.3 whereas they are −32.5, −32, −7.5, −32.3, −30, and −7 for group b.

Consequently, the behavior of solvent systems belonging to group a is characterized by a positive effect of the interaction between the choice of the mobile phase and the elution mode on the retention of the stationary phase. It means that the highest values of the retention of the stationary phase are obtained when the mobile phase is the heavier one (factor 2 at its high level) and is pumped from the head to the tail of the column (factor 7 at its high level). Solvent systems belonging to group b exhibit a negative effect of this interaction on the retention of the stationary phase, whereas systems belonging to group c gather very low values for this interaction either positive or negative. These rules are in fact the mathematical transcription of the recommended conditions given by Ito to obtain the highest retention of the stationary phase according to which group the solvent system belongs.

Interaction 27 was further calculated for 400 and 600 rpm and the values are gathered in Tables 12 and 13 for, respectively, 20 and 10 cm of the revolution radius. For all of the solvent systems, all β values, and both revolution radii, the rotational speed ω increases the absolute value of interaction **27**. At 800 rpm, the average value for interaction 27 is 40, which means than the suitable change of its level increases by 80% the retention of the stationary phase. For instance, a +40 value means interaction **27** should be set at its high level, which stands for a lighter mobile phase pumped from tail to the head of the column or a heavier mobile phase pumped from the head to the tail. This is a characteristic of group a.

C. Interpretations; Definition of β_{lim}

The behavior of solvent systems belonging to group c depends on the β value. According to Ito, a β_{lim} value can be defined: when the β of the apparatus is smaller than β_{lim}, the solvent system behaves as if it belongs to group b (interaction **27** negative), whereas a β higher than β_{lim} leads to a behavior of group a (interaction **27** positive).

This β_{lim} was observed to be within a 0.25–0.3 range. Using EDM, β_{lim} value can be obtained when the plot of interaction **27** vs. β crosses the 0 horizontal axis. Such a methodology was applied to all of the solvent systems studied by Ito for three rotational speeds, two revolution radii, and the three resulting β values. Tables 14–28 list the computed values of interaction **27**.

Table 12 Values of Interaction **27** for 15 Solvent Systems Studied by Ito at R = 10 cm and ω = 400, 600, and 800 rpm R = 20 cm

		Interaction **27**								
		β = 0.125			β = 0.25			β = 0.375		
System	Group	400 rpm	600 rpm	800 rpm	400 rpm	600 rpm	800 rpm	400 rpm	600 rpm	800 rpm
1	a	32.5	38.75	39.75	35.75	41.5	42.75	36.75	42	45.25
2	c	−2.5	−12	−11.5	26.5	29.75	29.5	38	40.75	40.75
3	a	22.25	30.5	34	32.5	38	40.5	35.75	39.5	40.25
4	c	0.25	0.75	−6.25	11	27.75	30	28.75	33.25	35.5
5	a	−10.25	16.75	22.75	31.75	36.25	41	32.5	39	41.25
6	c	−12.75	−23.75	−27	−6	−4.5	−14.5	27.75	26.5	25.5
7	c	−6.25	−11	−31.5	7.75	−5.25	−26	27.75	36.75	39.25
7a	c	−5.5	−11.5	−30.5	10.75	9.25	−21.75	30.75	37	37.75
7b	c	−3.75	−12.5	−29.25	12.75	17.25	−10	34.5	39.5	40.5
8	b	−21.5	−24.25	−24.25	−27	−33.5	−36.25	−11.25	−27	−28.5
8a	b	−21.75	−28.75	−29.5	−13.75	−33.75	−39.5	−11.5	−24.5	−28.5
8b	b	−6.75	−29.25	−37.25	−3.25	−20.25	−32.5	6.75	12.5	9.75
9	b	−17	−19.25	−18	−25.25	−31.75	−34	−13.75	−26.25	−29
9a	b	−23.25	−25.5	−27.25	−25.75	−35	−38.75	−14.25	−28	−29.5
9b	b	−7.75	−29.25	−34.75	−2.5	−21.75	−32	9.25	3.25	−10

Table 13 Values of Interaction **27** for 15 Solvent Systems Studied by Ito at $R = 10$ cm and $\omega = 400$, 600, and 800 rpm $R = 10$ cm

		Interaction **27**								
		β = 0.25			β = 0.5			β = 0.75		
System	Group	400 rpm	600 rpm	800 rpm	400 rpm	600 rpm	800 rpm	400 rpm	600 rpm	800 rpm
1	a	7.25	31.25	35.5	31.75	42	43.5	31.5	40.75	43.75
2	c	−0.75	−3.75	−9.5	6.75	19.25	26.25	21.75	30.25	33.5
3	a	9.25	31.5	36.75	29	38.25	40.5	21	33.75	36.75
4	c	3.75	4.5	1.25	7.25	23.75	26.25	−3	6	25.75
5	a	5	23	29.25	26	34.75	37.75	29	37.5	40.5
6	c	−7.5	−9.75	−12	21.75	28.25	30	21.75	30.75	34.75
7	c	18	1.75	−3.25	14	24.5	36	0.25	18	26
7a	c	17.75	4	−3.25	25.5	33	39	1.75	17	29.25
7b	c	25	15	−5.75	29	34.75	38.25	12.5	24.25	31.75
8	b	−17	−29	−32.5	−4.75	−11.25	−29.25	−5.25	−9.25	−9.5
8a	b	−8.25	−29.25	−32	−3.25	−10.5	−20.5	−7.5	−5.5	1.5
8b	b	3.5	−4.75	−7.5	5.75	5.5	16.25	−1.5	16.5	25.25
9	b	−10.25	−26.75	−32.25	−7	−14	−25.75	−6.75	−12.25	−12.75
9a	b	−4.25	−27.25	−30	−8	−11.75	−26	−3.25	−12	−7.75
9b	b	3.75	−2	−7	3.75	8.75	12.25	−0.75	15	25

Table 14 Solvent System 1

Revolution radius (cm)	Resulting β	Rotational speed (rpm) 400	600	800
10	0.25	7.25	31.25	35.5
	0.5	31.75	42	43.5
	0.75	31.5	40.75	43.75
20	0.125	32.5	38.75	39.75
	0.25	35.75	41.5	42.75
	0.375	36.75	42	45.25

Table 15 Solvent System 2

Revolution radius (cm)	Resulting β	Rotational speed (rpm) 400	600	800
10	0.25	−0.75	−3.75	−9.5
	0.5	6.75	19.25	26.25
	0.75	21.75	30.25	33.5
20	0.125	−2.5	−12	−11.5
	0.25	26.5	29.75	29.5
	0.375	38	40.75	40.75

Table 16 Solvent System 3

Revolution radius (cm)	Resulting β	Rotational speed (rpm) 400	600	800
10	0.25	9.25	31.5	36.75
	0.5	29	38.25	40.5
	0.75	21	33.75	36.75
20	0.125	22.25	30.5	34
	0.25	32.5	38	40.5
	0.375	35.75	39.5	40.25

Table 17 Solvent System 4

Revolution radius (cm)	Resulting β	Rotational speed (rpm) 400	600	800
10	0.25	3.75	4.5	1.25
	0.5	7.25	23.75	26.25
	0.75	−3	6	25.75
20	0.125	0.25	0.75	−6.25
	0.25	11	27.75	30
	0.375	28.75	33.25	35.5

Table 18 Solvent System 5

Revolution radius (cm)	Resulting β	Rotational speed (rpm) 400	600	800
10	0.25	5	23	29.25
	0.5	34.75	37.75	38.25
	0.75	37.5	40.5	40.5
20	0.125	−10.25	16.75	22.75
	0.25	31.75	36.25	41
	0.375	32.5	39	41.25

Table 19 Solvent System 6

Revolution radius (cm)	Resulting β	Rotational speed (rpm) 400	600	800
10	0.25	−7.5	−9.75	−12
	0.5	21.75	28.25	30
	0.75	21.75	30.75	34.75
20	0.125	−12.75	−23.75	−27
	0.25	−6	−4.5	−14.5
	0.375	27.75	26.5	25.5

Table 20 Solvent System 7

Revolution radius (cm)	Resulting β	Rotational speed (rpm) 400	600	800
10	0.25	18	1.75	−3.25
	0.5	14	24.5	36
	0.75	0.25	18	26
20	0.125	−6.25	−11	−31.5
	0.25	7.75	−5.25	−26
	0.375	27.75	36.75	39.25

Table 21 Solvent System 7a

Revolution radius (cm)	Resulting β	Rotational speed (rpm) 400	600	800
10	0.25	17.75	4	−3.25
	0.5	25.5	33	39
	0.75	1.75	17	29.25
20	0.125	−5.5	−11.5	−30.5
	0.25	7.25	10.75	9.25
	0.375	9.5	30.75	37

Table 22 Solvent System 7b

Revolution radius (cm)	Resulting β	Rotational speed (rpm) 400	600	800
10	0.25	25	15	−5.75
	0.5	29	34.75	38.25
	0.75	12.5	24.25	31.75
20	0.125	−3.75	−12.5	−29.25
	0.25	12.75	17.25	−10
	0.375	34.5	39.5	40.5

Table 23 Solvent System 8

Revolution radius (cm)	Resulting β	Rotational speed (rpm) 400	600	800
10	0.25	−17	−29	−32.5
	0.5	−4.75	−11.25	−29.25
	0.75	−5.25	−9.25	−9.5
20	0.125	−21.5	−24.25	−24.25
	0.25	−27	−33.5	−36.25
	0.375	−11.25	−27	−28.5

Table 24 Solvent System 8a

Revolution radius (cm)	Resulting β	Rotational speed (rpm) 400	600	800
10	0.25	−8.25	−29.25	−32
	0.5	−3.25	−10.5	−20.5
	0.75	−7.5	−5.5	1.5
20	0.125	−21.75	−28.75	−29.5
	0.25	−13.75	−33.75	−39.5
	0.375	−11.5	−24.5	−28.5

Table 25 Solvent System 8b

Revolution radius (cm)	Resulting β	Rotational speed (rpm) 400	600	800
10	0.25	3.5	−4.75	−7.5
	0.5	5.75	5.5	16.25
	0.75	−1.5	16.5	25.25
20	0.125	−6.75	−29.25	−37.25
	0.25	−3.25	−20.25	−32.5
	0.375	6.75	12.5	9.75

Table 26 Solvent System 9

Revolution radius (cm)	Resulting β	Rotational speed (rpm) 400	600	800
10	0.25	−10.25	−26.75	−32.25
	0.5	−7	−14	−25.75
	0.75	−6.75	−12.25	−12.75
20	0.125	−17	−19.25	−18
	0.25	−25.25	−31.75	−34
	0.375	−13.75	−26.25	−29

Table 27 Solvent System 9a

Revolution radius (cm)	Resulting β	Rotational speed (rpm) 400	600	800
10	0.25	−4.25	−27.25	−30
	0.5	−8	−11.75	−26
	0.75	−3.25	−12	−7.75
20	0.125	−23.25	−25.5	−27.25
	0.25	−25.75	−35	−38.75
	0.375	−14.25	−28	−29.5

Table 28 Solvent System 9b

Revolution radius (cm)	Resulting β	Rotational speed (rpm) 400	600	800
10	0.25	3.75	−2	−7
	0.5	3.75	8.75	12.25
	0.75	−0.75	15	25
20	0.125	−7.75	−29.25	−34.75
	0.25	−2.5	−21.75	−32
	0.375	9.25	3.25	−10

Using a simplified linear approximation between experimental β values, the intersection of this line with the horizontal axis (related to a null **27** interaction) gives a β_{lim} value. These computed data are gathered in Table 29.

Solvent systems belonging to group c, i.e., numbers 2, 4, 6, 7, 7a, and 7b, lead to β_{lim} values between 0.15 and 0.32 for 800 rpm. Such values are in agreement with the range of values given by Ito to these solvent systems. Surprisingly, systems from group a also show a β_{lim}, which can, however, be determined only at low rotational speeds. The approximate value is 0.2. Systems belonging to group b lead to values either close to 0.3 or close to 1.

The previous results of all solvent systems, whichever the group to which they belong, can be characterized by a β_{lim} value. For β values higher than this limit value, solvent systems behave as systems from group a: the lighter mobile phase has to be pumped from the tail to the head of the column, or the heavier mobile phase has to pumped from the head to the tail. On the contrary, β values smaller than β_{lim} require the heavier mobile phase to be pumped from the tail to the head of the column, or the lighter mobile phase to be pumped from the head to the tail, as for systems belonging to group b.

Figure 13 shows the relative average position of β_{lim} along the β axis compared to the β of the CCC apparatus for the three groups of solvent systems.

When the β of the column is smaller than the β_{lim} of the solvent system, the latter behaves like a system belonging to group b. On the contrary, when the β of the column is greater than the β_{lim} of the solvent system, then the latter behaves as if it belongs to group a.

Usual CCC devices are based on geometrical dimensions that lead to 0.4–0.7 β values. It explains why the reversal of behavior related to a change in β was observed only to solvent systems belonging to group c, which are characterized by 0.3 β_{lim} value. Such a reversal is expected for groups a and b, but they need a 0.1 β_{lim} or a 0.8 β_{lim}, respectively, which are not common values for CCC devices.

All of these studies have finally shown that the solvent systems can be classified within three groups for a given range of experimental conditions. They include a 0.125 to 0.75 β range, 10–20 cm for the revolution radius, 400–800 rpm rotational speed, and room temperature.

From a practical point of view, the knowledge of β_{lim} helps in two ways. First, it allows one to decide which pumping conditions will lead to satisfactory retention of the stationary phase, when the β of the column is known. Second, it allows one to decide if an increase of the rotation speed will or will not increase the retention of the stationary phase to a significant extent; the larger the difference between β_{lim} and β of the column and the more important the positive influence of the rotation speed on stationary phase retention.

Table 29 Values of β_{lim} for 15 Solvent Systems Studied by Ito, vs. Revolution Radius R and Rotational Speed ω

R (cm)	ω (rpm)	Solvent system 1	3	5	2	4	6	7	7a	7b	8	8a	8b	9	9a	9b
10	400	0.18	0.14	0.19	0.27	—	0.32	—	—	—	—	—	—	—	—	—
	600	—	—	—	0.29	0.19	0.32	0.23	0.21	0.06	—	1.02	0.37	—	—	0.30
	800	—	—	—	0.32	0.24	0.32	0.27	0.27	0.28	0.87	0.73	0.33	—	0.86	0.34
20	400	—	—	0.16	0.27	0.12	0.27	0.18	0.17	0.15	—	—	0.29	—	—	0.28
	600	—	—	—	0.27	0.12	0.27	0.27	0.19	0.18	—	—	0.33	—	—	0.38
	800	—	—	—	0.30	0.15	0.30	0.30	0.30	0.28	—	—	0.35	—	—	0.43
Group			a					c						b		

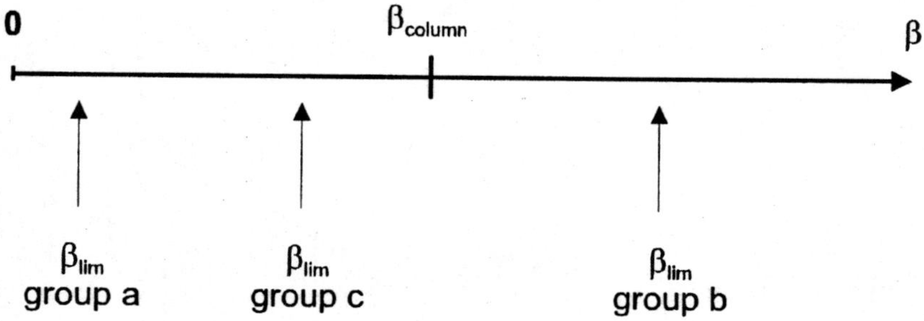

Figure 13 Average β_{lim} for each of the three solvent systems, relative to the β of a given apparatus.

III. EXPERIMENTAL DESIGNS APPLIED TO CROSS-AXIS COUNTERCURRENT CHROMATOGRAPHY

A. Objectives

Interpreting the values of the retention of the stationary phase obtained from a cross-axis apparatus raised an interesting problem. As many parameters may influence the retention of the stationary phase, a specific approach was needed to interpret the results when varying these parameters.

The first approach was developed by Ito et al. In front of many running parameters (or factors), he used a graphical method to determine which factors were influent. The method consisted of plotting the retention of the stationary phase as one parameter is changed between the horizontal and vertical axes. It was applied to the fourth prototype of cross-axis device (type X-LL) [19], to the fifth one (type X-LLL) [20], and to the sixth one (types X-1.5L and L) [21]. Solvent systems used with the fourth prototype included hexane/water, hexane/methanol, hexane/ethylacetate/methanol/water (1:1:1:1, v/v/v/v), ethyl acetate/water, ethyl acetate/acetic acid/water (4:1:5, v/v/v), chloroform/water, chloroform/acetic acid/water (2:2:1, v/v/v), *n*-butanol/water, *n*-butanol/acetic acid/water (4:1:5, v/v/v), and *sec*-butanol/water. Only one solvent based on an aqueous polymer mixture polyethylene glycol (PEG)–8000 4.4% (w/w)/dextran T500 7.0% (w/w) was used with the sixth prototype. Four solvent systems were finally used with the sixth prototype—*n*-butanol/0.13 M NaCl (1:1, v/v) + 1.5% (w/w) hexadecyl pyridinium chloride—and three polymer-based aqueous solvent systems—PEG-1000/K_2HPO_4, PEG-8000 4.4% (w/w)/dextran T500 7.0% (w/w) and PEG-8000 4.0% (w/w)/dextran T500 5.0% (w/w). The conclusions of his studies were that (1) the stationary phase is better retained with a shifted column relative to the central axis of rotation and (2) that the lighter mobile phase had to be pumped from the outside of the column toward its inside (closer to the central axis of rotation) and the heavier mobile phase had to be pumped from the inside of the column toward its outside (closer to the central axis of rotation). However, such an interpretation proved to be not accurate. Therefore, Ito et al. were not satisfied with the original conclusions. The drawbacks of such an approach were that many experiments were needed to detect influent factors and that only important effects could be visualized. Consequently, EDM has emerged as the choice method to be used in this case. It was applied in two steps.

First, all results obtained by Ito were analyzed through EDM calculations. The calculation results and the resulting choice of the factors were published by Goupy et al. [22]. They showed that the choice of optimal conditions for the retention of the stationary phase relies heavily on interactions between running parameters. Consequently, EDM gives simple choices that are always the best settings because it takes into account all the interactions.

The second step was to completely design the experiments according to EDM in order to fully use the power of the experimental designs. The following paragraphs will describe the implementation of this methodology, giving the experimental results, the computed values of the main effects and their interactions, and precise optimal conditions. Some interpretations will also be proposed.

B. "Cross-Axis" Apparatus

The characteristic features of this planetary centrifuge system are described in Chapter 4. More details about our apparatus are given in the following paragraphs.

1. Our Apparatus

The latest cross-axis apparatus (X-1.5L and L types) has been commercialized by Countercurrent Technologies, Inc. (see the list of manufacturers of CCC devices at the end of the book); it is similar to the sixth prototype used by Ito in Ref. 21. It was delivered in August 1993 to the Laboratoire de Chimie Analytique de l'ESPCI. All of its parts are made of stainless steel except the gears and the cylindrical holders, which are in Delrin. Its external dimensions are 60 × 60 × 35 cm. The two columns, made of several layers of Teflon tube wound on a cylindrical holder, are mounted in series. The internal diameter of the tube used for the columns is 2.6 mm. The rotational speed is regulated up to 1000 rpm, and the speeds were calibrated using a stroboscope at 200, 400, 600, 800, and 1000 rpm. The configurations of the stationary and planetary miter gears and the pulleys and belts force the column to rotate around its own axis at the same speed as its revolution speed around the central axis. Two counteraxes, whose rotation is attributable to the plastic gears, are required to prevent the twisting of the inlet and outlet Teflon tubes. The prototype was built to allow the counteraxis to be exchanged with the column holder. In that case, the type would be L.

2. Running Parameters

The cross-axis CPC involves the highest number of running parameters (named factors), compared to other CCC devices. They have been classified in two groups and are described hereafter.

Mechanical Characteristics

Many running parameters are related to mechanical characteristics that are set before any experiment. The six factors are specific to CCC and only one, which is the position of the column, is characteristic of a cross-axis chromatograph. All of them are related to the coil, which is made of a Teflon tubing wound around a cylinder. The name column relates to one or two coils in series.

Coil radius (*r*)
Internal diameter of the tubing (usually 0.8, 1.6, or 2.6 mm i.d. Teflon tubing)
Winding direction of the tubing on the cylindrical holder (right- or left-handed)
Column type (I or II)
Number of tubing layers
Column holder position (for our apparatus L or X-1.5L, as indicated in Figure 14)

Choices Left to the User

The six parameters are related to the design of the separation carried out by the experimenter. Their characteristics are given below:

Choice of mobile phase (heavier or lighter liquid phase from the solvent system)
Flow rate of the mobile phase
Rotation direction (clockwise or anticlockwise around the central vertical axis)
Rotational speed (around the central vertical axis, commonly from 400 to 1000 rpm)
Pumping mode (from the head to the tail or from the tail to the head of the column)
Pumping direction (inward, opposed to the direction of the centrifugal force, or outward, same direction as the centrifugal force, as indicated in Figure 15)

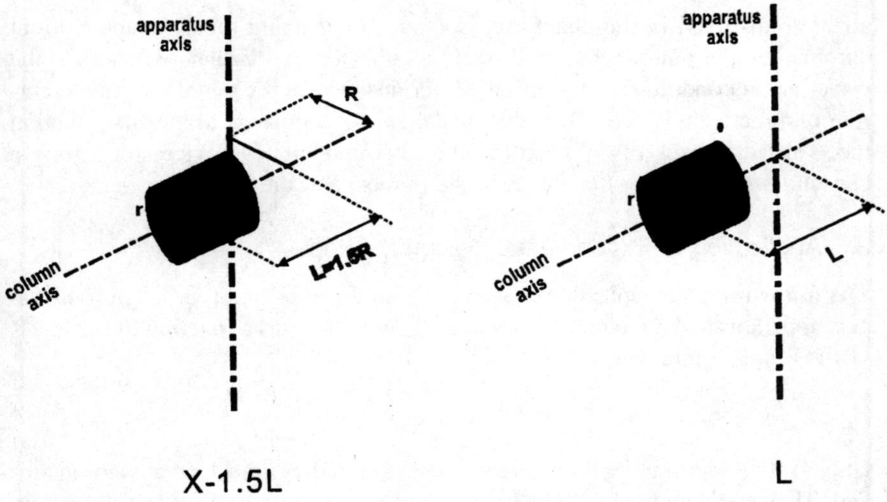

Figure 14 Positions X-1.5L and L.

Note: The temperature represents a major but implicit parameter of each CCC experiment. To ensure the reliability of an optimized purification or separation, the temperature has to be precisely set. Two devices may be used for that purpose. The best one is a thermoregulated room, in which the whole CCC chain is installed. The main advantage of such a device lies in the thermoregulation of

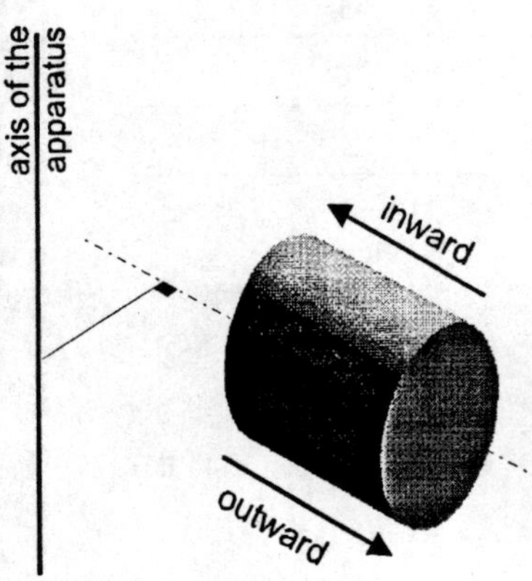

Figure 15 Definition of the pumping directions.

all units involved in the chain, i.e., solvents, connecting tubes, countercurrent chromatograph, pump heads, and fraction collector. A "lighter" solution is the use of an air conditioning unit installed, for instance, at the top of the countercurrent chromatograph. However, the smaller price compared to the first solution meets the drawback of the need of other thermoregulated devices for setting a chosen temperature for the solvents, the pumps, and the fraction collector.

3. Application of Experimental Design Method

The following paragraphs thoroughly describe the application of EDM to cross-axis apparatus and the retention of the stationary phase as a function of the levels of the running parameters.

Choice of Factors

Only 10 factors (running parameters) were studied from the 12 that were identified. They are gathered in Table 30, which indicates their numbering used during the study. Some factors are named according to Ito's names, e.g., P_{I} and P_{II} for the direction of rotation. The low and high levels of the factors were carefully chosen. For example, the lower rotational speed is 400 rpm, as 200 rpm would have led to too low retention of the stationary phase and 1000 rpm would have increased the wear of the rotating parts and of the connecting tubes. In the same way, the low 7°C temperature is the temperature that was measured when the apparatus was installed in a cold room at 5°C and the high level at 22°C was the measured running temperature without temperature control.

Table 30 Definition and Numbering of the 10 Studied Factors[a]

		Level	
Factor	No.	−	+
Coiling up	**1**	Right	Left
Mobile phase	**2**	Lighter	Heavier
Elution direction	**3**	I ("inward")	O ("outward")
Direction of rotation	**4**	P_{II} (clockwise)	P_{I} (anticlockwise)
Coil position	**5**	L	X-1.5L
Coil diameter	**6**	5 cm	7.5 cm
Elution mode	**7**	T ("tail-to-head")	H ("head-to-tail")
Number of layers	**8**	1	6
Rotation speed	**9**	400 rpm	800 rpm
Temperature	**10**	7°C	22°C

[a] Definition of the experimental domain.

Choice of Response

The same response as type J apparatus was chosen, i.e., the retention of the stationary phase, which is calculated in percent as the ratio of the volume of the stationary phase inside the column on the volume of the column.

The leak of the stationary phase is also indicated in order to eliminate any condition that would lead to such leaks with a satisfactory initial retention of the stationary phase.

Experimental Domain

The experimental domain is a 10-dimension space, whose limits are given by the low and high levels of the 10 factors. Factors **1**, **2**, **3**, **4**, and **7** are Boolean whereas the five others are continuous. The defined experimental domain applies only to a single solvent system.

The optimization for the best retention of the stationary phase has to take into account the 10 previously defined running parameters (factors). As each factor shows two levels, the number of experiments would be $2^{10} = 1024$! This represents a challenge for the EDM, which is expected to reduce the number of experiments to carry out and to help in interpreting the measurement results.

Chosen Solvent Systems

Only three solvent systems were studied, each belonging to a group defined by Ito on a type J device (see Chapter 2). The heptane/water system belongs to group a, "hydrophobic," *n*-butanol/water to group c, and *sec*-butanol/water to group b. These systems have quite distinct physical properties and hydrodynamic behaviors, so that any potential classification existing on the cross-axis device could be underlined by EDM results.

Original Features of the Apparatus

All the previous experiments carried out on a cross-axis apparatus have used a direct belt transmission from the base gear to the counteraxis gear, as shown on the left in Figure 16. Therefore, the direction of rotation of the central axis imposed that of the coil. However, the knowledge of the coiling up (factor **1**) and of the direction of rotation (factor **4**) determines the head and the tail of the column (factor **7**). Consequently, setting the elution mode (factor **7**) determines the global motion of the mobile phase along the coil axis; the elution direction (factor **3**) is then fixed. Taking into account the levels of the previous factors, the use of a belt transmission with the counteraxis leads to the aliase relation: **7** = **−134**. The existence of such an aliase complicates the interpretation of the results of EDM, as the value for the effect of a factor also includes other interactions.

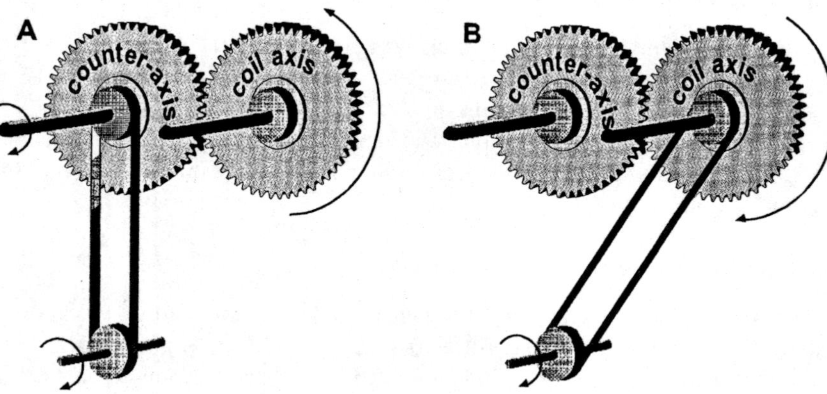

Figure 16 Principle of the belt transmission of the coil or the counteraxis; the coil is shown in the X-1.5L position.

We consequently modified the arrangement of the belts and of the connecting tubes to allow a direct transmission between the base gear and the coil itself, as shown at the right in Figure 16. In this case, the aliase can be written: **7** = **+134**. Use of both transmissions eliminates this aliase and eases the interpretations.

Results

Ten series of experiments were carried out with the three solvent systems. Their results are gathered in Tables 31–40. The values that are not underlined were read after the elution of a given volume of the mobile phase, which is indicated in the legend of the table. Underlined values were obtained just at the hydrodynamic equilibrium. When they are equal to the previous nonunderlined values, they are not reported. On the contrary, they are given and they reveal a leak of stationary phase.

Butanol-2/water system was used for more experiments than the two other solvent systems because the cross-axis apparatus permits high values of the retention of the stationary phase with important flow rates up to 2 mL/min. On the contrary, the Sanki apparatus (precisely described in Chapter 4) shows satisfactory retentions but the pressure drop limits the flow rates and type J apparatus leads to low values of retention.

Analysis

Experimental design methodology is the choice method when facing these 292 experiments. The results of its straight application to the 10 series have helped us in defining our strategy, by dividing original designs to study specific factors.

Table 31 Results of Series 1: System: Butanol-1/Water; Temperature: 22°C; 4 mL/min; 400 rpm; Type Ib Coil; Internal Volume: 625 mL; Coil Diameter: 7.5 cm; Left Coiling Up; i.d. = 2.6 mm; 6 Layers; Measurement of *Sf* After 500 mL Eluted

Position of coil	Phase mobile	Driven coil		Driven counteraxis	
		Conditions	*Sf* (%)	Conditions	*Sf* (%)
L	Lighter	P_I-T-I	36	P_I-H-I	39
		P_{II}-H-I	36	P_{II}-T-I	38
		P_{II}-T-O	14	P_{II}-H-O	17
		P_I-H-O	10	P_I-T-O	11
	Heavier	P_I-H-O	38	P_{II}-H-O	42
		P_{II}-T-O	35	P_I-T-O	38
		P_{II}-H-I	10	P_{II}-T-I	11
		P_I-T-I	7	P_I-H-I	10
X-1, 5L	Lighter	P_I-T-I	43	P_I-H-I	41
		P_I-H-O	33	P_{II}-T-I	31
		P_{II}-H-I	33	P_{II}-H-O	28
		P_{II}-T-O	7	P_I-T-O	8
	Heavier	P_I-H-O	52	P_{II}-H-O	42
		P_{II}-T-O	38	P_I-H-I	41
		P_{II}-H-I	35	P_I-T-O	27
		P_I-T-I	5	P_{II}-T-I	6

For each of the five series, five factors were always varied. They included the mobile phase (factor **2**), elution direction (factor **3**), direction of rotation (factor **4**), coil position (factor **5**), and elution mode (factor **7**). Then adequate factors were varied to obtain the various series; these are gathered in Table 41.

Computed results with JMP software from SAS Institute are gathered in Table 42. Effects lower than 3% were considered as not influent and therefore eliminated, and a 0 was then indicated in the table. Five factors (**2**, **3**, **4**, **5**, and **7**) were studied for a heptane/water system, leading to series 6. The same factors were also studied for series 2 with *n*-butanol/water and a 1.85-mm-i.d. tubing. Series 1 and 3 were different based on the number of layers. The coil diameter and the rotation speed were added factors for a *sec*-butanol/water system, so that four series were gathered with the introduction of factors 6 and 9.

Two points of view were then applied to interpret numerical results. The first is based on the analysis of the linear model between the retention of the stationary phase, *Sf*, and the studied factors, for each column of Table 42, i.e., for a given solvent system. The second is based on the analysis of the influence of selected factors.

Table 32 Results of Series 2: System: Butanol-1/Water; Temperature: 22°C; 4 mL/min; 800 rpm; Internal Volume, 21.9 mL; Coil Diameter, 7.5 cm; Right Coiling Up; i.d. = 1.85 mm; 1 Layer; Measurement of *Sf* After 30 mL Eluted

Position of Coil	Phase mobile	Driven coil		Driven counteraxis	
		Conditions	*Sf* (%)	Conditions	*Sf* (%)
L	Lighter	P_{II}-T-I	25	P_{II}-H-I	27
		P_I-H-I	18	P_I-T-I	21
		P_I-T-O	9	P_{II}-T-O	5
		P_{II}-H-O	7	P_I-H-O	2
	Heavier	P_{II}-H-O	39	P_I-H-O	34
		P_I-T-O	23	P_{II}-T-O	27
		P_I-H-I	16	P_{II}-H-I	11
		P_{II}-T-I	7	P_I-T-I	9
X-1, 5L	Lighter	P_I-H-I	41	P_I-T-I	43
		P_{II}-H-O	24	P_{II}-H-I	34
		P_{II}-T-I	23	P_I-H-O	27
		P_I-T-O	7	P_{II}-T-O	5
	Heavier	P_{II}-H-O	48	P_I-H-O	55
		P_I-H-I	47	P_{II}-H-I	37
		P_I-T-O	29	P_{II}-T-O	32
		P_{II}-T-I	7	P_I-T-I	0

Linear Model for Each Solvent System. Only influent factors are given in the following linear equations.

(1) Heptane/water. The results of series 6 lead to:

$$Sf = 51 - 9.7x_2 + 19.5x_2x_3 - 6.6x_2x_5 + 10.9x_2x_7 - 13.9x_2x_3x_5 + 11.2x_2x_5x_7$$

This model reveals that the direction of rotation (**4**) is influent neither by its direct effect nor by its interactions. Only the mobile phase is influent by its direct effect; other factors intervene as their interactions. The interaction between the mobile phase and the elution direction has an important effect on the retention of the stationary phase. In order to increase *Sf*, the heavier mobile phase should be pumped in the outward direction or the lighter mobile phase in the inward direction. In order to better interpret this model, the equation can be rewritten as:

$$Sf = 51 - x_2(9.7 + 6.6x_5) + x_2x_3(19.5 - 13.9x_5) + x_2x_7(10.9 + 11.2x_5)$$

Table 33 Results of Series 3: System: Butanol-1/Water; Temperature: 22°C; 4 mL/min; 400 rpm; Internal Volume, 43 mL; Coil Diameter, 7.5 cm; Left Coiling Up; i.d. = 2.6 mm; 1 Layer; Measurement of *Sf* After 50 mL Eluted

Position of coil	Phase mobile	Driven coil		Driven counteraxis	
		Conditions	*Sf* (%)	Conditions	*Sf* (%)
L	Lighter	P_I-T-I	37	P_{II}-T-I	42
		P_{II}-H-I	35	P_I-H-I	35
		P_I-H-O	8	P_{II}-H-O	7
		P_{II}-T-O	5	P_I-T-O	5
	Heavier	P_I-H-O	47	P_{II}-H-O	49
		P_{II}-T-O	44	P_I-T-O	42
		P_I-T-I	7	P_{II}-T-I	7
		P_{II}-H-I	5	P_I-H-I	7
X-1, 5L	Lighter	P_I-T-I	44	P_I-H-I	42
		P_{II}-H-I	36	P_{II}-H-O	28
		P_I-H-O	31	P_{II}-T-I	26
		P_{II}-T-O	5	P_I-T-O	6
	Heavier	P_I-H-O	51	P_{II}-H-O	45
		P_{II}-H-I	37	P_I-T-O	43
		P_{II}-T-O	35	P_I-H-I	40
		P_I-T-I	9	P_{II}-T-I	9

which shows the role of the position of the column on the influence of the mobile phase (**2**) and its interactions with the elution direction (**3**) and the elution mode (**7**).

A column in the L position ($x_5 = -1$) leads to

$$Sf = 51 - 3.1x_2 + 33.4x_2x_3$$

while a column in the X-1.5L position ($x_5 = +1$) leads to

$$Sf = 51 + 16.3x_2 + 5.6x_2x_3 + 22.1x_2x_7$$

It shows that the L position increases the influence of interaction **23** and decreases that of interaction **27**, while an X-1.5L position has the reverse effect. This model reveals that the interaction between the mobile phase and the elution mode (**27**), which is the basis of the behavior of solvent systems in a type J apparatus, is observed in the X-1.5L position. However, an important interaction between the mobile phase and the elution direction (**23**) is observed in the L position, but also to a lesser extent in the X-1.5L position.

Table 34 Results of Series 4: System: Butanol-2/Water; Temperature: 22°C; 4 mL/min; 400 rpm; Internal Volume, 43 mL; Coil Diameter, 7.5 cm; Left Coiling Up; i.d. = 2.6 mm; 1 Layer; Measurement of *Sf* After 50 mL Eluted

Position of coil	Phase mobile	Driven coil		Driven counteraxis	
		Conditions	*Sf* (%)	Conditions	*Sf* (%)
L	Lighter	P_I-T-I	31/40[a]	P_{II}-T-I	25/31
		P_{II}-H-I	21	P_I-H-I	23/26
		P_I-H-O	7	P_{II}-H-O	11
		P_{II}-T-O	4	P_I-T-O	7
	Heavier	P_I-H-O	28/34	P_{II}-H-O	32/37
		P_{II}-T-O	23/54	P_I-T-O	23/30
		P_I-T-I	12	P_{II}-T-I	12
		P_{II}-H-I	7	P_I-H-I	8
X-1,5L	Lighter	P_{II}-H-I	37	P_I-H-I	37
		P_I-T-I	23/42	P_{II}-H-O	26
		P_I-H-O	14	P_{II}-T-I	14/47
		P_{II}-T-O	2	P_I-T-O	5
	Heavier	P_I-H-O	44	P_{II}-H-O	33
		P_{II}-H-I	23/26	P_I-H-I	33
		P_{II}-T-O	23/54	P_I-T-O	11/30
		P_I-T-I	6	P_{II}-T-I	9/29

[a] Underscored values were obtained just at the hydrodynamic equilibrium.

(2) Butanol-1/water. The results of series 2 lead to

$$Sf = 23.1 + 3.2x_2 + 5.6x_5 + 6.1x_7 + 9.3x_2x_3 + 3.5x_2x_7 + 4.3x_5x_7$$

This can be rewritten:

$$Sf = 23.1 + 3.2x_2 + 5.6x_5 + x_7\,(6.1 + 4.3x_5) + 9.3x_2x_3 + 3.5x_2x_7$$

This model reveals that the direction of rotation (**4**) is influent neither by its direct effect nor by its interactions. The mobile phase (**2**), the position of the column (**5**), and the elution mode (**7**) directly intervene. The interaction between the mobile phase and the elution direction (**23**) has an important effect on the retention of the stationary phase. In order to increase *Sf*, the heavier mobile phase should be pumped in the outward direction or the lighter mobile phase in the inward direction. These conditions are similar to those for the heptane/water system. The interaction between the mobile phase and the elution mode (**27**) has a lower influence than that of its interaction with the elution direction (**23**). The positive interaction between the elution mode and the coil position (**57**) indicates that an X-1.5L position increases the beneficial influence of the elution mode on the retention of the stationary phase.

Table 35 Results of Series 5: System: Butanol-2/Water; temperature: 22°C; 4 mL/min; 800 rpm; Internal Volume, 43 mL; Coil Diameter, 7.5 cm; Left Coiling Up; i.d. = 2.6 mm; 1 Layer; Measurement of *Sf* After 50 mL Eluted

Position of coil	Phase mobile	Driven coil		Driven counteraxis	
		Conditions	*Sf* (%)	Conditions	*Sf* (%)
L	Lighter	P_{I}-T-I	51	P_{II}-T-I	54
		P_{II}-H-I	51	P_{I}-H-I	44
		P_{I}-H-O	16	P_{II}-H-O	11
		P_{II}-T-O	9	P_{I}-T-O	9
	Heavier	P_{I}-H-O	61	P_{II}-H-O	56
		P_{II}-T-O	55	P_{I}-T-O	52
		P_{I}-T-I	12	P_{II}-T-I	12
		P_{II}-H-I	8	P_{I}-H-I	7
X-1, 5L	Lighter	P_{II}-H-I	47	P_{I}-H-I	35
		P_{I}-T-I	38/56[a]	P_{II}-H-O	30
		P_{I}-H-O	19	P_{II}-T-I	24/49
		P_{II}-T-O	7	P_{I}-T-O	15/19
	Heavier	P_{I}-H-O	51	P_{II}-H-O	44
		P_{II}-H-I	30	P_{I}-H-I	34
		P_{II}-T-O	23/44	P_{I}-T-O	28/35
		P_{I}-T-I	5	P_{II}-T-I	9

[a] Underscored values were obtained just at the hydrodynamic equilibrium.

The results of series 1 and 3 leads to

$$Sf = 27 + 4.5x_7 + 12.4x_2x_3 + 4x_5x_7 - 3x_2x_3x_5$$

This model reveals that the direction of rotation (**4**) is influent neither by its direct effect nor by its interactions. Moreover, the number of layers (**8**) does not appear. Only the elution mode (**7**) directly intervenes. The interaction between the mobile phase and the elution direction (**23**) has an important effect on the retention of the stationary phase. In order to increase *Sf*, the heavier mobile phase should be pumped in the outward direction or the lighter mobile phase in the inward direction. In order to better interpret this model, the equation can be rewritten as:

$$Sf = 27 + x_7\,(4.5 + 4x_5) + x_2x_3(12.4 - 3x_5)$$

which shows the role of the position of the column (**5**) on the influence of the elution mode (**7**) and on the interaction between the mobile phase (**2**) and the elution direction (**3**).

A column in the L position ($x_5 = -1$) leads to

$$Sf = 27 + 15.4x_2x_3$$

Table 36 Results of Series 6: System: Heptane/Water; Temperature: 22°C; 4 mL/min; 800 rpm; Internal Volume, 43 mL; Coil Diameter, 7.5 cm; Left Coiling Up; i.d. = 2.6 mm; 1 Layer; Measurement of *Sf* After 50 mL Eluted

Position of coil	Phase mobile	Driven coil		Driven counteraxis	
		Conditions	*Sf* (%)	Conditions	*Sf* (%)
L	Lighter	P_{I}-T-I	88	P_{II}-T-I	91
		P_{II}-H-I	86	P_{I}-H-I	86
		P_{I}-H-O	23	P_{II}-H-O	26
		P_{II}-T-O	19	P_{I}-T-O	19
	Heavier	P_{II}-T-O	81	P_{I}-T-O	85
		P_{I}-H-O	79	P_{II}-H-O	85
		P_{II}-H-I	14	P_{I}-H-I	16
		P_{I}-T-I	13	P_{II}-T-I	16
X-1,5L	Lighter	P_{I}-T-I	95	P_{I}-T-O	95
		P_{II}-H-I	54	P_{II}-T-I	88
		P_{II}-T-O	42/83[a]	P_{I}-H-I	44
		P_{I}-H-O	44	P_{II}-H-O	30
	Heavier	P_{I}-H-O	58	P_{II}-H-O	61
		P_{II}-H-I	47	P_{I}-H-I	52
		P_{II}-T-O	19/47	P_{II}-T-I	2
		P_{I}-T-I	5	P_{I}-T-O	0

[a] Underscored values were obtained just at the hydrodynamic equilibrium.

while a column in the X-1.5L position ($x_5 = +1$) leads to

$$Sf = 27 + 8.5x_7 + 9.4x_2x_3$$

It shows that the L position increases the influence of interaction **23**, while an X-1.5L position slightly decreases the effect of this interaction while it increases the direct influence of the elution mode (**7**).

(3) Butanol-2/water. The results of series 4, 5, 7, and 8 leads to

$$Sf = 30.2 + 3.4x_5 + 3.2x_9 + 17.7x_2x_3 - 3.1\,x_2x_3x_6 - 4.3x_2x_3x_7 - 3.3x_2x_3x_5x_7$$

This model reveals that the direction of rotation (**4**) is influent neither by its direct effect nor by its interactions. The rotational speed (**9**) is influent only as a single factor, and it slightly increases the retention of the stationary phase. The coil diameter (**6**) is influent through its interaction with **2** and **3**, and the sign indicates that a bigger coil diameter decreases the influence of interaction **23**. The interaction between the mobile phase and the elution direction (**23**) has an important effect on the retention of the stationary phase. In order to increase *Sf*, the heavier mobile phase should be pumped in the outward direction or the lighter mobile

Table 37 Results of Series 7: System: Butanol-2/Water; Temperature: 22°C; 4 mL/min; 800 rpm; Internal Volume, 29,5 mL; Coil Diameter, 5 cm; Left Coiling Up; i.d. = 2.6 mm; 1 Layer; Measurement of *Sf* After 50 mL Eluted

Position of coil	Phase mobile	Driven coil		Driven counteraxis	
		Conditions	*Sf* (%)	Conditions	*Sf* (%)
L	Lighter	P_{II}-H-I	53	P_{II}-T-I	61
		P_{I}-T-I	48	P_{I}-H-I	51
		P_{I}-H-O	12	P_{I}-T-O	7
		P_{II}-T-O	3	P_{II}-H-O	3
	Heavier	P_{I}-H-O	61	P_{II}-H-O	64
		P_{II}-T-O	59	P_{I}-T-O	58
		P_{I}-T-I	7	P_{I}-H-I	14
		P_{II}-H-I	5	P_{II}-T-I	3
X-1, 5L	Lighter	P_{I}-T-I	56/<u>75</u>[a]	P_{I}-H-I	46
		P_{II}-H-I	44	P_{II}-T-I	36/<u>64</u>
		P_{I}-H-O	25	P_{II}-H-O	24
		P_{II}-T-O	3	P_{I}-T-O	5
	Heavier	P_{II}-T-O	58/<u>75</u>	P_{II}-H-O	51
		P_{I}-H-O	58	P_{I}-T-O	34/<u>68</u>
		P_{II}-H-I	25	P_{I}-H-I	34
		P_{I}-T-I	9	P_{II}-T-I	2

[a] Underscored values were obtained just at the hydrodynamic equilibrium.

phase in the inward direction. These conditions are similar to those for the heptane/water and butanol-1/water systems.

Influence of selected factors. The influencees of the number of layers (**8**), the coil diameter (**6**), the rotation speed (**9**), the temperature (**10**), and the nature of the solvent system are described in the following parts.

Number of layers. Series 1 and 3 achieved with the butanol-1/water system were different only by the number of layers (factor **8**). The latter factor was introduced as a direct factor and through its interactions with one or two other factors. Statistical examination of the results showed no influence of the number of layers itself and through its second- or third-order interactions. Consequently, the number of layers does not modify the hydrodynamic behavior of the solvent system inside the column. Then it is interesting to design and optimize a separation on a single layer small-volume column before scaling up the optimized separation to a multilayer high-volume column.

Coil diameter. The use of series 4, 5, 7, and 8 with butanol-2/water showed that the coil diameter (**6**) is influent only through its interaction with **2**

Table 38 Results of Series 8: System: Butanol-2/Water; Temperature: 22°C; 4 mL/min; 400 rpm; Internal Volume, 29.5 ml; Coil Diameter, 5 cm; Left Coiling Up; i.d. = 2.6 mm; 1 Layer; Measurement of *Sf* After 50 mL Eluted

Position of coil	Phase mobile	Driven coil		Driven counteraxis	
		Conditions	*Sf* (%)	Conditions	*Sf* (%)
L	Lighter	P_{II}-H-I	36	P_{II}-T-I	29/47
		P_I-T-I	30	P_I-H-I	29
		P_I-H-O	15	P_{II}-H-O	7
		P_{II}-T-O	7	P_I-T-O	5
	Heavier	P_IH-O	44	P_{II}-H-O	34/46
		P_{II}-T-O	37/44[a]	P_I-T-O	30/41
		P_{II}-H-I	3	P_I-H-I	3
		P_I-T-I	2	P_{II}-T-I	0
X-1, 5L	Lighter	P_{II}-H-I	42	P_I-H-I	34
		P_I-T-I	27/61	P_{II}-T-I	29/61
		P_I-H-O	24	P_{II}-H-O	22
		P_{II}-T-O	15	P_I-T-O	14
	Heavier	P_I-H-O	49	P_{II}-H-O	48
		P_{II}-T-O	41/71	P_I-H-I	30
		P_{II}-H-I	25	P_I-T-O	29/66
		P_I-T-I	3	P_{II}-T-I	0

[a] Underscored values were obtained just at the hydrodynamic equilibrium.

Table 39 Results of Series 9: System: Butanol-2/Water; Temperature: 7°C; 4 mL/min; 400 rpm; Internal Volume, 29.5 mL; Coil Diameter, 5 cm; Left Coiling Up; i.d. = 2.6 mm; 1 Layer; Measurement of *Sf* After 50 mL Eluted

Position of coil	Phase mobile	Driven coil		Driven counteraxis	
		Conditions	*Sf* (%)	Conditions	*Sf* (%)
L	Lighter	P_{II}-H-I	32	P_{II}-T-I	27/41
		P_I-T-I	24	P_I-H-I	12/20
	Heavier	P_I-H-O	24	P_I-T-O	25
		P_{II}-T-O	19/30[a]	P_{II}-H-O	24
X-1,5L	Lighter	P_{II}-H-I	25/32	P_{II}-T-I	22/44
		P_{II}-T-O	14	P_I-H-I	20/27
		P_I-H-O	12		
		P_I-T-I	7/17		
	Heavier	P_I-H-O	30	P_I-T-O	15/27
		P_{II}-T-O	3/24	P_{II}-H-O	15/27
		P_{II}-H-I	3		
		P_I-T-I	3		

[a] Underscored values were obtained just at the hydrodynamic equilibrium.

Table 40 Results of Series 10: System: Butanol-2/Water; Temperature: 7°C; 4 mL/min; 800 rpm; Internal Volume: 29.5 mL; Coil Diameter: 5 cm; Left Coiling Up; i.d. = 2.6 mm; 1 Layer; Measurement of *Sf* After 50 mL Eluted

Position of coil	Phase mobile	Driven coil		Driven counteraxis	
		Conditions	*Sf* (%)	Conditions	*Sf* (%)
L	Lighter	P_{II}-H-I	57	P_I-H-I	54/<u>74</u>
		P_I-T-I	51/<u>59</u>[a]	P_{II}-T-I	52/<u>68</u>
	Heavier	P_I-H-O	52	P_I-T-O	41
		P_{II}-T-O	41	P_{II}-H-O	39
X-1,5L	Lighter	P_I-T-I	37/<u>68</u>	P_{II}-T-I	19/<u>30</u>
		P_{II}-H-I	37	P_I-H-I	17/<u>24</u>
	Heavier	P_I-H-O	52	P_{II}-H-O	24
		P_{II}-T-O	20/<u>37</u>	P_I-T-O	19/<u>34</u>

[a] Underscored values were obtained just at the hydrodynamic equilibrium.

and **3**, and the sign indicates a bigger coil diameter decreases the influence of interaction **23**. As the latter has a positive effect on *Sf*, the smaller diameter coil is recommended.

Rotation speed. The use of series 4, 5, 7, and 8 carried out with butanol-2/water showed that the rotation speed (**9**) is influent only through its direct effect on *Sf*. A higher rotation speed increases the retention of the stationary phase. In the same way, series 7, 8, 9, and 10 carried out with

Table 41 Various Factors Used in EDM

System	Series	i.d. (mm)	Layers (no.)	Coiling up	Coil diameter (cm)	Speed (rpm)	Temp. (°C)
Heptane/water	6	2.6	1	Left	7.5	800	22
Butanol-1/water	1	2.6	6	Left	7.5	400	22
	2	1.85	1	Right	7.5	800	22
	3	2.6	1	Left	7.5	400	22
Butanol-2/water	4	2.6	1	Left	7.5	400	22
	5	2.6	1	Left	7.5	800	22
	7	2.6	1	Left	5	800	22
	8	2.6	1	Left	5	400	22
	9	2.6	1	Left	5	400	7
	10	2.6	1	Left	5	800	7

Table 42 Results Computed with JMP Software from SAS Institute

System	Heptane/water	*n*-Butanol/water		*sec*-Butanol/water
Series	*6*	*2*	*1 & 3*	*4, 5, 7, & 8*
Mean	**51**	**23.1**	**27**	**30.2**
2	−9.7	0	0	0
3	0	0	0	0
4	0	0	0	0
5	0	5.6	0	3.4
6	—	—	—	0
7	0	6.1	4.5	0
9	—	—	—	3.2
2*3	19.5	9.3	12.4	17.7
2*5	−6.6	0	0	0
2*7	10.9	0	0	0
5*7	0	0	4.0	0
2*3*5	−13.9	0	−3.0	0
2*3*6	—	—	—	−3.1
2*3*7	0	0	0	−4.3
2*5*7	11.2	0	0	0
2*3*5*7	0	0	0	−3.3

butanol-2/water were analyzed using a linear model based on factors 9 and 10 and their interaction:

$$Sf = 33.3 + 6.9x_9 - 3.9x_9x_{10} = 33.3 + x_9(6.9 - 3.9x_{10})$$

As for series 4, 5, 7, and 8, the rotation speed increases *Sf*, but its positive influence is diminished at higher temperatures.

Temperature. The above equation, based on the results of series 7–10, has shown that the temperature (**10**) intervenes only through its interaction with the rotation speed. A higher temperature means a lower benefic influence of the rotation speed on the retention of the stationary phase.

Solvent system. According to Table 41, butanol-2/water system and heptane/water system could be compared using series 5 and 6, as the internal diameter of the tubing, the number of layers, the direction of coiling up, the coil diameter, the speed, and the temperature were similar. In the same way, series 3 and 4 were used to compare butanol-1/water and butanol-2/water systems.

- Butanol-2/water (series 5, 800 rpm)

$$Sf = 31. + 17.1x_2x_3 - 4.1x_2x_3x_5$$

- Heptane/water (series 6, 800 rpm)

$$Sf = 51 - 9.7x_2 + 19.5x_2x_3 - 6.6x_2x_5 + 10.9x_2x_7 - 13.9x_2x_3x_5 + 11.2x_2x_5x_7$$

The signs of the common interactions, i.e., **23** and **235**, are the same for both solvent systems. The behavior of the heptane/water system involves three additional interactions that are based on the mobile phase (**2**) and on the coil position and the elution mode. For both systems, the influence of the coil position (**5**) on interaction **23** is the same, and its sign means a column installed in the L position increases the effect of interaction **23**. Interaction **27** is 2.1 for butanol-2/water and is not considered as influent. Its sign is similar to that of the heptane/water system, which indicates that a classification based on the sign of this interaction as defined on type J devices cannot be applied to cross-axis devices.

- Butanol-1/water (series 3, 400 rpm)

$$Sf = 27.2 + 3.3x_5 + 4.3x_7 + 13.7x_2x_3 + 4x_5x_7 - 3.8x_2x_3x_5$$

- Butanol-2/water (series 4, 400 rpm)

$$Sf = 25.2 + 3.9x_5 + 12.1x_2x_3 - 3.6x_2x_3x_7 + 3.4x_2x_4x_7$$

Common effects include the coil position (**5**) and interaction **23**, with the same signs. For the butanol-1/water system, the coil position (**5**) modifies the effect of the elution mode (**7**) and of interaction **23**. A coil in the L position increases the positive influence of interaction **23**, as was already observed for series 5 and 6. For the butanol-2/water system, the coil position does not intervene on the role of a factor or its interaction. Interaction **27** is not influent for both systems.

Conclusion of Experimental Design Methodology Analysis

Various interesting conclusions can be withdrawn from the previous analyses carried out by EDM. Of the 10 proposed factors, 9 were studied.

Factor **1**, the direction of winding, was not studied as it is related to 3, 4, and 7 according to the position of the belt (see Section "Original Features of the Apparatus").

Factor **2** (mobile phase) has a direct effect for two systems, either negative for heptane/water or positive for butanol-1/water. However, it mainly intervenes through an always positive interaction with factor **3** for all the studied solvent systems. It means that the choice of a heavier mobile phase requires the outward elution mode, whereas the use of a lighter one requires an inward elution mode, in order to increase the retention of the stationary phase. It may also play a role through its positive interaction **27** with the elution mode (**7**). A heavier mobile phase requires a head-to-tail elution mode, whereas a lighter one requires a tail-to-head mode to increase *Sf*. Its interactions with the position of the coil will be discussed in the factor **5** paragraph.

Factor **3** (elution direction) only intervenes through its positive interaction with the choice of the mobile phase (factor **2**).

Factor **4** (direction of rotation) was not observed to have any influence, either as a direct effect or as an interaction.

Factor **5** (coil position) can have a small positive effect for butanol-based solvent systems, but it mainly intervenes through its interactions with factors or interactions. It leads to a negative interaction with the choice of the mobile phase (**2**), the elution mode (**7**), and interaction **23**. On the contrary, its interaction with interaction **27** is positive. This implies a coil mounted in the L position (level −1 for factor **5**) decreases the influence of the mobile phase (factor **2**), of the elution mode (factor **7**), and of interaction **27**, while amplifying the role of interaction **23** on the retention of the stationary phase. On the contrary, a coil mounted in the X-1.5L position increases the effects of factors **2** and **7** and their interaction **27**, while interaction **23** plays a less important role on *Sf*.

Factor **6** (coil diameter) intervenes through its small negative interaction with interaction **23**. Consequently, a larger diameter coil slightly diminishes the positive influence of interaction **23** on *Sf*.

Factor **7** (elution mode) can intervene as a single factor or through its interactions, mainly with factor **2**. Its role is diminished by an L position of the coil.

Factor **8** (number of layers) plays no role. It reveals that a separation can be designed with a single-layer coil and easily extrapolated to a multilayer one.

Factor **9** (rotation speed) has a small but positive direct effect on *Sf*, which is altered by an increase of the temperature (**10**).

Factor **10** (temperature) shows no direct effect but intervenes through a small negative interaction with the rotation speed (**9**).

4. Recommended Conditions

The overall analysis carried out by the use of EDM has shown that a coil mounted in the L position simplifies the operation of the cross-axis apparatus. Indeed, the retention of the stationary phase is mainly related to the choice of the mobile phase and of the elution direction. A heavier mobile phase requires the outward elution direction while a lighter mobile phase requires an inward elution direction. For a further increase of *Sf*, the coil diameter may be reduced and the rotation speed increased, but the gain is likely to be in the order of a few percentage points. Moreover, it should be kept in mind that a higher temperature alters the benefits of a higher rotation speed. Other factors play nonsignificant roles. Moreover, the L position is less likely to lead to conditions that are submitted to leaks of the stationary phase (see Tables 34–40) than the X-1.5L position.

An L position of the coil consequently leads to a simple control of Sf *by the choice of the mobile phase and the elution direction, and does not depend on the solvent system. In case of a leak of the stationary phase, a change from the elution mode should solve the problem.*

5. Interpretations

The best retentions of the stationary phase require a lighter mobile phase to be pumped in the inward direction, while a heavier mobile phase has to be pumped in the outward direction. Figure 17A gives a simplified drawing of the column inside the apparatus. The behavior is simplified as the column is considered an empty cylinder. The chosen stationary phase is the heavier one.

The two upper drawings are related to the outward elution direction, in the same direction as the force field along the coil axis. The left-side drawing gives the distribution of both phases at a given time. The right-side drawing gives this distribution a few moments later; the heavier phase pushes away the lighter one toward the inside part of the coil, but as the lighter phase is continuously introduced at the inside part of the coil the heavier phase is expelled from the coil. Indeed, the lighter mobile phase cannot go through the stationary heavier one to exit by the outside part of the coil. The stationary heavier phase is consequently progressively expelled from the column, and the retention of the stationary phase is therefore very low.

The two lower drawings show on the contrary that a lighter mobile phase pumped from the outside part of the coil to the inside one (inward elution direction) is pushed away by the heavier stationary phase toward the inside part, from which it can exit the coil. The stationary phase is consequently well retained inside the coil. The satisfactory elution direction is consequently inward for a lighter mobile phase.

Similar explanations apply to the four drawings of Figure 17B. In order not to expel the lighter stationary phase, the heavier mobile phase has to be pumped from the inside to the outside part of the coil, which is the outward elution direction.

All of these drawings have shown that the recommended conditions are in fact based on a positive **23** interaction, which was observed for the three solvent systems under all conditions.

(A)

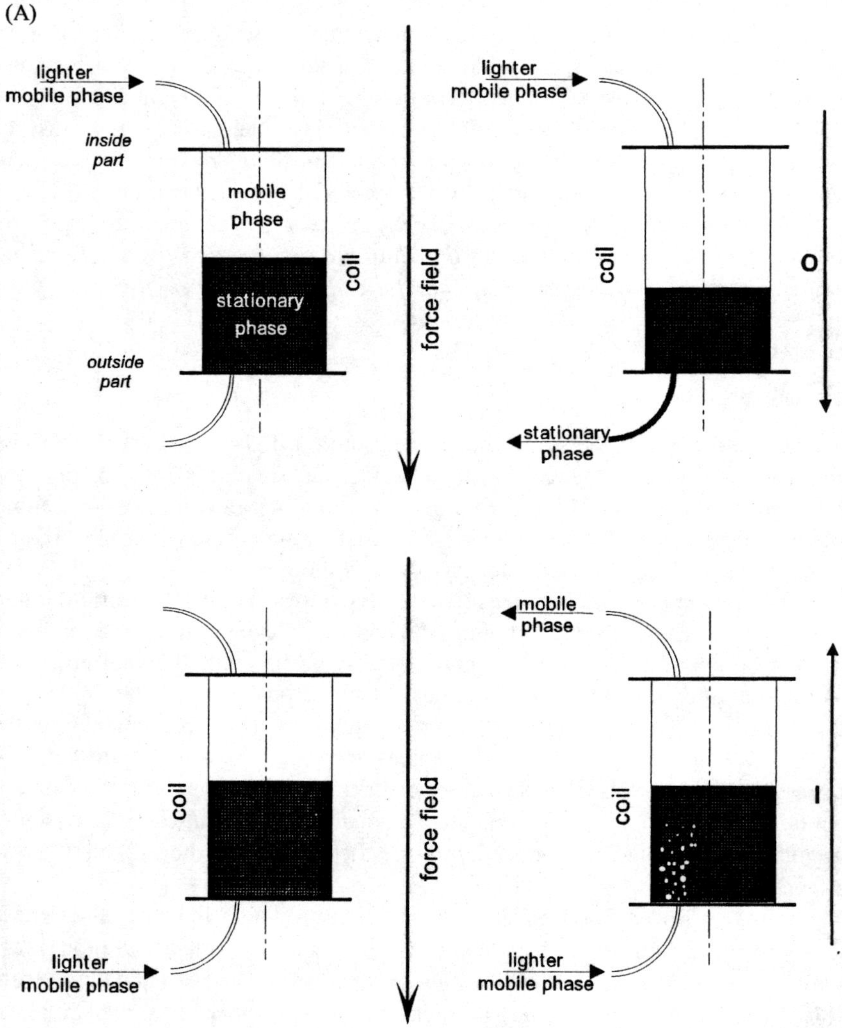

Figure 17 Influence of the elution direction (factor 3) on the retention of the stationary phase. (A) Lighter mobile phase; (B) heavier mobile phase.

(B)

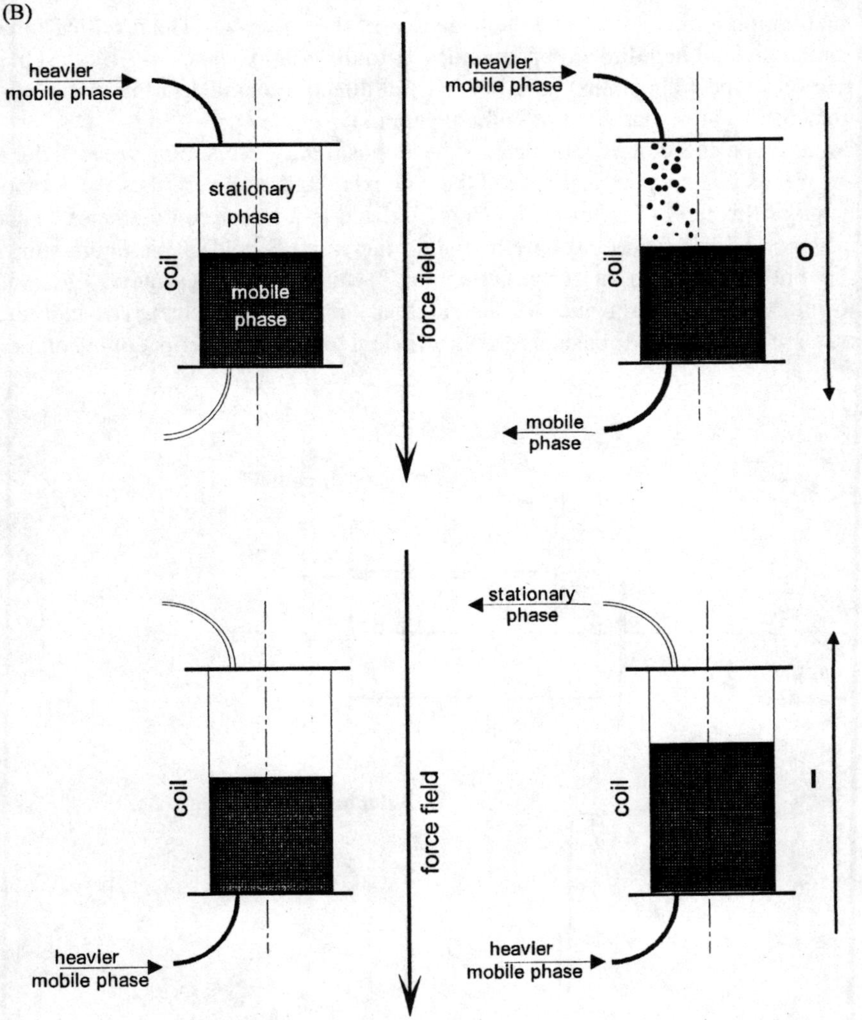

However, such an interpretation relies on the role of the force field along the coil axis, which enables the definition of the inside and outside parts of the coils, hence of the inward and outward elution directions. However, the coil rotates along its axis, which generates an additional force field. Indeed the cross-axis apparatus induces a three-dimensional force field, instead of the two-dimensional force fields of type J and Sanki devices. It is the combination of two force fields. One is generated by the rotation of the coil on its axis, the other one from the rotation of the coil around the main axis of the apparatus. The resulting force can be divided in a direction perpendicular to the coil axis, named J (representative of a type J apparatus), and in a second direction parallel to this same axis, named S (representative of a Sanki apparatus).

When the coil is mounted in the L position, the resulting force field is simple, as it is symmetrical around the coil axis. Figure 18 describes the components of this field. The lower drawing is related to a larger coil diameter, hence a bigger J force field (and only a slightly increased S field). This figure shows that both force fields intervene, but factor **7** (elution mode) was never observed to intervene on the retention of the stationary phase (see tables). The intensity of force field J can be considered as insufficient to reveal any effect of the elution

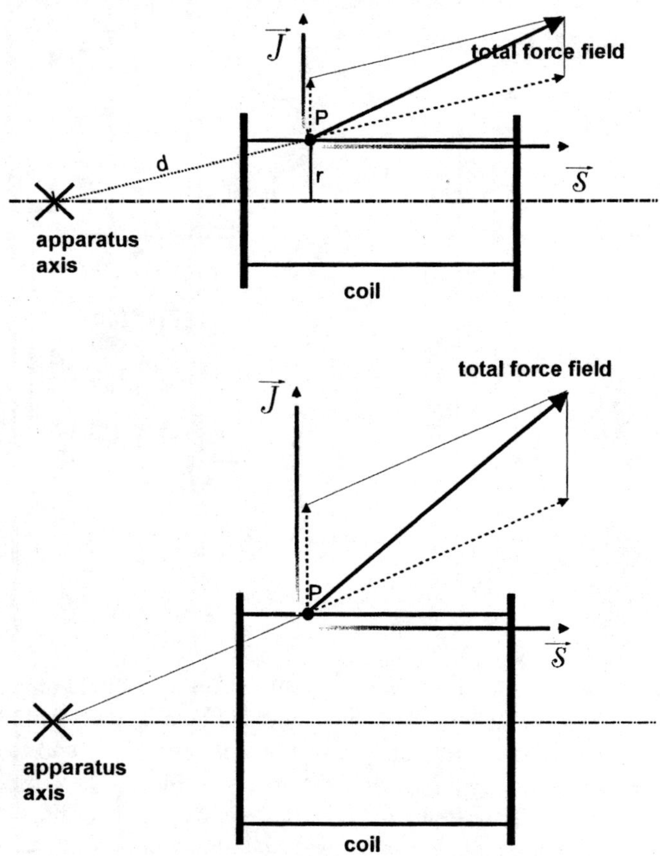

Figure 18 Diagram of the generated forces in the L position.

mode, whatever the coil radius. When the coil is mounted in the X-1.5L position, the geometry of the resulting force field evolves, and is no longer symmetrical around the coil axis. The maximum intensity of the J force field is increased at S field's expense. Consequently, the influence of the helix should be increased. This corresponds to the observed behaviors, as it was shown in the section "Conclusion of Experimental Design Methodology Analysis" that the coil position modifies the influences of interactions **23** and **27**. Indeed, its interaction with interaction **27** is positive. This implies that a coil mounted in the L position (level −1 for factor **5**) decreases the influence of the mobile phase (factor **2**), of the elution mode (factor **7**), and of interaction **27** (helix influence, related to J force field), while amplifying the role of interaction **23** on the retention of the stationary phase. On the contrary, a coil mounted in the X-1.5L position increases the effects of factors **2** and **7** and their interaction **27**, while interaction **23** has a less important influence on *Sf*.

These explanations are also consistent with the influence of coil diameter. An increase of this factor (**6**) increases the intensity of the J field while diminishing the intensity of the S field. The influence of interaction **23** should therefore decrease, as was shown in the above cited section.

REFERENCES

1. Y. Ito, *J. Biochem. Biophys. Meth. 5*:105 (1981).
2. M-C., Rolet, *La chromatographie de partage centrifuge en mode hydrostatique. Théorie et applications*, 1993. (Thesis Paris VI) Université Pierre et Marie Curie, Paris.
3. A. P. Foucault, *Centrifugal Partition Chromatography*, Chromatographic Science Series, Vol. 68, Marcel Dekker, New York, 1995.
4. Y. Ito, *Countercurrent Chromatography. Theory and Practice* (N. B. Mandava and Y. Ito, eds.), Chromatographic Science Series, Vol. 44, Marcel Dekker, New York. 1988, pp. 430–431.
5. Same as Ref. 4, except p. 369.
6. J.-M. Menet, D. Thiébaut, R. Rosset, J. E. Wesfreid, and M. Martin, *Anal. Chem. 66*:168–176 (1994).
7. J.-M. Menet, D. Thiébaut, and R. Rosset, *J. Chromatogr. 659*:3–13 (1994).
8. R. A. Fisher, *Statistical Methods for Research Workers*, Oliver and Boyd, 1925, p. 362.
9. R. A. Fisher, *The Design of Experiments*, Oliver and Boyd. 1935, p. 248.
10. F. Yates, *The Design and Analysis of Factorial Experiments*, 1937. 35, Imperial Bureau of Soil Science, Harpenden, Herts, England. Hafner (Macmillan).
11. W. J. Youden, *Statistical Methods for Chemists*, John Wiley and Sons, New York, 1951, p. 126.
12. R. L. Plackett and J. P. Burman, The design of optimum multifactorial experiments, *Biometrika 33*:305–325 (1946).

13. J. Goupy, *Methods for Experimental Design. Principles and Applications for Physicists and Chemists*, Elsevier, Amsterdam, 1993, p. 450.
14. J. Goupy, *La Méthode des Plans d'Expériences*, Dunod, Paris, 1988, p. 303.
15. G. E. P. Box, W. G. Hunter, and J. S. Hunter, *Statistics for Experimenters*, John Wiley and Sons, New York, 1971, p. 453.
16. J. L. Sandlin and Y. Ito, *J. Liq. Chromatogr. 8*:2153–2171 (1985).
17. H. Scheffe, *The Analysis of Variance*, John Wiley and Sons, New York, 1959.
18. Y. Ito, *Countercurrent Chromatography. Theory and Practice* (N. B. Mandava and Y. Ito, eds.), Chromatographic Science Series, Vol. 44, Marcel Dekker, New York. 1988, pp. 360–361 and 364–365.
19. Y. Ito, E. Kitazume, M. Bhatnagar, and F. D. Trimble, *J. Chromatogr. 538*:59–66 (1991).
20. Y. Shibuzawa and Y. Ito, *J. Liq. Chromatogr. 15*:2787–2800 (1992).
21. K. Shinomiya, J.-M. Menet, and Y. Ito, *J. Chromatogr. 44*:215–229 (1993).
22. J. Goupy, J.-M. Menet, K. Shinomiya, and Y. Ito, in *Modern Countercurrent Chromatography* (W. D. Conway and R. J. Petroski, eds.), ACS Symposium Series 593, 1995, pp. 47–61.

3
Coil Planet Centrifuges for High-Speed Countercurrent Chromatography

Yoichiro Ito
National Heart, Lung, and Blood Institute, National Institutes of Health, Bethesda, Maryland

Jean-Michel Menet
Rhône-Poulenc Rorer, Inc., Vitry-sur-Seine, France

I. INTRODUCTION

All existing systems for performing countercurrent chromatography (CCC) may be classified into two major categories: hydrostatic and hydrodynamic equilibrium systems [1,2]. In the hydrostatic system the mobile phase percolates through the column of the stationary phase on the effect of gravitational or centrifugal force where the mixing of the two phases relies entirely on the flow of the mobile phase. In contrast, the hydrodynamic system produces active phase mixing using a rotating force field induced by the planetary motion of the coil, thus yielding much more efficient solute partitioning.

In the 1970s, a series of planetary centrifuge systems had been developed for performing CCC based mainly on the hydrodynamic equilibrium system. All of these centrifuge systems were equipped with a rotary seal-free flow-through device so that the mobile phase could be eluted through the rotating column without the conventional rotary seal.

Figure 1 shows the synchronous planetary centrifuge systems wherein the rates of rotation and revolution are synchronized, i.e., the column rotates once about its own axis during one revolution around the central axis of the centrifuge.

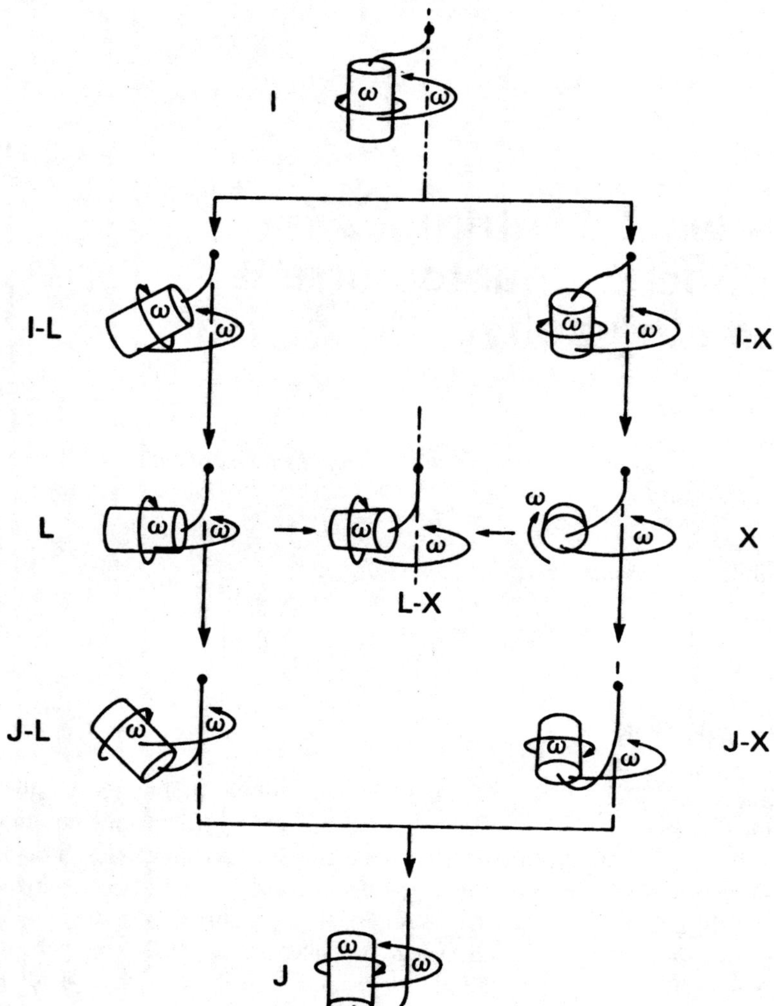

Figure 1 Series of synchronous coil planet centrifuge systems free of rotary seal.

Each system consists of a cylindrical column holder with a bundle of flow tubes, the end of which is tightly supported at the point marked "x" on the central axis of the revolution.

In type I, shown at the top of the diagram, the holder rotates around its vertical axis and simultaneously revolves around the central axis of the centrifuge at the same rate but in the reversed direction as indicated by a pair of arrows. This counterrotation of the holder unwinds the twist of the tube bundle caused by revolution, thus eliminating the need for the rotary seal. This seal-free mechanism can be extended to other systems with various orientations of the holder as shown in the diagram.

In the series shown in the left column, the holder axis is inclined toward the centrifuge axis to form types I-L, L, J-L, and J. In the right column, the holder axis is rotated while keeping the same distance to the central axis of the centrifuge, thus forming types I-X, X, J-X, and J as indicated in the diagram.

In the past, prototypes of these centrifuge systems, excluding types I-X and J-X, were constructed at the National Institutes of Health (NIH) machine shop to examine their performance for CCC. Among those, types J and X-L hybrid systems were found to be most useful in terms of stationary phase retention and partition efficiency. Below, the characteristic features of these two planetary centrifuge systems are described.

II. TYPE J COIL PLANET CENTRIFUGES

Since the early 1980s, the coil planet centrifuge system based on the type J planetary motion has been widely applied for the separation of natural and synthetic products. Various features of this system, including basic designs of the apparatus, mathematical analysis of acceleration, retention of the stationary phase, and so forth, are described in detail in the first monograph on CCC [1]. In this chapter, these features are briefly reviewed and a few selected examples of recent application are presented.

A. Basic Features of Type J Coil Planet Centrifuge

Figure 2 shows the planetary motion (A) and the resulting centrifugal force field (B) of the type J coil planet centrifuge. As shown in the force diagram, all centrifugal force vectors (counterforce against acceleration) are always confined in a plane perpendicular to the holder axis. As the holder rotates, both the direction and the net strength of the force vector fluctuate in such a way that the vector becomes longest at the point remote from the centrifuge axis and shortest at the point close to the central axis of the centrifuge. In most locations, the vectors are directed outwardly from the circle except for $\beta < 0.25$ where its direction is reversed. As indicated in the diagram, $\beta = r/R$, which is the ratio of the holder radius to the revolution radius. This characteristic fluctuation of the centrifugal force field produces different hydrodynamic distribution of the two solvent phases in the coiled column according to the orientation of the column on the holder.

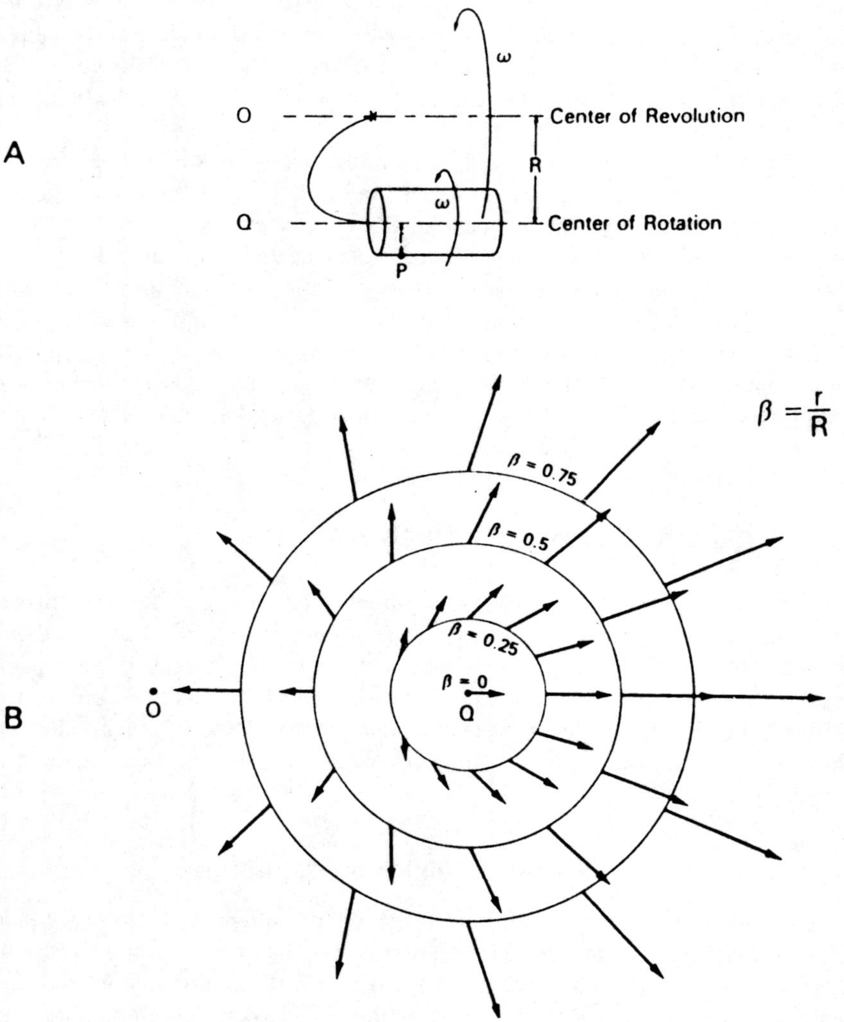

Figure 2 Planetary motion (A) and centrifugal force field (B) of type J planetary motion.

Figure 3 shows three different coil orientations on the holder. Among those both eccentric (B) and toroidal (C) forms ($\beta > 0.25$) have a common feature in that in each helical turn the two solvent phases are always subjected to an outwardly directed force field where the heavier phase is distributed toward the outer portion and the lighter phase toward the inner portion of each coiled turn, while the two phases are mixed back and forth by the fluctuating force field. This

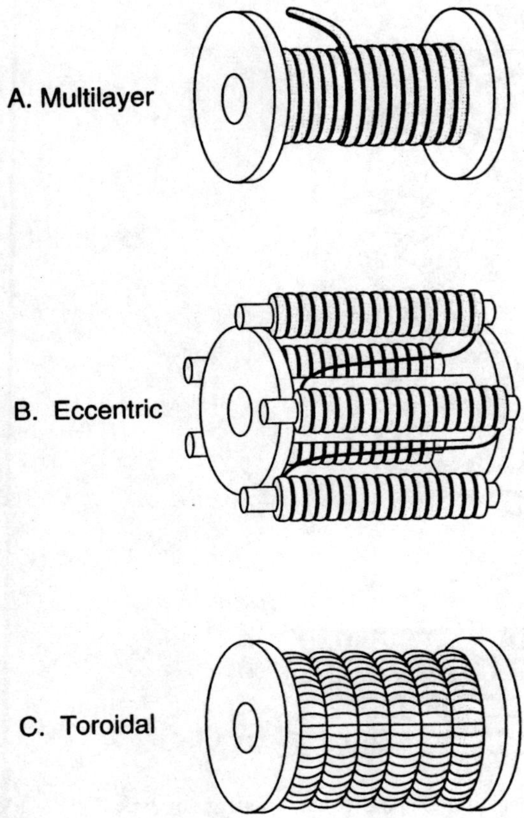

Figure 3 Three column orientations on the holder for type J coil planet centrifuge system: (A) multilayer, (B) eccentric, and (C) toroidal.

system, which is considered as a hybrid between the hydrostatic and hydrodynamic equilibrium systems, can provide a universal application of the solvent system including the polymer phase systems. However, the retention of the stationary phase in these columns is substantially lower than 50% of the total column capacity. While both columns are used for separating polar analytes such as peptides and proteins, the eccentric coil orientation (Figure 3B) is suitable for preparative scale separations and the toroidal coil orientation (Figure 3C) for analytical scale separations.

When the coil is directly wound around the holder hub as shown in Figure 3A (coaxial orientation), the resulting hydrodynamic phase distribution displays a completely different pattern as shown in Figure 4. The top diagram was drawn

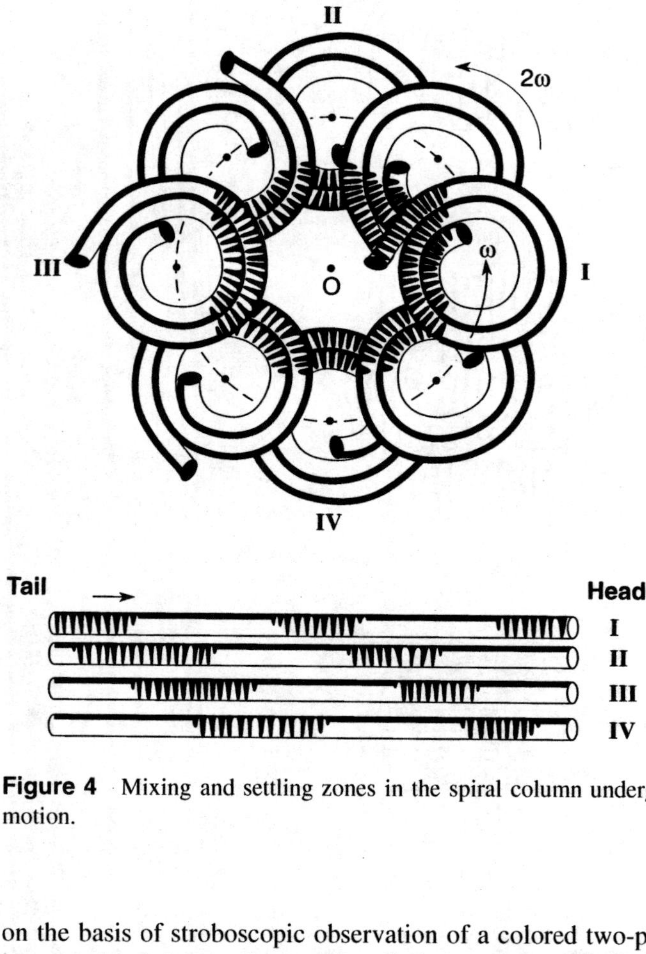

Figure 4 Mixing and settling zones in the spiral column undergoing type J planetary motion.

on the basis of stroboscopic observation of a colored two-phase solvent system in a rotating column where each coiled turn of the spiral column is divided into mixing and settling zones [3]. In about one-fourth of the area close to the center of revolution, the two solvent phases are vigorously mixed (mixing zone), whereas in the rest of the area they are separated by the strong centrifugal force showing a clear interface along the coil. In the bottom diagram, the spiral column at positions I, II, III, and IV was stretched and serially arranged in parallel so as to visualize the motion of the mixing zone along the length of the tube. Each mixing zone travels through the coil at a rate of one turn per revolution. This indicates that at 800 rpm of the optimum coil rotation the two phases at every location along the coil are subjected to a repetitive mixing and settling motion at a high frequency of 13 times per second. This explains an extremely high partition efficiency of the present system. In addition the retention of the stationary phase usually exceeds 60% of the total column capacity even at a high flow rate of the mobile phase, provided that the two-phase solvent system has a reasonable settling time of less than 20 sec [4]. However, this coaxial coil orientation on the type J coil planet centrifuge fails to retain viscous polymer phase systems that are useful for the separation of macromolecules and cell particles [5].

B. Design of Type J Coil Planet Centrifuge

Figure 5 shows a photograph of the original high-speed CCC centrifuge equipped with a multilayer coil separation column [6]. A commercial model based on this

Figure 5 The original high-speed CCC centrifuge [6].

design is available from PC Inc. (Potomac, Maryland), from Shimadzu Corporation (Kyoto, Japan), and from Kromaton (Mozé-sur-Loire, France). In this original design a single column holder is balanced by a counterweight mounted on the other side of the rotary frame. Consequently, the system requires adjustment of the counterweight according to the density of the two-phase solvent system used for separation. This problem was eliminated by a recent design illustrated in Figure 6 [7].

The upper diagram shows the original design, which requires a counterweight to maintain the centrifuge balance. The use of this counterweight is entirely eliminated by mounting a set of identical column holders around the rotary frame as shown in the lower diagram. There are single (right) and double (left) flow tube systems both requiring a set of tube supporters which counterrotate

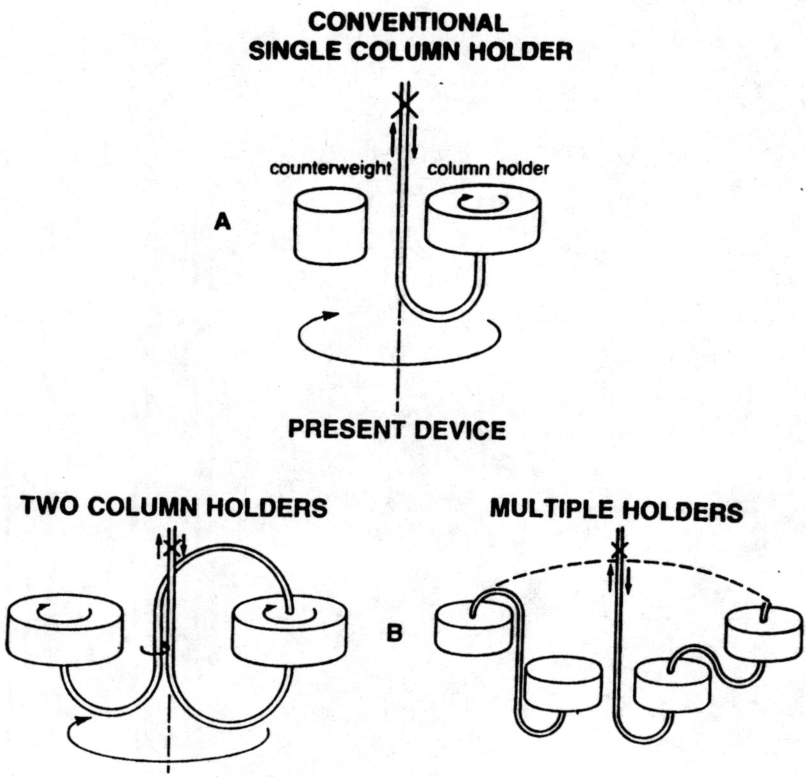

Figure 6 Improved design of the apparatus with multiple column holders [7]: (A) conventional device and (B) improved device.

Figure 7 Improved high-speed CCC centrifuge equipped with three column holders [7].

against the holder to prevent twisting of the flow tubes. The improved apparatus equipped with three column holders is illustrated in Figure 7. A similar apparatus is available from Pharma-Tech Research Corporation (Baltimore, Maryland) and from Kromaton.

Figure 8 shows a modified design of the apparatus with a pair of holders each equipped with eight double-layer coils. This eccentric coil planet centrifuge has been used for the separation of proteins with polymer phase systems [8].

C. Recent Applications of Type J Coil Planet Centrifuge

1. Improved High-Speed CCC Centrifuge

The improved design of the type J coil planet centrifuge with three column holders provides the following advantages over the original design with a single holder: (1) the centrifuge system is automatically balanced after the mobile phase emerges and the hydrodynamic equilibrium is established in the column, ensuring

Figure 8 Horizontal coil planet centrifuge with a pair of holders each equipped with eight eccentric double-layer coils [8].

minimum vibration and stable retention of the stationary phase; (2) both sample loading capacity and partition efficiency are increased.

Figure 9 shows typical examples of semianalytical separations of various compounds obtained by this improved high-speed CCC centrifuge, including (dinitrophenyl) (DNP) amino acids (Figure 9A), indole auxins (Figure 9B), flavonoids from *Hippophae rhamnoides* (Figure 9C), and rare earth elements by ligand affinity separation [9]. All of these separations were performed with a set of three multilayer coils of 1.07 mm i.d. with a total capacity of 270 mL (Figure 7).

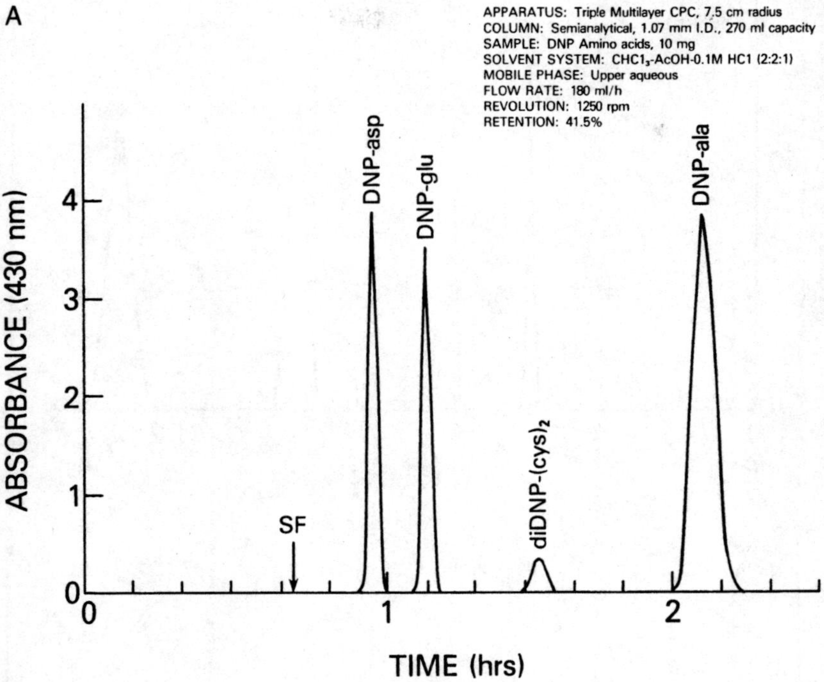

Figure 9 Small-scale preparative chromatograms obtained by the improved high-speed CCC. (A) *DNP-Amino acids.* Sample: DNP (dinitrophenyl)-L-aspartic acid (DNP-glu) (2.4mg), DNP-D,L-glutamic acid (DNP-glu) (2.4mg), diDNP-L-cystine [di(cys)$_2$] (0.5 mg), and DNP-L-alanine (DNP-ala) (4.8 mg); solvent system: chloroform/acetic acid/0.1 M HCl (2:2:1, v/v), upper aqueous phase mobile; flow rate: 180 mL/hr; revolution: 1250 rpm; retention of stationary phase: 41.5%. (B) *Indole auxins.* Sample: indole-3-acetamide (IA) (10 mg), indole-3-acetic acid (IAA) (30 mg), indole-3-butyric acid (IBA) (30 mg), and indole-3-acetonitrile (IAN) (30 mg); solvent system: *n*-hexane/ethyl acetate/methanol/water (1:1:1:1), lower aqueous phase mobile; flow rate: 150 mL/hr; revolution: 1250 rpm; retention of stationary phase: 53.3%. (C) *Flavonoids.* Sample: sea buckthorn extract 100 mg; solvent system: chloroform/methanol/water (4:3:2, v/v), lower nonaqueous phase mobile; flow rate: 180 mL/hr; revolution: 1250 rpm; retention of stationary phase: 60.0%. (D) *Rare earth elements.* Sample: lanthanum (La), praseodymium (Pr), and neodymium (Nd) each 25 mg; solvent system: 0.02 M DEHPA [di(2-ethylhexyl)phosphoric acid] in *n*-hexane/0.02 M HCl (1:1), lower aqueous phase mobile; flow rate: 300 mL/hr; revolution: 900 rpm; retention of stationary phase: 36.2%.

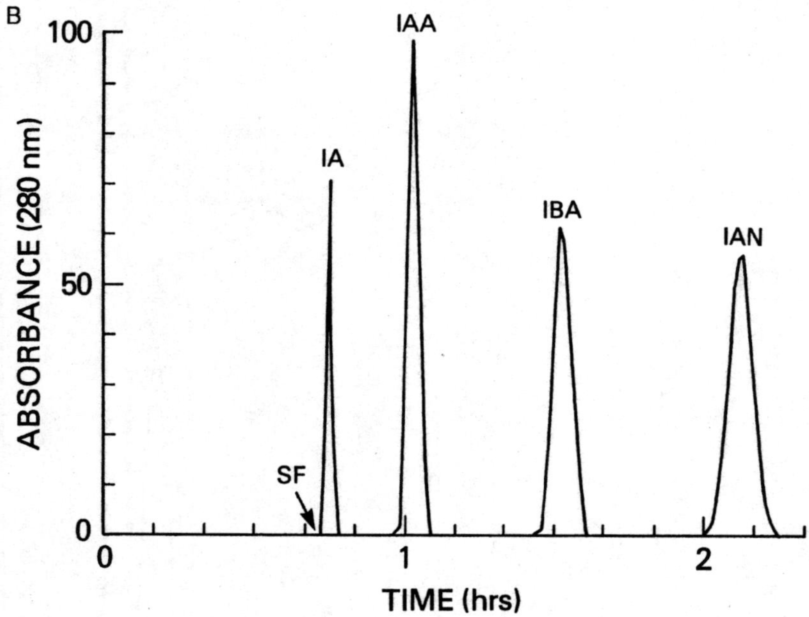

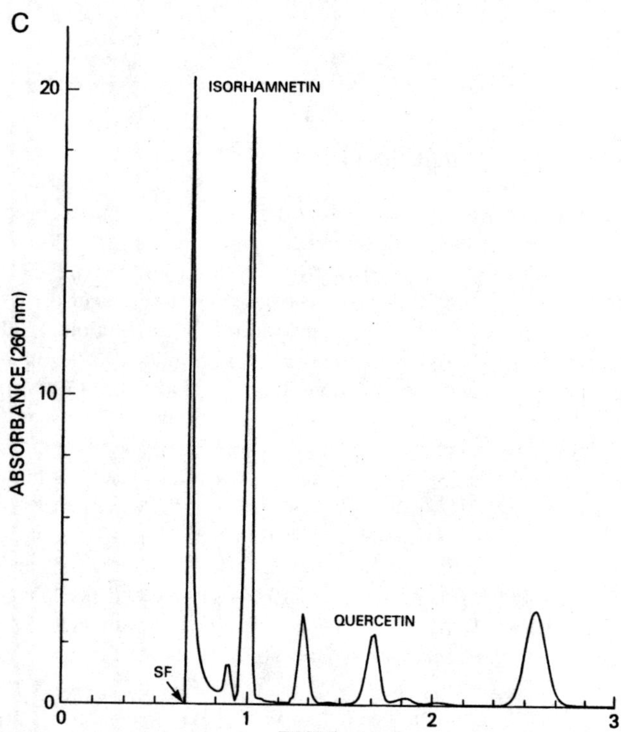

Figure 9 Continued

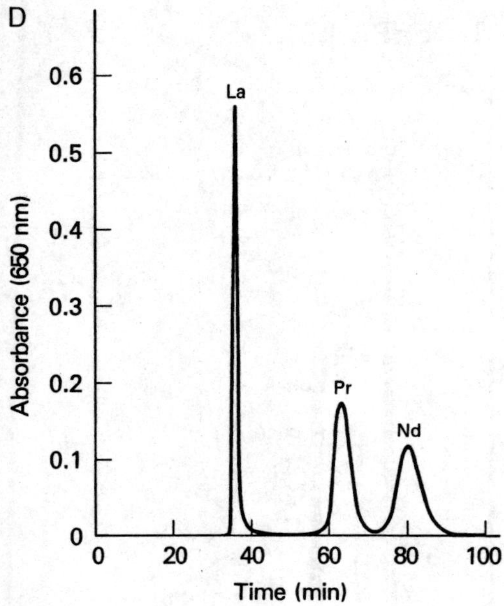

This semianalytical multilayer coil yielded high partition efficiencies of several thousand theoretical plates.

2. Affinity CCC

In liquid chromatography it is a common practice to perform affinity separation using a ligand-bound solid stationary phase. Recently, this affinity chromatographic technique has been successfully applied to CCC simply by dissolving a suitable ligand in the stationary phase [10]. Figure 10B shows an affinity chromatogram of five polar catecholamines using an affinity ligand, di-(2-ethylhexyl)-phosphoric acid (DEHPA). The separation was performed with a two-phase solvent system composed of methyl *tert*-butyl ether and the same volume of an aqueous solution containing ammonium acetate (0.1 M) and hydrochloric acid (0.05 M). Without a ligand all components elute at the solvent front as a single peak (Figure 10A). When a ligand was added to the stationary phase at 1.5% (v/v), all five components were completely resolved (Figure 10B). Although these catecholamines are unstable in a basic solution, the present mobile phase elutes them as stable HCl salts at an acidic pH.

Figure 11 shows chromatograms of four (dinitrobenzoyl) (DNB) amino acid racemates with and without ligand, *N*-dodecanoyl-L-proline-3,5-dimethylanilide (DPA). The separations were performed with a solvent system composed

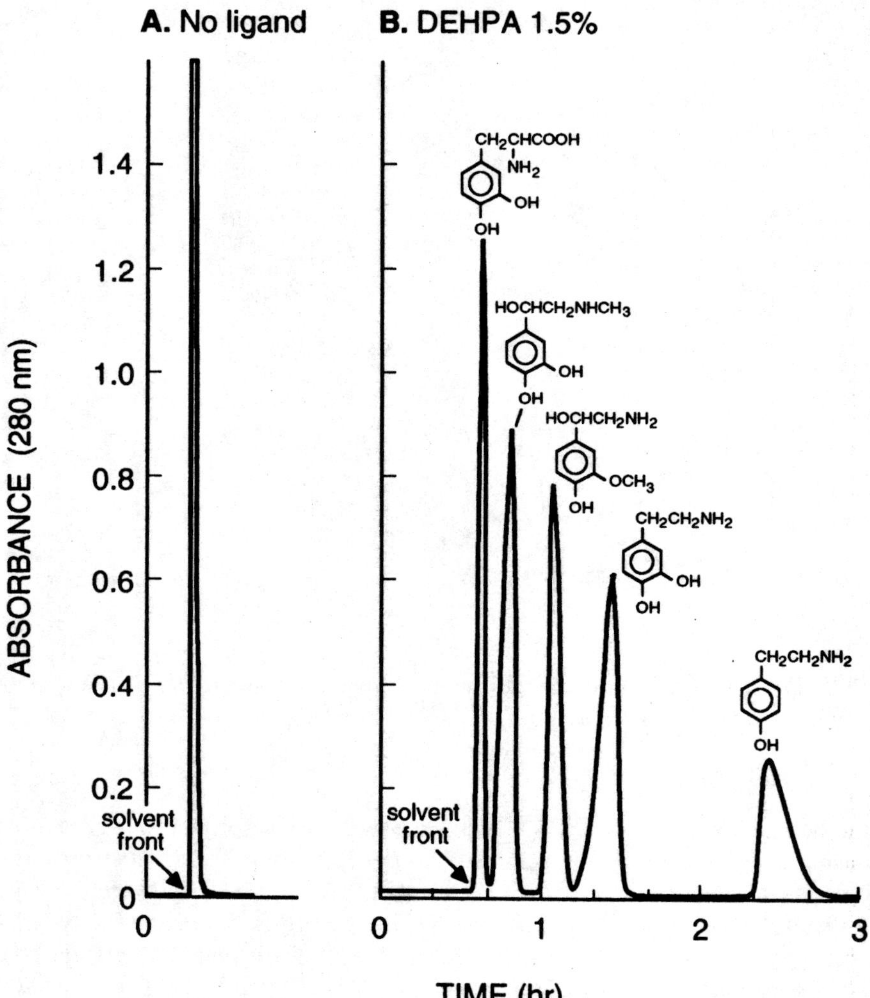

Figure 10 Separation of five catecholamines by affinity CCC. Chromatogram without ligand (A) and with 1.5% DEHPA in the organic stationary phase (B). Experimental conditions: apparatus: commercial semipreparative high-speed CCC centrifuge with 10-cm revolution radius; column: multilayer coil of 1.6 mm i.d. Tefzel tubing with a total capacity of 325 mL; solvent system: methyl *t*-butyl ether/water containing ammonium acetate (0.1 M) and HCl (0.05 M) (1:1) (A) and 1.5% DEHPA was added to the organic phase (B); sample: 5 mg each of five catecholamines indicated in the chromatogram B (also see Table 1); flow rate: 3.3 mL/min; revolution: 800 rpm; detection: 280 nm.

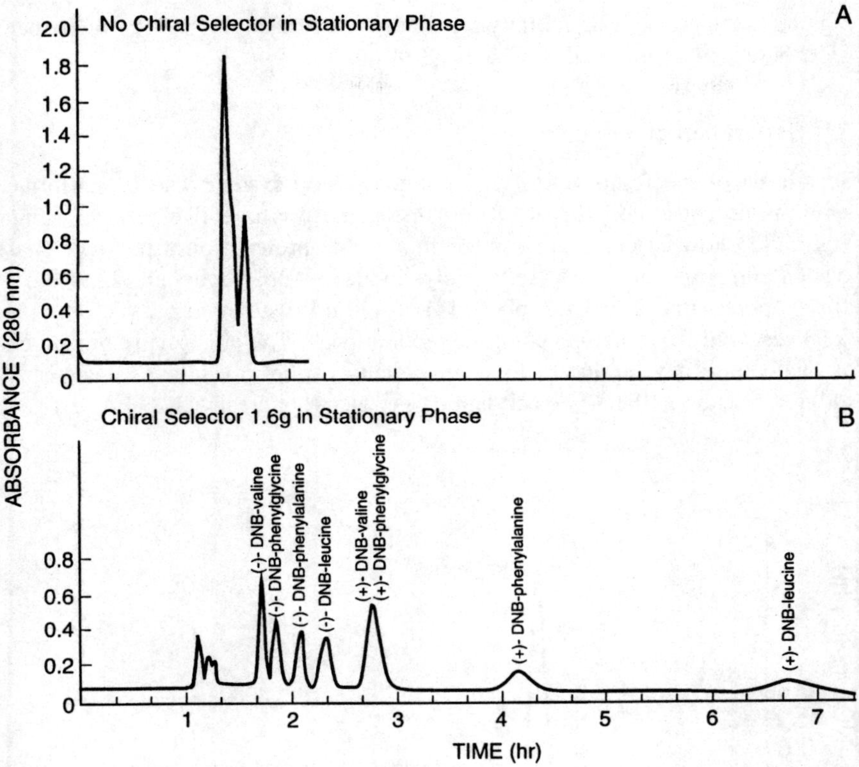

Figure 11 Separation of 4 DNB-amino acid racemates by affinity CCC: (A) Separation without ligand and (B) with ligand, DPA, in the stationary phase. Experimental conditions: apparatus: semianalytical high-speed CCC centrifuge equipped with three column holders at 7.6-cm orbital radius; columns: three multilayer coils consisting of 0.85 mm i.d. PTFE tubing connected in series with a total capacity of 180 mL; solvent system: hexane/ethyl acetate/methanol/10 mM HCl (8:2:5:5) without ligand (A) and DPA 1.6 g added to 200 mL organic stationary phase (B); sample: 10 mg each of (±)-DNB-amino acids indicated in the chromatogram B; flow rate: 1 mL/min; revolution: 1000 rpm; detection: 280 nm.

of hexane/ethyl acetate/methanol/10 mM hydrochloric acid (8:2:5:5, v/v) using a semianalytical high-speed CCC centrifuge. With ligand-free solvent system, all components were eluted together close to the solvent front forming two partially resolved peaks (Figure 11A). When a ligand was added to the stationary phase at 1.6 g/80 mL (Figure 11B), all components were retained longer in the column and well separated from impurities eluting near the solvent front. Eight components consisting of four pairs of enantiomers are resolved into seven peaks including one overlapping peak. Affinity separation of chiral compounds by high-speed CCC is described in detail in Chapter 8 of this book.

3. Separation of Proteins

Separation of macromolecules with polymer phase systems can be performed using an eccentric coil (Figure 8) mounted on a J-type coil planet centrifuge. Figure 12 shows a chromatogram of four stable proteins on a polymer phase system composed of 12.5% (w/w) polyethylene glycol 1000 and 12.5% (w/w) dibasic potassium phosphate (pH 9) [8]. The third ovalbumin peak is only partially resolved from the preceding myoglobin peak. The abnormally broad peak of ovalbumin may be due to its heterogeneity of the molecule as detected by sodium dodecyl sulfate–polyacrylamide gel electrophoresis (SDS-PAGE) [14].

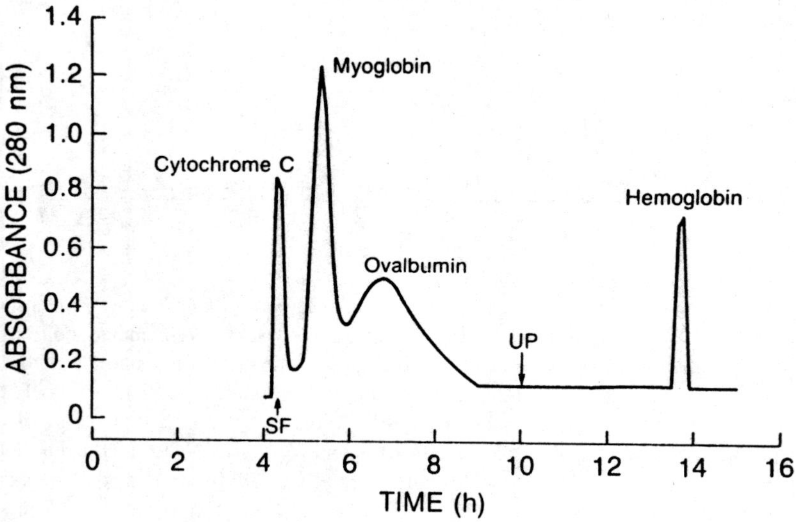

Figure 12 Chromatograms of four stable proteins with a polymer phase system by the horizontal coil planet centrifuge equipped with eccentric double-layer coils [8]. Experimental conditions: apparatus: horizontal eccentric CPC with 10-cm revolution radius; column: a set of eight double-layer coils, 1.6 mm i.d. with 220-mL capacity; sample: cytochrome *c* (horse heart), myoglobin (horse heart), ovalbumin (chicken egg), and hemoglobin (bovine), each 10–200 mg in 4 mL of the two-phase solvent system; solvent system: PEG 1000 12.5% (w/w), dibasic potassium phosphate 12.5% (w/w) in distilled water; mobile phase; lower phase; flow rate; 0.65 mL/min; revolution 800 rpm; retention of stationary phase: 19.3%. SF, solvent front; UP, upper phase eluted in the reverse direction.

The retention of the stationary phase was only 19% of the total column capacity. The cross-axis coil planet centrifuge described in the following section may be more suitable for protein separation because it provides much higher retention of the stationary phase.

D. Conclusion

Since the 1980s, the type J apparatus is the highest selling CCC device in the world, with about 300 units sold in the USA by PC Inc. and about 100 units sold in Japan by the Shimadzu Company. This somewhat crude apparatus has undergone almost no modification since the design of the first prototype by Ito. The design is simple, leading on the one hand to a restricted mechanical reliability and on the other to an easy maintenance. For several years, a thermoregulation unit has been optionally sold along with the CCC apparatus.

III. CROSS-AXIS COIL PLANET CENTRIFUGES

As mentioned in the Introduction, type X-L hybrid systems have been the second most studied hydrodynamic devices. The first prototype, type X, was designed by Ito in 1987 and constructed at the NIH machine shop [15].

A. Principle of the Apparatus

1. Design of the Apparatus

The vertical axis of the apparatus and the horizontal axis of the coil are always kept perpendicular to each other at a fixed distance. The cylindrical column revolves around the central axis at the same rotational speed with which it rotates on its own axis. As the Teflon inlet and outlet flow tubes can rotate on themselves without any twisting, the device is rotary seal-free. Three parameters displayed in Figure 13A explain the various versions of the cross-axis prototypes: r is the radius of the column holder, R the distance between the two axis, and L the measure of the lateral shift of the column holder along its axis. The name of a cross-axis device is based on the ratio L/R, when $R \neq 0$. Types X and L represent the limits for the column positions; the first type involves no shifting of the column holder while the second one corresponds to an infinite shifting. Some examples of Ito's prototypes are shown in Figure 13B. Table 1 gives the characteristics of the six cross-axis prototypes built by Ito [14–19]. They have different L/R ratios, hence their various names.

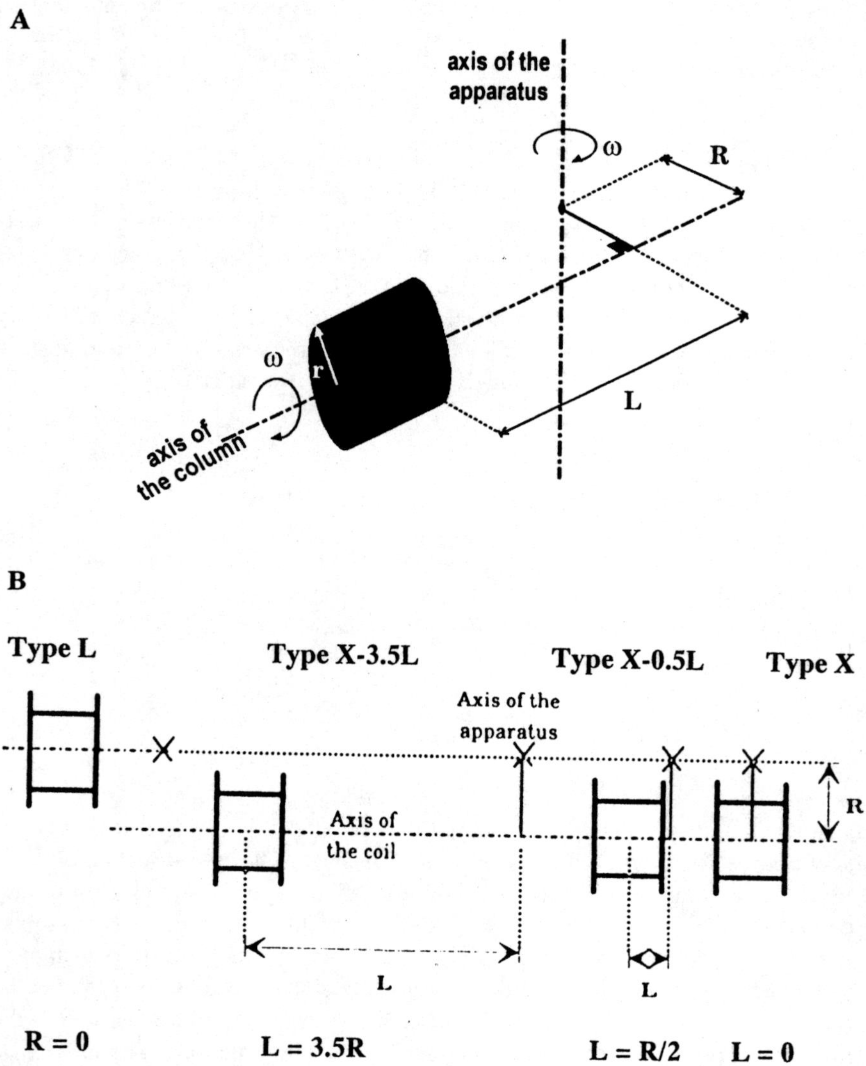

Figure 13 General principle of various cross-axis CPCs. (A) Geometrical parameters. (B) Some examples of Ito's prototypes: X type, built in 1987; X-0.5L type, built in 1988; X-3.5L type, built in 1991; and L type, built in 1992.

Table 1 Characteristics of Cross-Axis Coil Planet Centrifuges

Year	Column position	*L* (cm)	*R* (cm)	β[a]	Column volume (mL)
1987 [15]	X	0	10	0.25	15
				0.50	15
				0.75	15
				0.5 to 0.8	400
				0.19 to 0.9	?
1988 [16]	X or X-0.5L	0 or 10	20	0.125	15 or 28
				0.375	15 or 28
				0.625	15 or 28
				0.375 to 0.625	800
1989 [17]	X-1.25L	12.5	10	0.5	28
				0.75 to 1.15	750
1991 [18]	X-LL	15.2	7.6	0.5	20–30
				1	20–30
				1.6	20–30
				0.25 to 0.60	280
				0.50 to 1.00	250
				1.00 to 1.20	450
1991 [19]	X-3.5L	13.5	3.8	0.50 to 1.30	150
				0.44 to 1.50	220
1992 [14]	X-1.5L or L	16.85	10.4 or 0	0.26 or 0.16	20
				0.48 or 0.30	41
				0.26 to 0.48 or	
				0.16 to 0.30	287
				0.10 or 0.06	
				0.02 or 0.01	18
					17

[a] β is calculated as defined in the text ($\beta = r/R$), except for the L position: $\beta = r/L$.

2. Types of Columns

Studies on the β ratio (r/R) led to the creation of various columns by varying the radius of the column holder. However, another idea was to change the holder configuration. Figure 3 has shown the types of column holders used on type J and cross-axis coil planet centrifuges. The multilayer coil (Figure 3A) is the best known and simplest as the Teflon tube is directly wound on a cylinder. It is also called a type I column [20]. This standard type (Ia) consists of alternative layers of right- and left-handed coils. A second subtype (Ib), mainly used for cross-axis coil planet centrifuges, consists of entirely left- or right-handed coils with interconnection tubes between each layer.

The eccentric coil assembly, shown in Figure 3B, is also called a type II column [20]. For this type of column, β is defined as the ratio of the radius of a cylinder forming a column unit on R. The purpose is the decrease in β, which can be as low as 0.01. The internal volume of such columns is small (a few tens of mL) so that they have an analytical purpose. Moreover, using small β values increases the efficiency (number of theoretical plates) of the column [14].

Except for the first prototype, it was possible to assemble two columns in series. The advantage is to continuously equilibrate the whole apparatus from a mechanical point of view. Another advantage is to set the internal total volume; the use of one eccentric coil leads to a few tenths of mL, whereas two multilayer coils in series may lead to more than 0.6-L capacity.

3. The Latest Prototype

The latest cross-axis coil planet centrifuge (X-1.5L and L types [14]) has been commercialized by Countercurrent Technologies, Inc. (see the list of manufacturers of CCC devices at the end of the book); a photograph is shown in Figure 14. All parts are made from stainless steel, except for the gears and the cylindrical holders which are in Delrin. Its external dimensions are 60 × 60 × 35 cm. A horizontal section is shown in Figure 15, perpendicular to the central axis. The two columns, made of several layers of Teflon tube wound on the cylindrical holder, are mounted in series. The internal diameter of the tube used for the columns is 2.6 mm. The rotational speed is regulated up to 1000 rpm, the usual operating speed range being 400–800 rpm. The configurations of the stationary and planetary miter gears and the pulleys and belts force the column to rotate around its own axis at the same speed as its revolution speed around the central axis. Two counteraxes, rotating thanks to the plastic gears, are required to prevent the twisting of the inlet and outlet Teflon tubes. The prototype was built to allow the counteraxis to be exchanged with the column holder. In that case, the type would be L.

B. Use of the Apparatus

Before running the apparatus, some parameters must be set to ensure the best retention of the stationary phase and, to a lesser extent, high efficiency.

For centrifugal partition chromatography and type J hydrodynamic apparatus, only two parameters are involved. The choice of the mobile phase should be based on the upper (lighter) liquid phase or the lower (heavier) liquid phase of the solvent system and its pumping direction. The hydrostatic system has been the simplest to use as the best retentions of the stationary phase are obtained for only two combinations, i.e., lower mobile phase pumped in the descending mode

Figure 14 Photograph of X-1.5L and L prototypes. All of the rotating parts are enclosed in a stainless steel box. The pulley driven by a belt from the electric engine is at the bottom of the central vertical axis. Two cylindrical holders (in white) are mounted in the X-1.5L position. The inlet and the outlet Teflon tubes go through the upper plate of the apparatus inside the emptied part of the central vertical axis. On counteraxis is installed for each coil to prevent the twisting of the Teflon connections. A thin circular metallic plate around the main holder of the columns decreases the torque by an average of 30%. Owing to the sufficient distance between the axis of a column holder and its counteraxis, the connecting tubes are reliable and do not need to be often replaced.

or upper mobile phase pumped in the ascending mode (see Chapter 5). Type J hydrodynamic systems require a more complicated choice of the two parameters because the best combinations depend mainly on the nature of the solvent system as well as on the geometrical dimensions of the apparatus and on the temperature (see Chapter 1).

1. Running Parameters of the Cross-Axis Coil Planet Centrifuge

The cross-axis coil planet centrifuge (CPC) involves the highest number of running parameters, compared to the other CCC devices. They have been classified into two groups and are described hereafter.

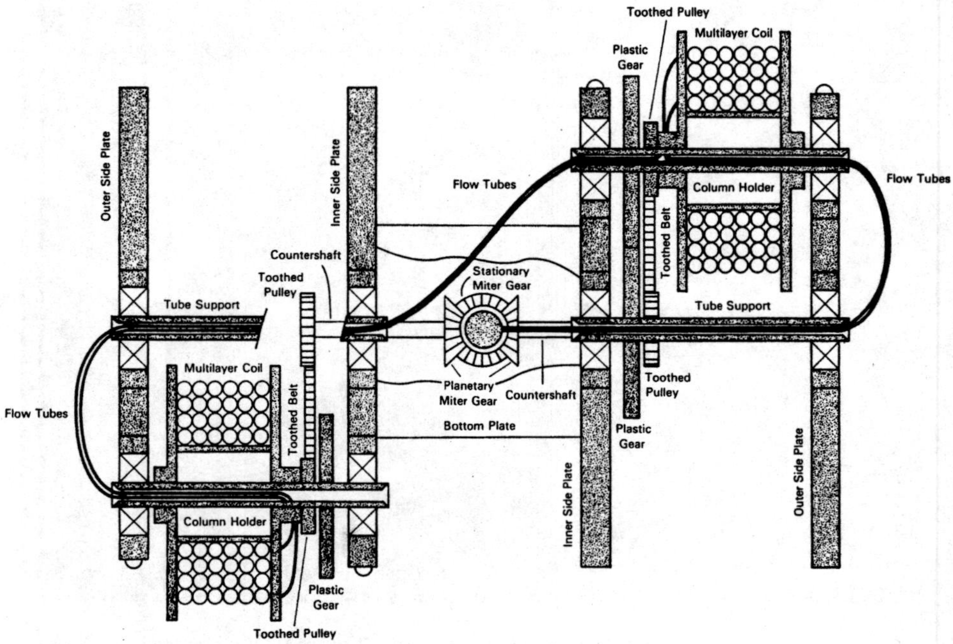

Figure 15 Section of the latest cross-axis CPC. The plane of the section is perpendicular to the central vertical axis of the apparatus. The two column holders are shown in the X-1.5L position. The paths of the connecting tubes are shown. The inlet tube comes from the central axis, enters the counteraxis, makes a U turn to enter the column holder, then continues toward the second counteraxis, makes another U turn to enter the second column axis. To come back, the path has to be exactly the same to prevent any twisting of the Teflon tubes.

Mechanical Characteristics

For this group, the six parameters are set before a series of experiments and are not changed very often. Their descriptions are given below:

1. Coil radius (*r*, shown in Figure 13A)
2. Internal diameter of the tubing (usually 0.8, 1.6, or 2.6 mm i.d. Teflon tubing)
3. Winding direction of the tubing on the cylindrical holder (right- or left-handed)
4. Column type (type I or II)
5. Number of tubing layers
6. Column holder position (e.g., L or X-1.5L)

Choices Left to the User

The six parameters are related to the design of the separation carried out by the experimenter. Their characteristics are given below:

7. Choice of the mobile phase (heavier or lighter liquid phase from the solvent system)
8. Flow rate of the mobile phase
9. Rotation direction (clockwise or anticlockwise around the central vertical axis)
10. Rotational speed (around the central vertical axis, commonly from 400 to 1000 rpm)
11. Pumping mode (from head to tail or from tail to head)
12. Pumping direction (Inward, opposed to the direction of the centrifugal force or outward, same direction as the centrifugal force)

Temperature

Temperature represents a major but implicit parameter of each CCC experiment. To ensure the reliability of an optimized purification or separation, the temperature has to be precisely set. Two devices may be used for that purpose.

The best one is the use of a thermoregulated room, in which the whole CCC chain is installed. The main advantage of such a device lies in the thermoregulation of all units involved in the chain, i.e., solvents, connecting tubes, countercurrent chromatograph, pump heads, fraction collector.

A ''lighter'' solution is the use of an air conditioning unit installed, for instance, at the top of the countercurrent chromatograph. However, the lower price compared to that of the first solution balances the drawback of the other thermoregulated devices needing a chosen temperature for the solvents, pumps, and fraction collector.

2. Optimization for Retention of Stationary Phase

As for the Sanki devices and the type J coil planet centrifuges, the cross-axis CPC needed first to be optimized for the retention of the stationary phase. Two methods have been employed to investigate the role of each parameter and to consequently give guidelines for choosing the parameters in order to achieve a more than satisfactory retention of the stationary phase.

The first method is based on graphics and has been introduced and explained by Ito [16]. However, it gives interesting guidelines only if the effect of the studied parameter is high or if the effects of the interactions are important.

Goupy et al. have introduced the experimental design method [21,22] which has proved to be very efficient in giving precise guidelines for optimizing the retention of the stationary phase. An extensive study has been carried out on the latest version of the cross-axis prototype [23] in order to give precise and simple guidelines for good retention of the stationary phase, along with a better understanding of the role of each factor and of the multiple interactions.

Only the main conclusion is given below and we refer the reader to Refs. 21 and 22 for more details on experimental designs applied to CCC and the experiments we have carried out.

For an organic phase/aqueous phase solvent system, the following guidelines have to be followed in order to obtain a retention of the stationary phase among the best attainable values:

1. The mobile lighter phase has to be pumped from the tail to the head of the column, in the inward direction while the rotation is clockwise, *or* the mobile heavier phase has to be pumped from the tail to the head of the column in the outward direction.
2. In case of some leaks of the stationary phase, the tail-to-head pumping mode has to be replaced by the head-to-tail pumping mode.
3. The coils must be in the eccentric position (e.g., X-1.5L).
4. The rotational speed must be high (800 rpm for instance).
5. The diameter of the coil must be high.

The first two guidelines are of paramount importance in order to ensure the retention of the stationary phase and to obtain a high value. The last three guidelines may help in increasing by a few percents the retention of the stationary phase.

3. Separation Procedure

The procedure is similar to that used for a type J unit. The liquid phase chosen as stationary is pumped in the column while the apparatus is stopped. When the column is completely filled, the cross-axis CPC is rotated at the desired speed and direction. The mobile phase is then pumped and the hydrodynamic equilibrium of the phases is achieved when the mobile phase emerges from the outlet tube. The sample can be introduced using an injection valve. The solutes are dissolved in the stationary phase, the mobile phase, or a mixture of both.

The above guidelines help the user to choose the running conditions. He or she has to keep in mind that a higher flow rate of the mobile phase means a lower retention of the stationary phase in most cases.

For on-line solute monitoring, the outlet tube is connected to detectors that are readily available for liquid chromatography, such as a UV detector, a fluorimeter, or an evaporating light scattering detector (ELSD) [24].

C. Features

1. Retention of Stationary Phase

As we saw in Chapter 1, Ito defined three solvent system groups according to the hydrophobicity of the nonaqueous phase. The first one, "hydrophobic," gathers solvent systems containing a hydrophobic organic phase, such as heptane/water or chloroform/water. Such systems are easily retained by the type J [25] and cross-axis CPCs [26] and by the CDCCC [27], with a particularly high value for the x axis: average values are shown in Table 2.

''Intermediate'' solvent systems involve a more hydrophilic organic phase; examples are the chloroform/acetic acid/water or *n*-butanol/water systems. Their tendency to evolve after mixing to a more stable emulsion than hydrophobic systems reduces the retention of stationary phase. As indicated in Table 2, type J CPC undergoes the highest decrease [25], whereas the two other devices show only a small decrease [26,27].

The ''hydrophilic'' group encompasses biphasic systems containing a polar phase, such as the *n*-butanol/acetic acid/water or *sec*-butanol/water systems. Their emulsions are nearly stable. Consequently, their stationary phase is even less retained in the column; the type J device leads to such low values that it can hardly be used with such systems [25]. The CDCCC allows good retention of stationary phase [27] at a low flow rate limited by the internal pressure drop due to the viscosity of the organic phase. The cross-axis apparatus does not encounter such a limitation [14,26]; its internal pressure drop is similar to that observed inside the type J CPC, which is much smaller than that obtained on a CDCCC.

This advantage is magnified when systems containing at least one polymer phase are used, as demonstrated in Table 2. The retention of stationary phase ranges from 30% to 70% at an average 3 mL/min flow rate [14] for the cross-axis CPC, whereas the CDCCC enables a smaller range but at a lower than 1 mL/min flow rate [27]. Type J CPC using type I columns does not retain these solvent systems [1], whereas the use of type II columns allows a small retention of stationary phase (typically 20%) [28].

Precise examples have been discussed in the thesis of Menet [23]. The solvent system made of heptane/acetic acid/methanol 1:1:1 (v/v/v), introduced for the separation of fatty acids by Murayama et al. [29] and used as a test separa-

Table 2 Maximum Retentions of Stationary Phase for Type J and Cross-Axis Coil Planet Centrifuges and CDCCC

Solvent system group	Type J	Cross-axis	CDCCC
Hydrophobic	90	>95	75
Intermediate	50	>80	75
Hydrophilic	20 to 30	50–90	50–70
Containing a polymer phase	0	30–70	<50

The retention of stationary phase is given as the percentage of the total volume of the column. For hydrophilic systems and those containing a polymer phase, the average values are obtained on the cross-axis coil planet centrifuge at a 3 mL/min flow rate, while they are typically obtained at a 1 mL/min flow rate on the CDCCC.

tion by our laboratory, has been used to compare various CPCs. Three devices have been compared: a hydrostatic-type CPC, named by Sanki whose description and features are given in Chapter 5; a type J apparatus, named by PC Inc.; and a cross-axis type of unit. Two figures illustrates the features of the cross-axis device relative to the two other devices. Figure 16 shows that the retentions of the stationary phase all increase with the rotational speed, whatever the apparatus. However, for the same flow rate, the values are 20–25% lower for the Sanki than for the cross-axis CPC. The use of flow-rates up to 8 mL/min does not prevent the latter from leading to satisfactory retentions of the stationary phase. Figure 17 shows that for the hydrodynamic-type CPCs, the retention of the stationary phase is less sensitive to the flow rate than the hydrostatic-type device. Below 6 mL/min, the retention of the stationary phase is at least 20% lower for the Sanki unit than the values for the PC Inc. or the cross-axis device. These two figures have shown that the cross-axis presents a unique possibility to retain the stationary phase compared to a type J CPC and a Sanki unit. An increase in

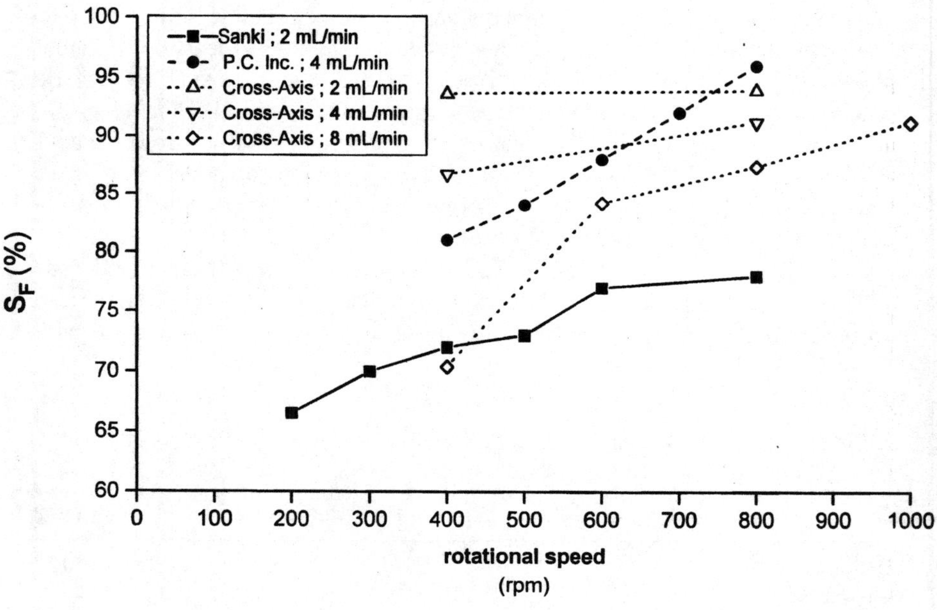

Figure 16 Variation of the retention of the stationary phase vs. the rotational speed; solvent system: heptane/acetic acid/methanol (1 : 1 : 1, v/v/v), lighter mobile phase; Sanki LLN unit (six cartridges, ascending mode), PC Inc. (one coil of 310 mL of internal volume), and a cross-axis unit (two coils of a total 625 mL of internal volume, P_I-T-I pumping mode).

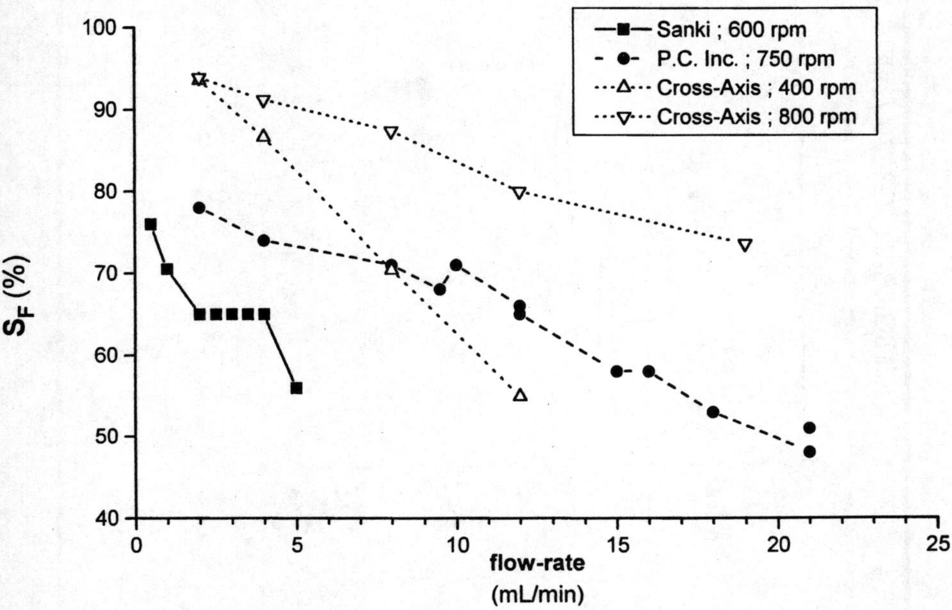

Figure 17 Variation of the retention of the stationary phase vs. the flow rate; solvent system: heptane/acetic acid/methanol (1 : 1 : 1, v/v/v), lighter mobile phase; Sanki LLN unit (six cartridges, ascending mode), PC Inc. (one coil of 310 mL of internal volume) and a cross-axis unit (two coils of a total 625 mL of internal volume, P_I-T-I pumping mode).

rotational speed induces only a small increase in the retention in the stationary phase but ensures a better stability when the flow rate is increased.

2. Efficiency and Resolution

According to the following formula:

$$R_S = 2V_S\,(K_2 - K_1)/(w_2 + w_1) \tag{1}$$

where R_S is the resolution between two adjacent peaks, K_2 and K_1 the partition coefficients of the separated solutes, and w_2 and w_1 the peak base width in volume units, the resolution between two peaks is proportional to the volume of stationary phase inside the column. As a result, the higher the retention of stationary phase, the higher the resolution. The high retentions of stationary phase obtained with the cross-axis CPC are consequently revealed as being particularly helpful to achieve good separations.

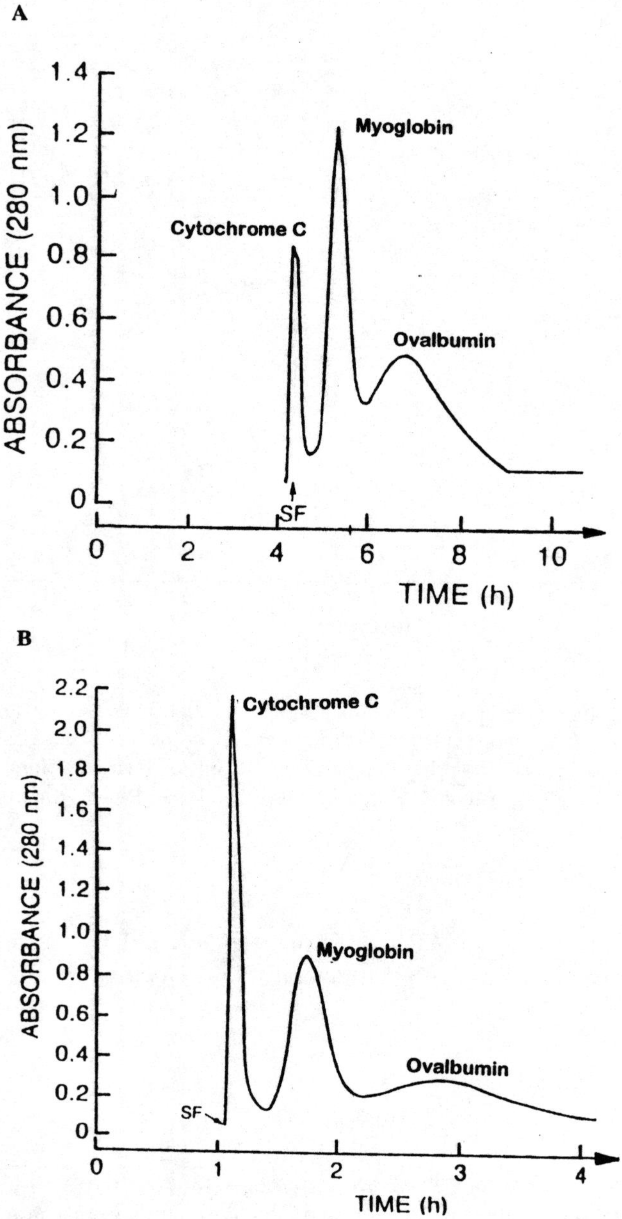

Figure 18 Comparison of separations of a mixture of three proteins carried out on a type J CPC and on a cross-axis CPC. (A) Type J CPC: equipped with type II columns giving a 220-mL capacity; sample: 10–100 mg of each protein; solvent system: PEG-1000/K_2HPO_4/water (12.5, 12.5, 75%, w/w/w); heavier mobile phase, flow rate: 0.65 mL/min; rotational speed: 800 rpm. (B) Type X-LL cross-axis CPC: equipped with type Ib columns giving a 250-mL capacity; same conditions as above except flow rate 0.65 mL/min and rotational speed 800 rpm.

Using the same solvent system, a separation of three proteins (e.g., cytochrome *c*, myoglobin, and ovalbumin) allows comparison of type J and cross-axis CPCs [8]. To obtain a sufficient retention of the polyethylene glycol (PEG)–rich stationary phase, type II columns were used with type J CPC. However, a limited 0.65 mL/min flow rate leads to 19% of retention of stationary phase, which is low compared to the 49% obtained with the cross-axis at a 2 mL/min flow rate. The low flow rate for type J CPC increases the separation time to 15 hr as shown in Fig. 18A, three times longer than the experiments carried on the cross-axis CPC (Fig. 18B). Moreover, the latter shows a higher resolution, partly due to the increase in stationary phase retention. Efficiencies, measured as the number of theoretical plates using the myoglobin and ovalbumin peaks, are similar between the two devices.

D. Selected Applications

Table 3 shows separations or purifications achieved with the various cross-axis prototypes. They have analytical or preparative purposes and can be classified

Table 3 Examples of Separations and Purifications

Separated compounds	Quantities	Solvent systems
DNP amino acids [15,28,30,31]	100 mg to 10 g	Chloroform/acetic acid/0.1 M HCl
Dipeptides (containing a tyrosine moiety) [15,28]	100 mg to 2.5 g	*n*-Butanol/dichloroacetic acid/ 0.1 M ammonium formate (100:1:100 to 100:0:100, v/v) or *n*-butanol/acetic acid/water
Indole auxins [30,31]	3 g	*n*-Hexane/ethyl acetate/ methanol/water
Proteins (containing a heme group, lipoproteins, globulines, histones, recombinant enzyme) [19,8,32,33]	10 mg to 1 g	PEG-1000/potassium phosphate buffer or PEG-8000/Dextran T500 + potassium phosphate buffer
Polysaccharides [34]	[a]	*n*-Butanol/0.13 M NaCl + HPC (15 g/L)
Steroids (crude synthetic mixture) [30]	2.4 g	*n*-Hexane/ethyl acetate/ methanol/water
Flavonoids (from a crude extract of sea buckthorn) [30]	100 mg	Chloroform/methanol/water
Antibiotics (bacitracin) [31]	5 g	Chloroform/95% ethanol/water

HPC, hexadecyl pyridinium chloride.
[a] Data not available.

into three groups. One encompasses the experiments carried with biphasic polar solvent systems; the second group deals with more ''classical'' solvent systems as they are based on a low-polarity organic phase; and the third group gathers aqueous two-phase polymer systems (ATPSs).

1. Polar Organic Phase/Aqueous Phase Systems

Figure 19 shows a chromatogram of the analytical separation of six dipeptides containing a tyrosine moiety [15] that was obtained on the first prototype (X type). The solvent system was made of two polar phases, i.e., a butanol-rich phase and an aqueous one. A gradient of dichloroacetic acid from 0.01 to 0 (v/v) was used. The retention of stationary phase was 55%, which is a high value for such a solvent system.

2. Low-Polarity Organic Phase/Aqueous Phase Systems

Figure 20 shows a chromatogram of 2.4 g of a crude reaction mixture of synthetic steroids [29]. The semipreparative separation was carried out at a high 4 mL/

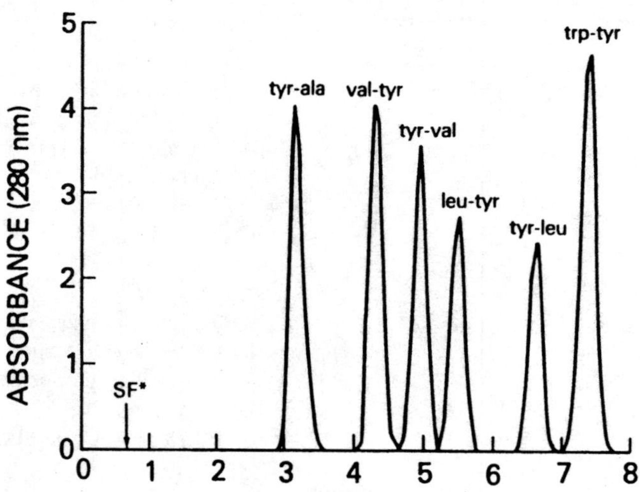

Figure 19 Chromatogram of a dipeptide separation on a type X cross-axis CPC [15]. Amounts of a 100 mg of tyrosylalanine, valyltyrosine, tyrosylvaline, leucyltyrosine, tyrosylleucine, and tryptophyltyrosine. Solvent system: *n*-butanol/dichloroacetic acid/0.1 M ammonium formate; gradient 100:1:100 to 100:0:100 (v/v/v) with a heavier mobile aqueous mobile phase; flow rate: 2 mL/min; rotational speed: 800 rpm; single type Ia column, total volume: 400 mL. Retention of the stationary phase: 55%.

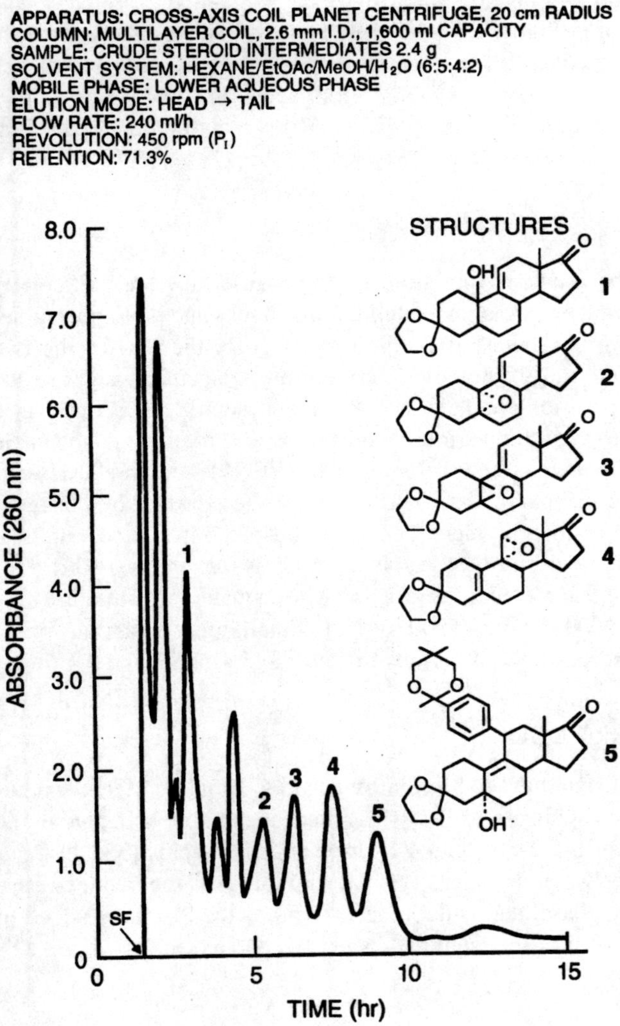

Figure 20 Chromatogram of a steroid reaction mixture on a type X-0.5L cross-axis coil planet centrifuge [30]. A 2.4-g amount of a mixture of synthetic steroids was used. Solvent system: hexane/ethyl acetate/methanol/water (6:5:4:2, v/v/v/v) with a heavier aqueous mobile phase; flow rate 4 ML/min; rotational speed: 450 rpm; two type Ia columns in series for a total volume of 1600 mL. Retention of the stationary phase: 71%.

min flow rate. However, the retention of stationary phase remained high with a 71% value. Five products were identified by nuclear magnetic resonance and their formula are given in Figure 20. The same solvent system also separates a 3-g mixture of indole auxins [30]. Solvent systems based on a chloroform-rich organic phase were used to separate DNP-amino acids and flavonoids from a crude extract of Sea buckthorn (*H. Rhamnoides*) [30]. An antibiotic, i.e., bacitracin, was purified with the same system [31].

3. Aqueous Two-Phase Polymer Systems

These systems have been extensively studied [5] because they are particularly suitable for the separation or extraction of cellular organelles and biological molecules. PEG/potassium phosphate buffer allows one to fix the pH for the two liquid phases between 4 and 9. Moreover, varying the molecular weight of the polymer modifies the partition coefficient of the solute. Such a system was used on the cross-axis prototypes for various applications to proteins, as shown in Table 3. Another well-known system is based on the PEG/dextran biphasic mixture. The partition coefficients of the solutes can also be adjusted by changing the molecular weights of the polymers; adding a phosphate buffer sets the pH. The very low interfacial tension of these solvent systems (as low as 0.0001 dyn/cm) are suitable to fragile molecules, as proteins whose quaternary structure may be broken. The PEG-8000/dextran T500 system containing a potassium phosphate buffer enabled separations of various histones and globulines [19].

4. "pH zone Refining" CCC

The pH zone refining technique has been introduced by Ito in the 1990s and has raised a wave of various applications [11–13]. A recent example is related to the separation of isomers of the 2-, 3-, and 4-hydroxycinnamic acids [35]. Using a *tert*-butyl methyl ether/water biphasic system, ammonia as the retainer base added to the stationary phase, and trifluoroacetic acid as the displacer added to the organic phase, 100 mg of the isomers has been separated in 2.5 hr.

REFERENCES

1. Y. Ito, in *Countercurrent Chromatography: Theory and Practice* (N. B. Mandava and Y. Ito, eds.), Marcel Dekker, New York, 1988, pp. 79–442.
2. Y. Ito, *J. Biochem. Biophys. Methods* 5(2):105 (1981).
3. W. D. Conway, *Countercurrent Chromatography: Apparatus, Theory and Applications*, VCH, New York, 1990.
4. Y. Ito and W. D. Conway, *J. Chromatogr. 301*:405 (1984).

5. P. Å. Albertsson, *Partition of Cell Particles and Macromolecules*, Wiley-Interscience, New York, 1986.
6. Y. Ito, J. Sandlin, and W. G. Bowers. *J. Chromatogr. 244*:247 (1982).
7. Y. Ito, H. Oka, and J. L. Slemp. *J. Chromatogr. 475*:219 (1989).
8. Y. Shibusawa and Y. Ito. *J. Chromatogr. 550*:695 (1991).
9. Y. Ito, H. Oka, E. Kitazume, M. Bhatnagar, and Y. W. Lee. *J. Liq. Chromatogr. 13*:2329 (1990).
10. Y. Ma and Y. Ito, *Anal. Chem. 68*:1207 (1996).
11. A. Weisz, A. Scher, K. Shinomiya, H. M. Fales, and Y. Ito. *J. Am. Chem. Soc. 116*: 704 (1994).
12. Y. Ito, K. Shinomiya, H. M. Fales, A. Weisz, and A. Scher, in *Countercurrent Chromatography* (W. D. Conway and R. Petroski, eds.), ACS Symposium Series, Am. Chem. Soc, Washington, DC, 1995.
13. Y. Ito, in *High-Speed Countercurrent Chromatography* (Y. Ito and W. D. Conway, eds.), Wiley-Interscience, New York, 1996.
14. K. Shinomiya, J.-M. Menet, H. M. Fales, and Y. Ito, *J. Chromatogr. 644*:215 (1993).
15. Y. Ito, *Sep. Sci. Technol. 22*:1971 (1987).
16. Y. Ito and T.-Y. Zhang, *J. Chromatogr. 449*:135–151 (1988).
17. Y. Ito, H. Oka, and J. Slemp, *J. Chromatogr. 463*:305 (1989).
18. Y. Ito, E. Kitazume, M. Bhatnagar, and F. Trimble, *J. Chromatogr. 538*:59 (1991).
19. Y. Shibusawa and Y. Ito, *J. Liq. Chromatogr. 15*:2787 (1992).
20. J.-M. Menet, D. Thiébaut, and R. Rosset, *J. Chromatogr. 659*:3–13 (1994).
21. J. Goupy, J.-M. Menet, K. Shinomyia, and Y. Ito, in *Countercurrent Chromatography* (W. D. Conway and R. Petroski, eds.), ACS Symposium Series, Am. Chem. Soc, Washington, DC, 1995.
22. J. Goupy, J.-M. Menet, and D. Thiébaut, submitted to *Analytical Chemistry*.
23. J.-M. Menet, Thèse de Doctorat de l'Université Pierre et Marie Curie, Paris VI, 1995.
24. S. Drogue, M.-C. Rolet, D. Thiébaut, and R. Rosset, *J. Chromatogr. 626*:41–52 (1992).
25. Y. Ito, *J. Chromatogr. 301*:387 (1984).
26. Y. Ito, *J. Chromatogr. 538*:67 (1991).
27. M.-C. Rolet, Thèse de Doctorat de l'Université Pierre et Marie Curie, Paris VI, 1993.
28. Y. Ito and T.-Y. Zhang, *J. Chromatogr. 449*:153 (1988).
29. W. Murayama, T. Kobayashi, Y. Kosuge, H. Yano, Y. Nunogaki, and K. Nunogaki, *J. Chromatogr. 239*:643 (1982).
30. T.-Y. Zhang, Y.-W. Lee, Q.C. Fang, R. Xiao, and Y. Ito, *J. Chromatogr. 454*:185 (1988).
31. M. Bhatnagar, H. Oka, and Y. Ito, *J. Chromatogr. 463*:317 (1989).
32. Y. Shibusawa, Y. Ito, K. Ikewaki, D.J. Rader, and H.B. Brewer, *J. Chromatogr. 596*:118 (1992).
33. Y. Shibusawa, M. Yamaguchi, and Y. Ito, *J. Liq. Chrom. & Rel. Technol. 21*:121 (1998).
34. Y. Ito, personal communication.
35. Y. Shibuzawa, Y. Hagiwara, C. Zhimao, Y. Ma, and Y. Ito, *J. Chromatogr. A 579*: 47 (1997).

4

Centrifugal Partition Chromatography: Operating Parameters and Partition Coefficient Determination

Alain Berthod and Karine Talabardon
Centre National de la Recherche Scientifique, University of Lyon 1, Villeurbanne, France

I. INTRODUCTION

The very name of the technique, countercurrent chromatography (CCC), may be misleading for the nonspecialist in that there is no countercurrent motion in CCC. CCC is a chromatography technique in which both the stationary phase and the mobile phase are liquid. As detailed in previous pages of this book, the solvent biphasic liquid system is the heart of CCC. Concerning the "hardware," two kinds of CCC chromatographs are marketed: the hydrodynamic machines and the hydrostatic machines.

The hydrodynamic machines were developed by Yoishiro Ito [1–3]. They are described in other chapters of this book. The hydrostatic machines are developed and marketed by a single company: Sanki Engineering Limited (2-16-10, Imazato, Nagaokakyo, Kyoto 617, Japan). The technique name as chosen and registered by Sanki is centrifugal partition chromatography, and the corresponding acronym, CPC, may be confusing. The CPC acronym was used in CCC to name the coil planet centrifuge. In this chapter, the CPC acronym will not be used in order to avoid confusion.

The first part of the chapter compares the two CCC modes: hydrostatic and hydrodynamic. It also describes the centrifugal partition chromatographs. The second part describes the principle and running parameters of centrifugal partition chromatography. The third part emphasizes an important application of the technique: the partition coefficient determination. The last part presents other applications of the original features of centrifugal partition chromatography.

II. HYDRODYNAMIC VS. HYDROSTATIC CCC DEVICES

A. Principle

Most of the hydrodynamic CCC devices developed by Yoishiro Ito contain open Teflon tubes coiled in spools with various geometry [1–3]. A combined rotation

Table 1 Comparison of the Features of the Two Main High-Speed CCC Machines

Type	Hydrodynamic	Hydrostatic
Liquid retained in	Coiled Teflon tubes	Channels
Rotating connections	None	Two rotary seals
Gyration axis	Two or more	One axis
Centrifugal field	Cyclic	Constant
Liquid retention quantity (% of V_c)	Between 0% and 96%	Between 40% and 80%
Stability[a]	Low	High
Efficiency	Up to 4 plates per tube turn or 50 plates per mL or less	Up to 1 plate per channel or 20 plates per mL or less
Internal volume adjustment	Changing spool or tubing	Sanki LLN: changing the cartridge number
Pressure[b]	Low, 0.01–0.8 MPa (0.1–8 kg/cm^2 or 1–110 psi)	Medium, 0.2–7 MPa (2–70 kg/cm^2 or 30–1000 psi)
Maintenance	Connecting tubing to change every ~50 hr	Rotary seals to lubricate every ~50 h
Other	Noisy gear assembly generating heat; no control on temperature, which increases during the run	Quiet centrifuge; low-temperature change during a run; very good temperature control on the Sanki LLN
Device name and acronym	Multilayer coil separator extractor, high-speed countercurrent chromatography, (HSCCC)	Centrifugal partition chromatograph (CPC)

[a] The liquid retention is stable when a small disruption (e.g., solute injection) does not cause a collapse of the mobile/phase stationary phase equilibrium.

[b] 1000 psi ≈ 70 kg/cm^2 ≈ 7 MPa ≈ 70 bars.

motion is applied with two or more axes of gyration. It creates a changing and cyclic centrifugal field that maintains the liquid stationary phase when the mobile phase is pushed through it. The hydrostatic machines derived from the droplet CCC (DCCC) units with serially connected glass tubes and only the gravitational field to maintain the liquid stationary phase. Basically, the hydrostatic machines were developed to increase the gravitational field by putting the DCCC tubes in the rotor of a centrifuge. In a centrifugal partition chromatograph, there is a single gyration axis producing a constant centrifugal field that maintains the liquid stationary phase in little chambers called *channels* serially connected by *ducts*. Table 1 compares the features of both modes.

The main difference between the two modes is illustrated by Figure 1. In

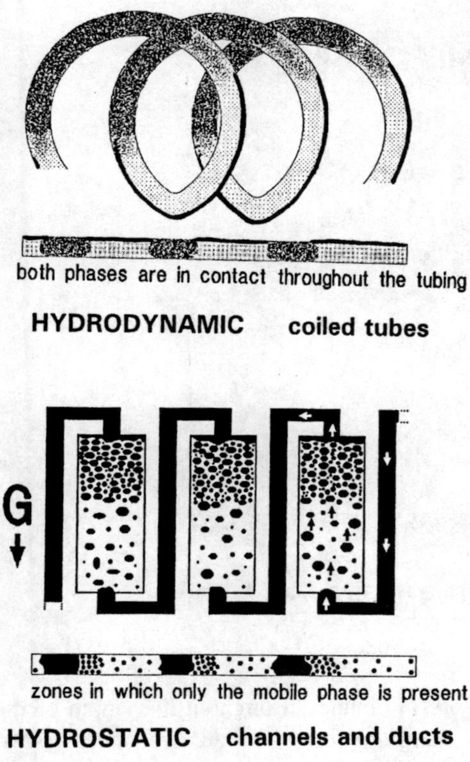

Figure 1 Oversimplified view of the hydrostatic and hydrodynamic principle of CCC apparatuses.

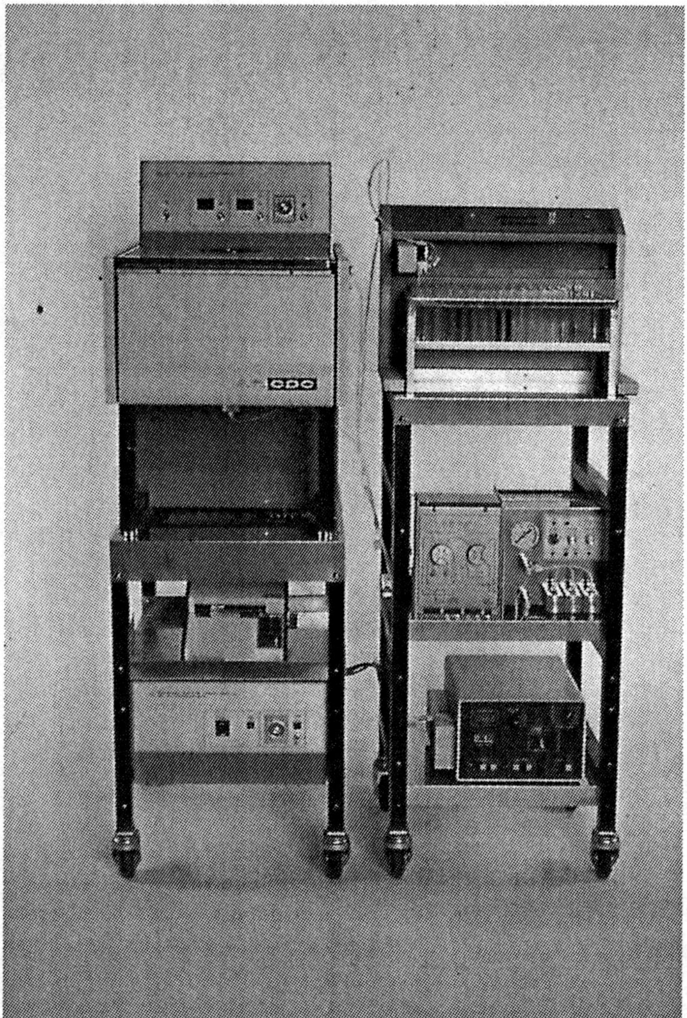

Figure 2 The complete CPC-LLN apparatuses from Sanki Engineering.

hydrodynamic devices, both phases are in contact throughout the length of the coiled tubes. In hydrostatic devices, there are zones, the ducts connecting two adjacent channels, in which only the mobile phase is present. This difference explains why the efficiency of hydrostatic machines is lower than that of hydrodynamic units at comparable apparatus volumes: in the latter, there is no unused

Figure 3 The rotor of the CPC-LLN apparatus. The upper rotary seal and six 250-W cartridges can be seen. (From Ref. 4, p. 358.)

Figure 4 The channels engraved in a disk mounted inside the rotor of a HPCPC 1000 apparatus. (From Ref. 4, p. 361.)

Figure 5 The rotor of the HPCPC 1000 apparatus (cf. Figure 3).

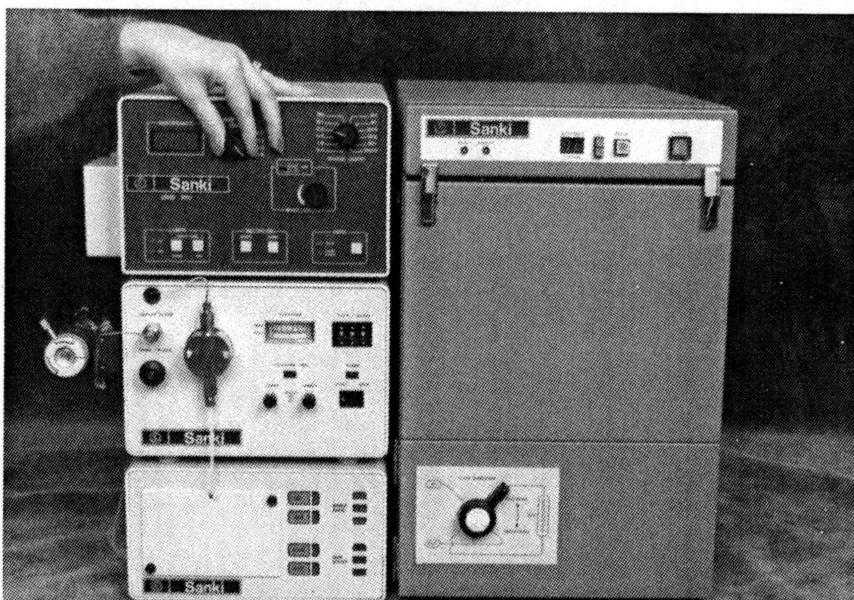

Figure 6 The complete HPCPC 1000 from Sanki (cf. Figure 2).

Table 2 Technical Data on Centrifugal Partition Chromatographs [4]

	CPC-LLN 250 W	CPC-LLN 1000 W	HPCPC 1000	HPCPC 2000	LLI 005	LLI 030
Channel						
Length (cm)	1.24	3	1.5	2.9	4.6	8
Section (mm^2)	2.7	44	5.9	30	70	370
Volume (μL)	36	1350	88	860	3300	30,000
Number	From 400 to 4800, 400 per cartridge	From 40 to 480, 40 per cartridge	2136	1332	1144	792
Duct						
Volume (mL)	5.5 mL/cartr.	21 mL/cartr.	33	250	1200	6200
%[a]	27%	28%	15%	18%	24%	21%
Internal volume	20 mL/cartr.	75 mL/cartr.	220	1400	5000	30,000
rotor radius (cm)	15	15	10	15	15	25
Max rpm	2000	2000	2000	1500	1300	1000
Max field G[b]	550	550	400	320	230	230
Flow rate range (mL/min)	1–10	1–10	1–20	5–40	10–140	100–700

[a] This percentage indicates the geometrical fraction of the internal volume occupied by the ducts.
[b] Centrifugal field inside the rotor rotating at maximum speed expressed in $G = 9.8$ m/sec^2, the earth gravitational field.

volume. Also, it explains why there is little pressure buildup in hydrodynamic CCC; and why the maximum stationary phase retention volume cannot be 96% in a hydrostatic apparatus. The maximum liquid retention volume is the total channel volume, which is the machine volume minus the volume of the ducts. A complete book was recently dedicated to centrifugal partition chromatography [4].

B. Centrifugal Partition Chromatographs

The Sanki Engineering Company produced the centrifugal partition chromatograph Model CPC-LLN from 1985 to 1993. Figure 2 is a general view of the complete setup. The CPC-LLN machine has a centrifuge rotor that holds up to 12 cartridges (Figure 3). There are 400 channels per cartridge with a total internal volume of 20 mL. A complete description of all features of this apparatus can be found in Refs. 4, 5, and 15. The CPC-LLN has a temperature-regulated rotor. The working temperature could be set between 10°C and 50°C. The internal V_C volume could be adjusted between 20 mL up to 240 mL, working with 1 up to 12 cartridges, respectively.

The production of the LLN model was discontinued in 1993 when the HPCPC series 1000 was introduced. The cartridge arrangement was abandoned. The stationary phase is retained in channels engraved in disks (Figure 4). Several disks are piled in the centrifuge rotor shown by Figure 5. The channel volume was redesigned with a trapezoidal form to enhance the mixing efficiency. The depth of the connecting ducts was reduced to decrease the volume in which there is no phase contact. The HPCPC machine is smaller in size than the CPC-LLN (Figure 6). Larger centrifugal partition chromatographs can be custom-made to fit the need of a specific application. Table 2 lists the technical data of the several machines of the Sanki Company.

III. CHROMATOGRAPHIC PARAMETERS

The chromatographic parameters used for hydrostatic devices are not different from those used for hydrodynamic machines and other liquid chromatography (LC) apparatuses [6].

A. Solute Retention

The fundamental equation of partition chromatography is

$$V_R = V_M + \mathbf{P}V_S \tag{1}$$

where V_R, V_M, and V_S are the solute retention volume, the mobile phase volume, and the stationary phase volume, respectively. **P** is the solute distribution coefficient between the two phases. This equation is difficult to use in LC because partitioning is seldom the only cause of solute retention. Silanol interaction, adsorption, and pore size exclusion can be other factors participating in the LC solute retention. Also the V_S volume is difficult to know accurately in LC. There are no such problems in CCC. The solute retention depends on one parameter only: the liquid–liquid partition coefficient, **P**. The stationary phase volume inside the CCC "column," V_S, is exactly known because the sum

$$V_M + V_S = V_C \tag{2}$$

is the internal volume, V_C, of the CCC "column" or instrument.

The retention factor, k', is the parameter most commonly used in LC to study solute retention. It is defined as

$$k' = (V_R - V_M)/V_M \tag{3}$$

The partition coefficient, **P**, is defined referring to the stationary phase:

$$\mathbf{P} = \frac{\text{[solute concentration in the stationary phase]}}{\text{[solute concentration in the mobile phase]}} \tag{4}$$

The partition coefficient depends on which solvent or solvent mixture is used as the stationary phase. If the phase role is reversed, the mobile phase becomes the stationary phase and the partition coefficient **P** becomes $\mathbf{P}' = 1/\mathbf{P}$. This leads to one of the big advantages of CCC over LC: no solute can be irreversibly retained in the column. A high retention volume means a high partition coefficient **P**. Reversing the phases will produce a low **P′** value ($\mathbf{P}' = 1/\mathbf{P}$), then a low retention volume.

The retention factor, k', can be related to the solute partition coefficient by

$$k' = \mathbf{P}V_S/V_M \tag{5}$$

The difference in retention between two solutes is measured by the separation factor, α:

$$\alpha = k'_2/k'_1 = \mathbf{P}_2/\mathbf{P}_1 \tag{6}$$

where α is always higher than or equal to 1.

B. Peak Efficiency

The efficiency, N, is related to the peak width. Efficiency is measured by plate numbers. A classical equation for a Gaussian-shaped peak is

$$N = 4(V_R/W_{0.6h})^2 \tag{7}$$

in which $W_{0.6h}$ is the peak width at 60% of the peak height. $W_{0.6h}$ corresponds exactly to 2σ, the variance, if the peak is perfectly Gaussian. The Gaussian peak base width, W_b, is equal to 4σ or $2W_{0.6h}$.

Efficiency is related to the solute mass transfer between the two liquid phases. In general, it is observed that the efficiency increases when the viscosity of the liquid phases decreases. Also, the efficiency is directly proportional to the number of channels. Consequently, more channels, i.e., more cartridges, give more plates and sharper peaks, but also higher pressure drops. Efficiency is also related to the mobile phase flow rate and the stationary phase volume retention in a more complex manner as described in the first chapter of this volume.

C. Solute Resolution

The resolution, *Rs*, is a parameter measuring the quality of a separation. A 1.5 resolution value means that two adjacent peaks are separated with a baseline return in between [6]. A resolution factor lower than 1.5 means there is some peak overlap. A resolution factor higher than 1.5 means that there is space between the two peaks. *Rs* is expressed as the ratio of the distance between the two peak maxima to the mean value of the peak width, *W*, at the baseline:

$$Rs = \frac{V_2 - V_1}{(W_2 + W_1)/2} \tag{8}$$

Using Eqs. (1) and (6), and assuming the efficiency, *N*, is constant, *Rs* can be rewritten as

$$Rs = \frac{1}{4}\sqrt{N}\,\frac{P_2 - P_1}{\dfrac{V_M}{V_S} + \dfrac{P_2 + P_1}{2}} \tag{9}$$

The resolution factor can be expressed with retention factors instead of partition coefficients. Substituting Eq. (5), one obtains:

$$Rs = \frac{1}{4}\sqrt{N}\,\frac{k_2' - k_1'}{1 + \dfrac{k_2' + k_1'}{2}} \tag{10}$$

Using the separation factor [Eq. (6)] and making the approximation $k_1' \approx (k_2' + k_1')/2$, the classical resolution equation used in LC [6] is obtained:

$$Rs = \frac{1}{4}(\alpha - 1)\sqrt{N}\,\frac{k_1'}{1 + k_1'} \tag{11}$$

IV. RUNNING PARAMETERS AND THEORY

A. Stationary Phase Retention

The liquid stationary phase retention is of paramount importance in CCC. It was shown that the maximum stationary phase volume, V_S^{max}, that can be retained in a centrifugal partition chromatograph depends linearly on the flow rate. The operating mode to obtain the V_S^{max} volume is described as follows [7]:

Filling Procedure:

1. The centrifugal partition chromatograph is filled with the liquid stationary phase.
2. The rotor of the centrifuge is started.
3. The liquid mobile phase is pushed at the desired flow rate, entering the top of the instrument (for obvious reasons, it is also called the *head*) and moving in a *descending* way if it is heavier than the liquid stationary phase. If the mobile phase is the lightest liquid phase, it should enter the centrifugal partition chromatograph through the bottom (or *tail*) and move toward the top or head of the machine in an *ascending* way. The equilibrium shown by Figure 1 is established channel after channel inside the machine.
4. The exiting phase is collected in a graduated cylinder. As long as the liquid stationary phase leaves the machine, the pressure increases. When the first drops of the mobile phase appear in the cylinder, the pressure stabilizes.
5. The volume of stationary phase collected in the cylinder corresponds to the volume of mobile phase inside the machine. This volume is V_M^{min}, the minimum volume possible to have inside the machine at the selected rotation speed and flow rate, and with the biphasic liquid system used [8,9].

The V_S^{max} stationary phase volume is directly related to the V_M^{min} minimum mobile phase volume by

$$V_S^{max} = V_C - V_M^{min} \tag{12}$$

The retention factor, *Sf*, expressed in percentage of the instrument volume, is defined as:

$$Sf = 100 \times V_S/V_C \tag{13}$$

It was shown by Foucault and Bousquet [10–12] that the V_M^{min} volume increased linearly with the flow rate. That means that the V_S^{max} volume decreases linearly with the flow rate. The slope and intercept of the V_M^{min} vs. F lines depends on the liquid system used but not on the rotation speed above 200 rpm. These results have produced interesting theoretical observations, discussed later in Section IV.B.

Between the 0–V_S^{max} limits, it is possible to obtain any volume of stationary phase retained inside the centrifugal partition chromatograph by filling the machine differently [13]. It is possible to use two pumps—one for the liquid stationary phase and the second for the mobile phase. Setting correctly the flow rates of each pump, the centrifugal partition chromatograph is filled, not rotating, by the desired V_M and V_S volumes. With only one pump, it is possible to fill the machine at a higher flow rate than the operating flow rate. The stationary phase volume retained will be lower than the V_S^{max} volume corresponding to this flow rate.

In a particular case when a minimum volume of stationary phase was desired, the *underload* mode was developed [14]. The filling and equilibrating of the centrifugal partition chromatograph was done in the way opposite that given in step 3 of the filling procedure listed above. Once the mobile phase was seen exiting the device, the normal way was resumed with a very low volume of stationary phase.

B. How Does the Mobile Phase Move Through the Stationary Liquid Phase?

It was shown by Menet et al. [15] that in a constant centrifugal field a liquid droplet forms and moves in a liquid phase at a velocity that depends only on the interfacial tension and on the two liquid phase viscosities. The droplet velocity does not depend *on the strength of the field or on the flow rate* [9–16]. Only centrifugal partition chromatographs work with constant centrifugal fields (Table 1); as a result, the theoretical work of Menet et al. [15,16] can be used to understand the fundamentals of centrifugal partition chromatography. This work has two implications:

1. In a constant field, the droplet velocity depends on the droplet size. This is the driving principle of centrifugation. If droplet velocity is independent of field strength, then a higher field, i.e., higher centrifuge rpm, must produce smaller droplets. The droplet size depends on the field strength.
2. If the flow rate increases and the droplet velocity does not, then it is requisite that the mobile phase forms more droplets of equal size. All droplets move at the same speed in the stationary liquid phase.

The first point indicates that the efficiency should increase with the rotation speed of the centrifuge. A decrease in droplet size is linked to an increase of the liquid–liquid interphase area. Then, the mass transfer between phases should be easier, producing a higher efficiency and sharper peaks at high rpm values. The second point explains why the stationary phase volume retained decreases linearly when the flow rate increases: if there are more droplets of mobile phase, there is less space for the stationary phase.

These theoretical points were observed experimentally by Foucault et al. [10–12,17]. They established that for any centrifugal partition chromatograph the linear relationship between V_M^{min} and the flow rate, F, is expressed by:

$$\frac{V_M^{min}}{V_C} = \frac{1-d}{\bar{u}}\frac{F}{S} + d \tag{14}$$

where d is the relative volume of the ducts expressed in percentage of the column volume (Table 2), $\bar{u}$ is the average mobile phase velocity inside the channel, and S is the channel section (Table 2). The slope of the V_M^{min} vs. F lines produces the average velocity $\bar{u}$. The velocity $\bar{u}$ was only depending on the biphasic liquid system used, not depending on the rotation speed or on the flow rate. The $\bar{u}$ value allowed classification of the liquid systems as very stable ($\bar{u} > 10$ cm/sec), stable ($10 > \bar{u} > 6$ cm/sec), and less stable ($\bar{u} < 6$ cm/sec) [17]. The product $S \times \bar{u}$ corresponds to the flow rate at which V_M^{min} is equal to V_C (i.e., $V_S = 0$, no stationary phase retention). It can be related to the flooding flow rate introduced by Berthod et al. [18]. The flooding flow rate was defined as the mobile phase flow rate high enough to push the stationary phase out of the rotor of the centrifugal partition chromatograph, eventually leading to $V_S = 0$. The intercept of Eq. (14) is d. It was found that the relative duct volume, d, was slightly dependent on the biphasic liquid system used.

Table 3 lists the theoretical droplet velocity described by Menet et al. [15], the experimental average mobile phase velocity measured by Foucault [17], the corresponding $S \times \bar{u}$ maximum flow rate, and the experimental relative duct volume for the HPCPC 1000 centrifugal partition chromatograph and several liquid systems. For the methanol heptane liquid system, one set of experiments was done with the CPC-LLN 250W centrifugal partition chromatograph and another with the HPCPC 1000 apparatus; the results differ significantly. The "flooding" flow rate, $S \times \bar{u}$, is 14 mL/min and 34 mL/min for the CPC-LLN and the HPCPC machine, respectively. It shows that the stability of the phase retention is much better in the HPCPC machine than in the older CPC-LLN chromatograph. The channel design is also better in the HPCPC machine, the d value is in the 20% range, one-fifth of the internal volume is useless. The CPC-LLN d value is in the 30% range, one-third of the machine volume is unused for interphase exchange.

Figure 7 shows two chromatograms obtained with an HPCPC 1000 unit in the same conditions but the rotational speed. The peak efficiency is higher when the centrifugal field is higher [9]. It is a proof that the solute mass transfer is faster at high rotation speed, which speaks for an increase of the interphase, i.e., a decrease of the droplet size at constant flow rate.

Table 3 Theoretical and Experimental Mobile Phase Velocities

Liquid system	Mobile phase[a] (cm/sec)	V^b (cm/sec)	S^c (mL/min)	d^d vol%
Chloroform-water				
Water (ascending)	28.1	63.4	99	23
Chloroform (descending)	28.8	42.0	102	20
Ethylacetate-water				
Water (descending)	19.0	16.3	67	26.5
Ethylacetate (ascending)	14.2	9.8	50	15
1-Butanol-water				
Water (descending)	13.6	1.3	48	21
1-Butanol (ascending)	11.8	1.9	41	15.5
2-Butanol-water				
Water (descending)	5.1	0.47	18	16
2-Butanol (ascending)	5.3	0.69	19	28
DMSO-heptane				
DMSO (descending)	23	28.5	70	22
Heptane (ascending)	15	7.0	46	18
Methanol-heptane				
Methanol (descending)	9.6	2.8	34	23
Methanol (descending)[e]	9.2[e]	—	14[e]	31[e]
Heptane (ascending)	9	2.2	32	19
1-Butanol/acetic acid/water (4:1:5 v/v/v)				
Aqueous (descending)	9.8	0.75	35	21
Organic (ascending)	5.5	0.84	20	18

All experiments were done with a HPCPC 1000 instrument with $S = 5.9$ mm² and $V_c = 240$ mL, the flow rate changed from 1 to 10 mL/min [17].
[a] Average mobile phase velocity from Eq. (14) and experiments from Ref. 17.
[b] Theoretical mobile phase velocity from Ref. 15.
[c] Flow rate corresponding to a nil stationary phase retention.
[d] Intercept of Eq. (14) from Ref. 17 in percentage of the column volume.
[e] Experiments done with the CPC LLN 250W [$S = 2.6$ mm², $V_c = 120$ mL (6 cartridges)] [12].

This was visualized by Van Buel et al. [19]. They used a special experimental centrifugal partition chromatograph with a glass plus Teflon transparent cartridge and made stroboscopic photographs. Figure 8 was drawn from the photographs they took at constant flow rate (6.6 mL/min) and different centrifugal fields. Large droplets, with diameters in the 2-mm range, are seen at 670 rpm (field 26 g). At 1000 rpm (field 56 g), there is no droplet but a continuous jet of mobile phase. At higher rotational speed (1400 rpm and higher) and higher field (100 g and up), the jet breaks in small droplets. Figure 8 shows that the stationary phase retention volume is similar at different fields. However, the Figure 8 arrows

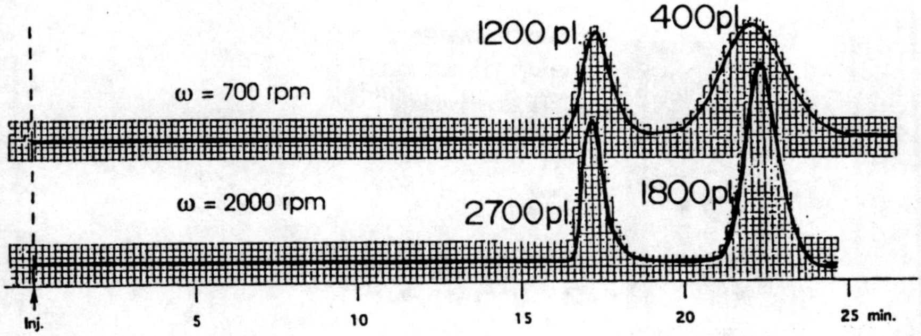

Figure 7 Effect of the rotational speed on efficiency. The higher rotational speed induces smaller droplet size and higher efficiency. HPCPC 1000, methanol descending mobile phase 7 mL/min, heptane stationary phase, *Sf* = 50%, phenyl acetate and butyl phenyl acetate. (From Ref. 17.)

point out that the stationary phase is partly forced in the duct by the high centrifugal field [19]. Also, a part of the mobile phase stays in the channel again due to the high centrifugal field. This explains the variations of the relative duct volume observed experimentally (Table 3) [17]. For example, the high 28% *d* value obtained for 2-butanol in the ascending mode (Table 3) might indicate that the water stationary phase penetrates deeply into the 2-butanol-filled ducts leaving a significant volume of 2-butanol in the channels. In the opposite mode, the *d* value is 16%, close to the 15% geometrical value (Table 2). It means that the 2-butanol stationary phase does not penetrate into the water-filled ducts in the descending mode.

These observations lead us to propose a different channel geometry (Figure 9) that would reduce useless volumes and band broadening. The channel volume is reduced at both the upper left and lower right side of the rectangle design. The duct volume is also minimized so that the *d* relative duct volume remains close to 10–15%. With the Figure 9 channel design, the centrifugal partition chromatograph efficiency would be better at comparable machine volume.

C. Pressure Drop

The experimental inlet pressure is much higher with hydrostatic devices than with hydrodynamic ones. Berthod et al. proposed that the observed pressure drop, *Pd*, in operating a centrifugal partition chromatograph was expressed as the sum of two terms [7]:

$$Pd = n[\Delta d\ \omega^2 Rh + \eta\psi F] \tag{15}$$

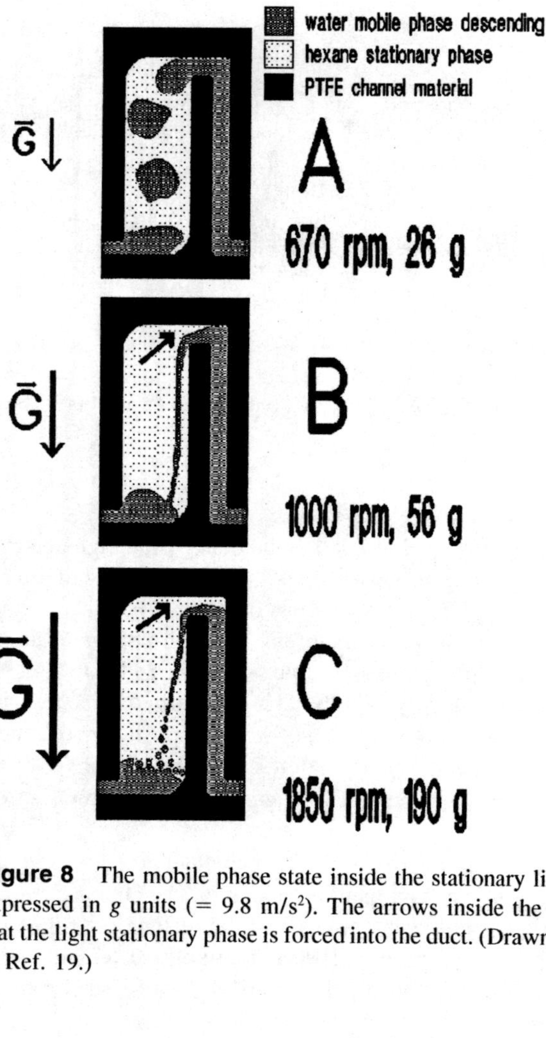

Figure 8 The mobile phase state inside the stationary liquid. G is the centrifugal field expressed in g units ($= 9.8\ m/s^2$). The arrows inside the channel in B and C point out that the light stationary phase is forced into the duct. (Drawn by Van Buel from photograph in Ref. 19.)

in which n is the number of channels, Δd is the phase density difference, ω is the rotor spin rate, R is the rotor radius, h is the stationary phase height in a channel that also contains the mobile phase, η is the mobile phase viscosity, ψ is a constant dependent for the type or geometry of the instrument used, and F is the mobile phase flow rate. Assuming the volume of the ducts is completely filled by the mobile phase, the stationary phase height, h, can be taken as the ratio:

$$h = V_S/(nS) = SfV_C/(nS) \tag{16}$$

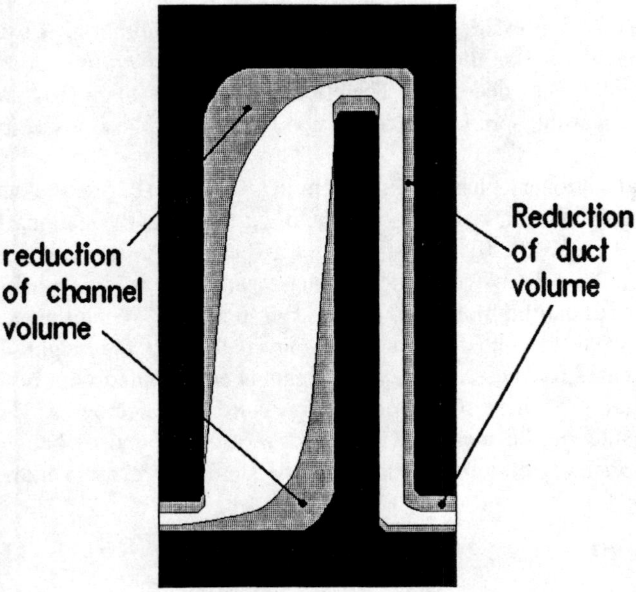

Figure 9 An improved design of for centrifugal partition chromatograph channels. The open area corresponds to the channel volume. The dotted area shows the difference between the existing design and the proposed design.

in which S is the channel section and Sf is the retention factor [eq. (13), ratio V_S/V_C]. Then, all terms of Eq. (15) can be obtained experimentally [19].

The first term of Eq. (15), $n\,\Delta d\,\omega^2 Rh$, or $\Delta d\,\omega^2 RSfV_C/S$, is the hydrostatic term, it increases *quadratically* with the centrifuge rotation speed. The second term of Eq. (15), $n\eta\psi F$, is the hydrodynamic or viscous term, it increases *linearly* with the flow rate. If the density difference between the two liquid phases is low ($\Delta d \leq 0.1$ g/cm^3), the maximum operating pressure, 70 kg/cm^2, will not be reached. With a density difference of 0.4 g/cm^3 (e.g., hexane-water, $\Delta d \leq 0.34$ g/cm^3), the pressure drop can reach the 70 kg/cm^2 upper pressure limit if the rotation speed is in the range of 800–2000 rpm, depending on n, the channel number (or V_C, the machine volume).

Equation (15) shows that the pressure drop depends on h, the average height of the stationary phase crossed by the mobile phase inside the channels. This means that at the beginning of the equilibration of the centrifugal partition chromatograph the pressure drop is always low, whatever liquid system is used, due only to the hydrodynamic viscous term. Then, it increases linearly with time as more channels are equilibrated [7]. With a centrifugal partition chromatograph, it is important to watch the pressure during the "column" equilibration; if it is noted that the pressure will pass the 70 kg/cm^2 limit, then *the rotation speed, and not the flow rate*, should be decreased. The effect of the flow rate on pressure drop is less dramatic than the spin rate except when the mobile phase viscosity is high.

Losses of liquid stationary phase is a problem in CCC [18,20]. Such losses can be due to bleeding, which is a small but continuous carryover of the stationary phase by the mobile phase [7,18,20]. Also, the pressure inside the unit can induce a stationary phase solubilization by the mobile phase saturated at atmospheric pressure [7]. Such losses can be monitored by pressure readings. When the stationary phase bleeds from the centrifugal partition chromatograph, the height, h, in each channel decreases (or the number, n, of channels equilibrated with both phases decreases, or both of these occur) and the pressure drop decreases. The stationary phase volume can be measured by injecting a compound of known partition coefficient, P. Its retention volume gives the stationary phase volume by:

$$V_S = (V_R - V_C)/(P - 1) \tag{17}$$

Van Buel et al. found experimentally that the hydrodynamic term of Eq. (15) was reliable. However, the hydrostatic term overestimates the experimental pressure at high rotation speeds [19]. They thought this is due to the part of the stationary phase that is pushed into the ducts at high rotation speeds (arrows in Figure 8). A third negative term, combining ω, the rotation speed, and F, the flow rate, should be introduced in Eq. (15) to take into account the phenomenon experimentally observed. Such a third negative term would be liquid system–dependent. Equation (15) is a good usable first prediction of the experimental pressure drop obtained with an actual set of parameters and a centrifugal partition chromatograph.

V. APPLICATIONS

Centrifugal partition chromatography is a CCC technique. CCC can be usefully applied to the purification of virtually any chemical sufficiently soluble in a biphasic solvent system that provides a partition coefficient in the 0.2–5 range (Eq. (1), [21]). Many CCC applications were described in various fields such as *biochemistry*: amino acids, peptides, proteins, antibiotics, drugs; *natural products*: flavonoids, digitonins, coumarins, carotenoids, retinals, fatty acids and esters, alkaloids, tannins and polyphenol compounds, saponins, indole auxins, gibberellins; *industrial organic and pharmaceutical chemistry*: dyes, pesticides and herbicides, catecholamines, drugs and metabolites, optical isomers; *inorganic chemistry*: alkali and rare earth cations, anions. They are described in several books besides this one [3–4,21]. The partition coefficient determination, especially octanolwater coefficient, and a single application of saponin separation will be exposed to illustrate the use of the technique.

A. Partition Coefficients Determination

Equations (1) and (17) show that the partition coefficient, **P**, of a solute is directly related to its retention volume. Then, CCC is a very useful tool for accurate liquid–liquid partition coefficient determination. In a recent work [22], we described the thermodynamics of solute partitioning and the effect of solvent polarity. The octanol water partition coefficient, P_{oct}, is used as the hydrophobicity scale in quantitative structure activity relationship. It explains the paramount importance of the P_{oct} coefficient in the prediction of the biological effects of chemicals. Three CCC methods for P_{oct} (and any **P**) coefficient determination are described: the direct method, the dual-mode method, and the cocurrent method.

1. Direct P_{oct} Measurements

When a CCC device is filled with octanol as the stationary phase and water is the mobile phase, the solute retention volume is directly related to its P_{oct} without any assumption. The big advantage is that there is no need to have highly pure solutes. Any impurity will probably have a differing P_{oct} and will be separated from the solute of interest. The P_{oct} range obtained by direct measurement is $0.01 < P_{oct} < 200$. The minimum retention volume should be 5 mL higher than the V_M volume [23] to produce an error on P_{oct} lower than 10% [23]. Very hydrophilic solutes have very low P_{oct} values, then very low V_R volumes that cannot be accurately differentiated from V_M, giving the lower limit of the P_{oct} range. The upper limit is due to the experiment duration. A solute with a 200 P_{oct} value has a retention volume higher than 200 times the octanol stationary phase volume. With $V_S = 30$ mL and $V_M = 10$ mL (CPC-LLN 250W with two cartridges) the retention volume of the 200 P_{oct} solute is 6010 mL [Eq. (1)]. At 6 mL/min, this volume corresponds to 16 hr 40 min.

The *underload* mode was proposed to decrease the retention volumes [14]. The V_S volume was reduced to a minimum. In the preceding example, the CPC-LLN device (two cartridges) can be underloaded with only 7 mL of octanol ($Sf = 17\%$). The retention volume of the 200 P_{oct} solute is reduced to only 1433 mL. The experiment duration at 6 mL/min is 4 hr. The drawback is that the efficiency is very low, which means that the peaks are broad and difficult to detect. Figure 10 shows the direct P_{oct} measurement of seven compounds with an underloaded CPC-LLN machine with six cartridges. It should be noted that the 167 P_{oct} measurement of 2-chloronitrobenzene took 13 hr ($V_R = 3750$ mL). The *Rs* factor between compounds 6 and 7 is about 0.3, the plate number is in

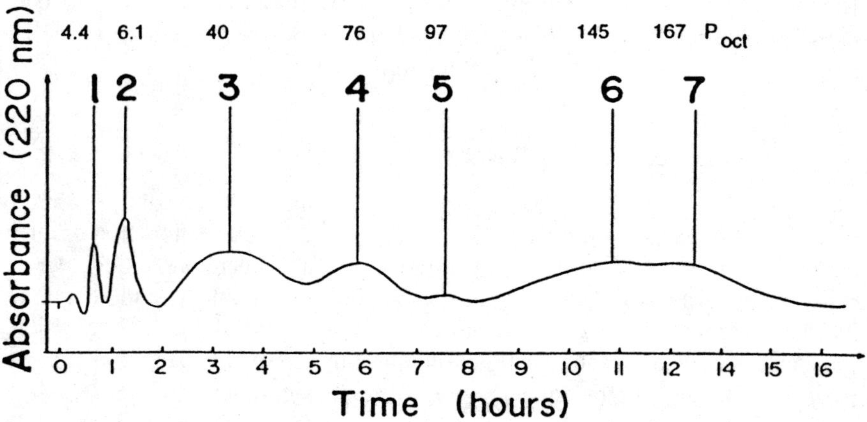

Figure 10 P_{oct} direct determination of seven compounds. CPC-LLN machine with six 250-W cartridges, V_T = 125 ml, 5 mL/min water pH 4, 21.7 < V_{oct} < 22.7 ml, 100 mL injection volume of about 2 mg of each solute, **1**: benzamide, **2**: 2-acetoxy benzoic acid, **3**: acetophenone, **4**: benzoic acid, **5**: 2-chlorobenzoic acid, **6**: 2-chlorophenol, **7**: 2-chloronitrobenzene. The small figures are the P_{oct} values. (From Ref. 7, p. 1451.)

the 60-plate range, i.e., 40 channels per plate. However, the high accuracy of the measurement should be pointed out: the absolute error on the 167 P_{oct} value is ±1 unit, producing the result 2.22 < log P_{oct} < 2.225 [22]. Testa et al. have had success using the octanol phase as the mobile phase for P_{oct} measurements of lipophilic solutes [24].

2. The Dual-Mode or Back-Flushing Technique

The similar dual-mode method and back-flushing method were proposed by Gluck and Martin [25] and Menges et al. [26], respectively. The idea is simple: solutes with very high P_{oct} values move very slowly in the octanol phase; they need a too long time to emerge outside the machine. To force them from the CCC apparatus, the roles of the aqueous and octanol phases and their flowing direction are reversed after some reasonable flowing time in the normal direction. Defining V_{aq}, the volume of aqueous phase pushed in the normal direction, and V_{oct}, the volume of octanol phase pushed in the reversed direction and needed to retrieve the studied solute, its P_{oct} value is simply expressed by [25,26]:

$$P_{oct} = V_{aq}/V_{oct} \tag{18}$$

This method allowed measurement of P_{oct} values as high as 10,000 (log P_{oct} = 4). The drawback is the low accuracy. A 10,000 P_{oct} value implies a V_{aq} volume

10,000 times greater than the V_{oct} volume [Eq. (18)]. It means that if the V_{aq} volume is 5 L (5 mL/min for 17 hr), the corresponding V_{oct} is only 0.5 mL. The mode reversal should be perfect with no change of the volumes of the octanol and aqueous phases. This is impossible to fulfill with a hydrodynamic machine that does not held the stationary phase enough tightly. Only centrifugal partition chromatographs can be used [27]. Anyway, a minimum volume of 4 mL for V_{oct} was recommended [25].

3. Cocurrent CCC: The Moving Stationary Phase

If a lipophilic solute stays too long inside the CCC unit, why isn't it pushed out, slowly with the liquid stationary phase in the same direction as the mobile phase? Defining F_{oct} and F_{aq} as the octanol "stationary phase" flow rate and the aqueous "mobile phase" flow rate, respectively, Berthod demonstrated that the retention time, t_R, of a solute was expressed by [28]:

$$t_R = (V_{aq} + P_{oct} V_{oct})/(F_{aq} + P_{oct} F_{oct}) \tag{19}$$

The global retention volume corresponds to the liquid flow rate ($F_{aq} + F_{oct}$) multiplied by the retention time expressed by Eq. (19):

$$V_R = (F_{aq} + F_{oct}) \left(\frac{V_{aq} + P_{oct} V_{oct}}{F_{aq} + P_{oct} F_{oct}} \right) \tag{20}$$

The retention volume, V_R, of lipophilic solutes decreases dramatically with F_{oct}. For example, with a CPC machine of 125 mL (CPC-LLN 250W with six cartridges), $V_{oct} = 40$ mL and $V_{aq} = 85$ mL and an aqueous flow rate of 6 mL/min, a $P_{oct} = 1000$ solute has a retention volume of 40 L and a retention time of 4 days and 15 hr. With the cocurrent method, a second pump is used to slowly push the octanol phase with an F_{oct} flow rate of only 0.1 mL/min. Equation (19) shows that the retention volume drops to 2.3 L and the retention time drops to 6 hr and 20 min, a 94% reduction in experiment duration. The P_{oct} value is computed using:

$$P_{oct} = \frac{t_R F_{aq} - V_{aq}}{V_{oct} - t_R F_{oct}} \tag{21}$$

The maximum retention volume change with P_{oct} corresponds to the selectivity maximum. It was shown that this maximum was located around the P_{oct} values corresponding to the F_{aq}/F_{oct} flow rate ratio [27,28]. P_{oct} values 10 times higher than that corresponding to the selectivity maximum can be determined with a ± 0.1 log unit error [22]. With a maximum flow rate of 10 mL/min for the aqueous phase and a minimum flow rate of 10 μL/min for the octanol phase, the best precision is obtained for the 1000 P_{oct} value. A 10,000 P_{oct} value could be measured in such experimental conditions. The retention time would be 40 hr [Eq. (19)] and the retention volume 24 L. The phenanthrene P_{oct} was measured as shown by Figure 11 with a CPC-LLN 250W (three cartridges) centrifugal partition chromatograph. The P_{oct} value of 20,000 ($\pm$8000) or log P_{oct} value of 4.3 $\pm$ 0.2 was obtained with the retention volume of 9800 mL and the retention time of 18 hr.

The interesting thing about the method is that there is no abrupt change; it is continuous. The octanol volume retained in the hydrostatic machine is very stable, more stable than with other methods because there is a constant input of octanol. The octanol volume changes due to dissolution as noted in the direct method [22,27] or to phase reversal in the back-flushing method [25,26] do not exist with the cocurrent CCC method. A hydrodynamic machine was not able to operate properly in cocurrent mode. It was retaining increasing amounts of octanol and then releasing bursts of stationary phase randomly [27]. Both the cocurrent method and the dual-mode method of P_{oct} measurement could not be performed with hydrodynamic instruments [27]. Table 4 compares the different P_{oct} measurement methods [22].

Both the dual-mode and cocurrent method can be performed only with hydrostatic units capable of tightly holding the octanol phase.

B. An Application Example: Saponin Purification

Saponins are a type of glycoside widely distributed in plants. The saponin family consists of a sapogenin, which constitutes the aglycon moiety of the molecule, and a sugar [30]. The sapogenin may be a steroid or a triterpene and the sugar moiety may be glucose, galactose, a pentose, or a methylpentose. The saponin name comes from the surfactant properties. The aglycon moiety is apolar and the sugar moiety is polar. Saponins can form emulsions that can disrupt the liquid–liquid equilibrium inside a CCC device.

1. Sample Origin and Description

Fenugreek, *Trigonella foenum graecum*, is a mediterranean plant of the Leguminosae Lotoideae family, also found in India. It was found that polar extracts of fenugreek, used as nutritional additives in dog food, increased the food intake and at the same time decreased the cholesterol level in the blood of the dogs [31]. The analysis of the fenugreek extract was rich in saponins. The aglycon moiety was a furostanic steroid. The different sugar moieties produced more than 10 different saponins. It is of paramount interest to identify which saponin molecule has the biological activity. Then it is necessary to purify milligram amounts of every saponin found in the fenugreek polar extract. CCC was used for this task.

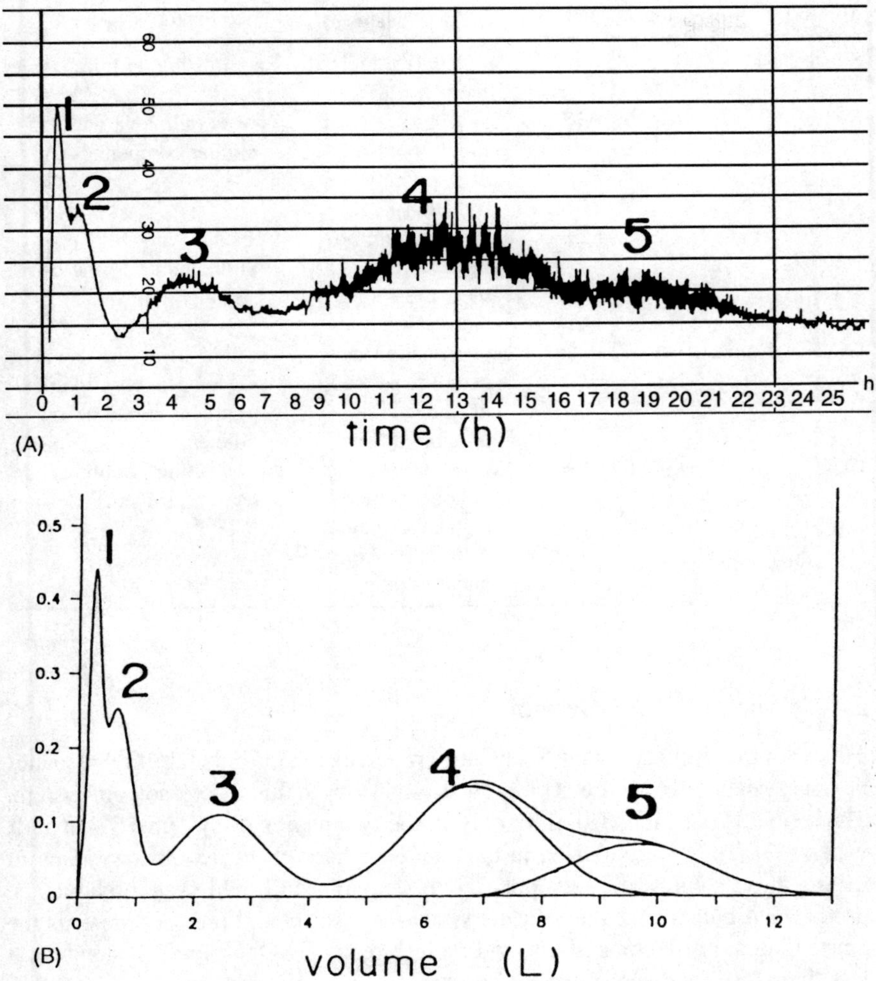

Figure 11 Actual cocurrent chromatogram (A) and the corresponding smoothed chromatogram (B) used for the retention volume measurements. Injected volume: 0.3 ml, **1** phthalimide (0.9 mg inj.), **2** *p*-hydroxybenzoic acid (3.1 mg), **3** *o*-chlorophenol (4 mg), **4** 1-naphthol (2.2 mg), **5** phenanthrene (1 mg). F_{aq} = 9 mL/min, F_{oct} = 0.02 mL/min, V_{oct} = 26.3 mL, detection UV, 210 nm, 0.16 aufs, postcolumn addition of 2-propanol at 2.7 mL/min. (From Ref. 22, p. 192.)

Table 4 Comparison of the Different P_{oct} Measurement Methods

Method	log P_{oct} range	Advantages	Drawbacks
Shake flask	−4; +4	Officially recognized method [28]	Requires highly pure solutes
HPLC	0; +5.5	Fast, does not need highly pure solutes, possible automation	Poor correlations with dissimilar compounds
CCC direct	−2; +2.3	No correlation, no need for pure solutes, possible automation	Does not work for lipophilic solutes, time consuming
CCC dual-mode	+1; +4	No correlation, no need for pure solutes, recycling possible	Possible error when switching modes, very long experiment duration, low efficiency
CCC cocurrent	−1; +4.6	No correlation, continuous method, high efficiency, Adjustable experiment duration	No recycling possibility, detection problems

2. Choice of Solvent System

The Oka-Ito approach that scans the water/methanol/butanol/ethyl acetate/hexane system in 16 mixtures [4,32] was used to try to find a good solvent system. The saponins partitioned slightly only in the two phases of system 16 (butanol/water 5-5). They were located in the aqueous phase of the 15 other systems. It means these solutes are very polar. A polar biphasic liquid system should be used. The *n*-butanol/methanol/water system was selected. Figure 12 presents the phase diagram in volume, mass, and mole percentage. Only liquid systems with less than 15% methanol present two phases.

The retention of the phases of the system 1-butanol/methanol/water 38:12:50 vol% (or 3:1:4 mixture) was studied with a hydrodynamic coil planet centrifuge CCC machine. Both the aqueous and the organic phases were poorly retained with *Sf* values lower than 20%. It was not possible to use the hydrodynamic machine. A hydrostatic centrifugal partition chromatograph (Sanki HPCPC 1000) was tested for phase retention with the same liquid system. The *Sf* values were in the 60% range for both phases. The centrifugal partition chromatograph was well suited to work with the polar butanol/methanol/water system when a hydrodynamic CCC was not.

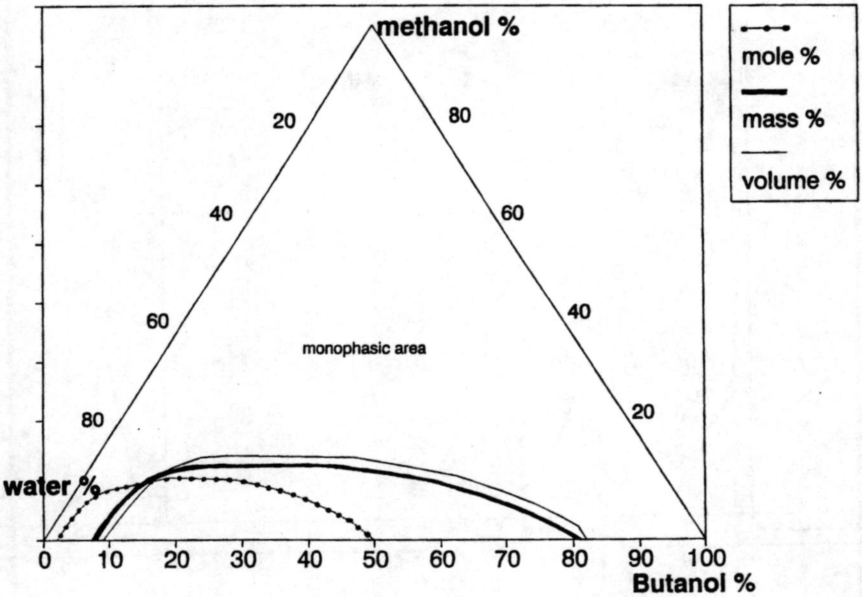

Figure 12 The *n*-butanol/methanol/water liquid system at 22°C. The binodial location and the biphasic area depends on the parameter used.

3. Saponin Separation

It was found that complete separation of the saponin sample was difficult in a simple run. We chose to use the back-flushing method [4,21] described in Section IV.A.2. The best results are obtained with *Sf* = 50%. The centrifugal partition chromatograph was loaded with the organic, butanol-rich phase. The aqueous mobile phase was pumped in the descending mode at 8 mL/min, the rotor being set at 200 rpm. The mobile aqueous phase volume was 118 mL at equilibrium (*Sf* = 43%). Fifty milligrams of the saponin sample dissolved in the aqueous phase was injected. The chromatogram was developed for several hours at 2 mL/min and 1100 rpm (Figure 13). With 500 mL of aqueous phase, four peaks were eluted. At this point, the mode was switched to ascending, and the organic phase was pumped at 2 mL/min for 1 hr. This is enough time to push out the stationary phase with the remaining saponins in it. The partition coefficients corresponding to the peaks obtained are indicated in Figure 13. Each peak was collected and analyzed by thin-layer chromatography. As shown by the plate in Figure 13, the A, B, and C peaks correspond to a purified saponin. The D peak in descending mode and the E and F peaks in ascending mode contain more than

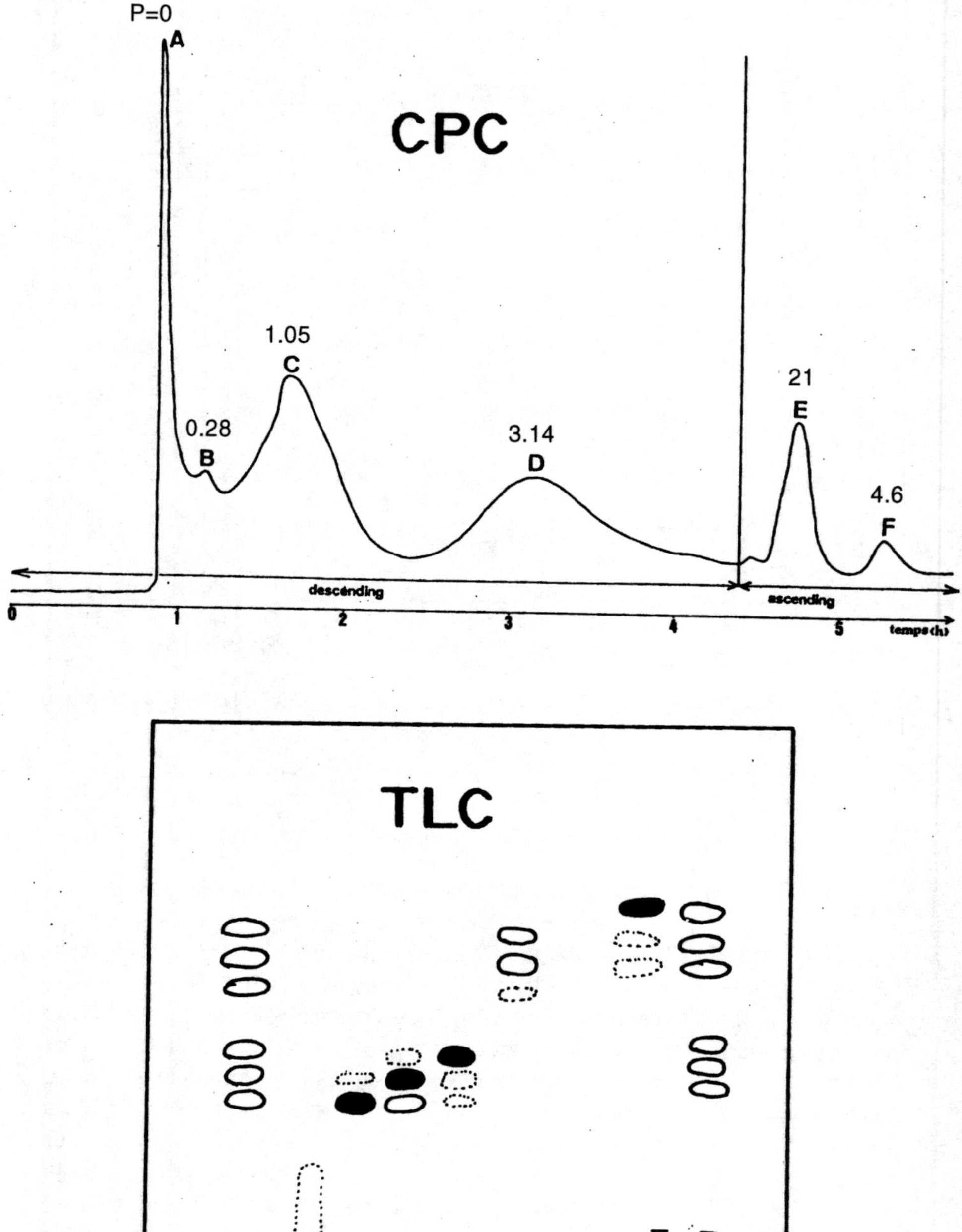

Figure 13 Saponin separation. (top) HPCPC centrifugal partition chromatography. *n*-butanol/methanol/water 3:1:4 (v/v/v) biphasic system, 1100 rpm, 2 MPa, *Sf* = 43%, 2 mL/min descending then ascending, injection of 1 mL of aqueous phase with 52 mg of saponin mixture, detection UV 230 nm, 0.08 aufs. (bottom) Thin-layer chromatography on silica, chloroform/methanol/water 65:42.5:10 (v/v/v) eluent.

one compound. However, the biological activity of every saponin fraction was assessed. Several fractions were found to be active. Different liquid systems are still under investigation to improve this separation of saponins.

V. CONCLUSION

An important point should be emphasized concerning the centrifugal partition chromatographs: At the moment, they are the most reliable CCC apparatuses, and they can work for days without technical problems and without noise. They are used with notable successes in industry [33]. The relatively low efficiency per machine volume could be technically improved. This drawback is well compensated by the stability of the retention of the stationary liquid phase. In several cases, hydrodynamic CCC instruments could not be used because they could not retain enough stationary phase when centrifugal partition chromatographs could work flawlessly. The Achilles heel of the technique was pointed out by Margraff [33]: The customer service and commercial practices of the sole centrifugal partition chromatograph manufacturer, Sanki, are not necessarily compatible with the introduction of the technique for industrial processes outside of Japan. There is an urgent need to change this situation because centrifugal partition chromatography is a very useful and promising tool for industrial chromatography.

REFERENCES

1. Y. Ito, in *Advances in Chromatography*, J. C. Giddings, E. Grushka, and J. Cazes (eds), Vol. 24, Marcel Dekker, New York, 1984. p. 181.
2. Y. Ito, *High Speed Countercurrent Chromatography, CRC Crit. Rev. Anal. Chem., 17*:65–143 (1986).
3. N. B. Mandava and Y. Ito, *Countercurrent Chromatography*, Chromatographic Science Series, Vol. 44, Marcel Dekker, New York, 1989.
4. A. P. Foucault, *Centrifugal Partition Chromatography*, Chromatographic Science Series, Vol. 68, Marcel Dekker, New York, 1995.
5. M. C. Rolet, D. Thiebaut, and R. Rosset, *Analusis, 20*:1–11 (1992).
6. L. R. Snyder and J. J. Kirkland, *Introduction to Modern High Performance Liquid Chromatography*, 2nd ed., John Wiley and Sons, New York, 1979, pp.36–42.
7. A. Berthod and D. W. Armstrong, *J. Liq. Chromatogr., 11*:547–566 (1988).
8. O. Bousquet, A. P. Foucault and F. Le Goffic, *J. Liq. Chromatogr., 14*:3343–3357 (1988).
9. M. C. Rolet, Thèse de Doctorat, Université de Paris 6 (1993).
10. A. P. Foucault, O. Bousquet, and F. Le Goffic, *J. Liq. Chromatogr., 15*:2691–2706 (1992).

11. A. P. Foucault, O. Bousquet, F. Le Goffic, and J. Cazes, *J. Liq. Chromatogr., 15*: 2721–2733 (1992).
12. O. Bousquet, Thèse de Doctorat, Université de Paris 6 (1994).
13. J. M. Menet, M. C. Rolet, D. Thiebaut, R. Rosset, and Y. Ito, *J. Liq. Chromatogr., 15*:2883–2908 (1992).
14. A. Berthod, Y. I. Han, and D. W. Armstrong, *J. Liq. Chromatogr. 11*:1441–1456 (1988).
15. J. M. Menet, D. Thiebaut, R. Rosset, J. E. Wesfreid, and M. Martin, *Anal. Chem., 66*:168–176 (1994).
16. J. M. Menet, Thèse de Doctorat, Université de Paris 6 (1995).
17. A. P. Foucault, in *Centrifugal Partition Chromatography* (A. Foucault ed.), Chromatographic Science Series, Vol. 68, Marcel Dekker, New York, 1995, pp. 25–49.
18. A. Berthod and D. W. Armstrong, *J. Liq. Chromatogr., 11*:567–583 (1988).
19. M. J. Van Buel, L. A. M. Van der Wielen and K. C. A. M. Luyben, in *Centrifugal Partition Chromatography* (A. Foucault, ed.), Chromatographic Science Series, Vol. 68, Marcel Dekker, New York, 1995, pp. 51–69.
20. A. Berthod and N. Schmitt, *Talanta, 40*:1489–1498 (1993).
21. W. D. Conway, *Countercurrent Chromatography: Apparatuses, Theory and Applications*, VCH, Weinheim, 1990.
22. A. Berthod, in *Centrifugal Partition Chromatography* (A. Foucault ed.), Chromatographic Science Series, Vol. 68, Marcel Dekker, New York, 1995, pp. 167–197.
23. A. Berthod and D. W. Armstrong, *J. Liq. Chromatogr., 11*:1187–1204 (1988).
24. R. S. Tsai, P. A. Carrupt and B. Testa, in *Modern Countercurrent Chromatography* (W. D. Conway and R. J. Petroski, eds.), ACS Symposium Series, Vol. 593, American Chemical Society, Washington, D.C., 1995, pp. 143–154.
25. S. J. Gluck and E. J. Martin, *J. Liq. Chromatogr. 13*:3559–3572 (1990).
26. R. A. Menges, G. L. Bertrand, and D. W. Armstrong, *J. Liq. Chromatogr. 13*:3061–3073 (1990).
27. A. Berthod and V. Dalaine, *Analusis, 20*:2508–2513 (1992).
28. A. Berthod, *Analusis, 18*:352–358 (1990).
29. Rules and Regulations, *Fed. Reg. 50, 188*:39252 (Sept 1985).
30. *The Merck Index*, 11th ed., Merck, Rahway, NJ, 1989.
31. Y. Sauvaire, O. Ribes, J. C. Baccou, and M. C. Mariani, *Lipids, 28*:191–195 (1991).
32. F. Oka, H. Oka, and Y. Ito, *J. Chromatogr., 538*:99–110 (1991).
33. R. Margraff, in *Centrifugal Partition Chromatography* (A. Foucault ed.), Chromatographic Science Series, Vol. 68, Marcel Dekker, New York, 1995, pp. 331–350.

5
Cross-Axis Countercurrent Chromatography: A Versatile Technique for Biotech Purification

Y. W. Lee
Research Triangle Institute, Research Triangle Park, North Carolina

I. INTRODUCTION

Recombinant DNA technology has advanced rapidly over the past decade. It is now possible to express a large number of important mammalian genes in bacteria or yeast. The overexpression of these genes allows a specific protein to be produced in large quantity—up to 30–40% of the total cellular proteins. Once a protein with commercial value is successfully overexpressed in the cellular system, a cost-effective recovery procedure must be selected. In most cases, the estimated cost of isolating and purifying a recombinant protein can be 80% of the total production cost [1]. Therefore, protein purification plays a pivotal role in the evolution of biotechnology.

Protein purification in preparative scale presents many difficulties. For instance, centrifugation is the most convenient bioprocess separation technique; however, it is ineffective where particles to be separated are close in density to the suspending liquid. In bacterial cell lysates, particles are extremely fine, and centrifugation cannot provide the degree of separation required for product clarity. In tangential flow filtration, fluid velocity is often increased to overcome the gel layer that builds up. This velocity increase often results in membrane fouling. Selective precipitation suffers from low yields and sometimes losses of activity.

The most popular technique for protein purification is ion exchange chromatography (IEC). However, the resolution is limited, especially in preparative scale separation. Several sophisticated chromatographic methods have been developed, such as capillary zone electrophoresis, reverse phase high-performance liquid chromatography (HPLC), and free-flow isoelectric focusing. However, the cost of operating these systems is high, and often sample size is limited. The development of a cost-effective and preparative method for protein purification is urgently needed.

Recently, high-speed countercurrent chromatography (HSCCC) based on the principle of liquid–liquid partition has emerged as a powerful method in the isolation of small organic molecules [2–4] and in the purification of synthetic polypeptides [5]. In general, a nonmiscible two-phase solvent system is required for performing HSCCC. A wide variety of two-phase solvent systems can be readily formed by mixing binary, ternary, and even quaternary organic solvents with water. This provides ample opportunity to create a particular two-phase solvent system that is suitable for the isolation of a specific component in a crude mixture. For instance, HSCCC has been applied in the isolation of antibiotics from actinimycete fermentation with minimal degradation and loss of bioactivity, phytoecdysteroids from the root bark of *Vitex madiensis*, and numerous bioactive substances from crude natural sources. However, most two-phase solvent systems containing organic solvents are not suitable for the purification of proteins because the protein molecules denature readily in organic solvents.

Aqueous polymer two-phase systems provide a gentle, non-denaturing milieu for enzymes, subcellular particles, and cells. Various configurations, typically created by mixing solutions of polyethylene glycol (PEG) and dextran or PEG and specific salts of phosphate and sulfate, have been developed by Albertsson and his collaborators for the large-scale extraction of macromolecules. These systems use either single-stage extraction or continuous countercurrent distribution [6–10]. Single-stage extraction with an aqueous two-phase solvent system appears to be well suited for the large-scale purification of proteins with markedly different partition coefficients, e.g., the extraction of pullulan-6-glucan hydrolase and 1,4-glucan phosphorylase from 5 kg of *Klebsiella* cell paste. However, protein components with similar partition coefficients can be resolved only by multiple discrete transfer or extraction techniques, such as countercurrent distribution. For instance, multistage countercurrent distribution has been employed in the fractionation of proteins from serum with an aqueous two-phase solvent system [10]. Continuous countercurrent distribution with an aqueous two-phase solvent system was elegantly demonstrated in the purification of organelle- and membrane-bound receptors with a toroidal coil centrifuge [11,12]. However, because of its limited capacity and number of transfers, these techniques have not been widely used in laboratories. One prime feature of the modern HSCCC apparatus including toroidal coil centrifuge is that it uses centrifugal force to hold the stationary phase and facilitates the mobile phase partitioning through the stationary phase. This is done because the general physical properties of the aqueous two-phase polymeric solvent system, such as low interfacial tension (three or four orders of magnitude less than in the aqueous/organic system) requires a long settling time, which is not compatible with an ordinary (type J) flow-through planetary HSCCC designed for the separation of small organic molecules. Thus, poor retention of the stationary phase (less than 10%) and, consequently, poor resolution occurred in the early study of the aqueous two-phase solvent systems.

Modifications of the orientation and position ($L/R = 2$) of the multilayer columns of the ordinary type J HSCCC apparatus has made a significant improvement in the retention of the polymeric stationary phase. The newly designed cross-axis planetary centrifuge system (X-CPC), detailed in Chapter 3, can produce a greater lateral force field relative to the radial centrifugal force field at a given speed, and enhances significantly the retention of the polymeric stationary phase [13,14]. The X-CPC has been applied successfully in the separation of cytochrome *c*, myoglobin, ovalbumin, hemoglobin [15], and lipoproteins [16], and in the purification of recombinant proteins including uridine phosphorylase (UrdPase) [17] and purine-nucleoside phosphorylase (PNPase). This chapter describes the progress made in applying the X-CPC with an aqueous two-phase solvent system in the purification of enzymes and recombinant proteins.

II. PREPARATION OF THE AQUEOUS TWO-PHASE SOLVENT SYSTEM

Aqueous two-phase solvent systems were well developed in the 1960s and were used extensively in the purification of enzymes, proteins, and cells. The large literature on countercurrent distribution of proteins in these aqueous two-phase solvent systems is an excellent source for the selection of appropriate aqueous two-phase solvent systems for purification of biotechnical products [6,7]. A list of commonly used aqueous two-phase solvent systems is given in Table 1.

The polymers used to make aqueous two-phase systems are available from many sources and in a range of molecular weights, e.g., glucose polymer dextran, predominantly poly(α-1,6-glucose) (Pharmacia Fine Chemicals, Piscataway, New Jersey); the poly(sucrose) Ficoll (Fi, Pharmacia); methylcellulose (Union Carbide); and the hydrocarbon ether polymers including PEG, a linear synthetic polymer available in a variety of molecular weights and from many sources including Union Carbide, and BASF. Systems composed of dextran and PEG are by far the most widely used.

There are two ways to prepare an aqueous two-phase solvent system: two hydrophilic polymers in sufficient concentrations may be added to an aqueous solution, or one polymer may be added to a buffer solution of suitable salts. The partition of a target protein in an aqueous two-phase system is governed by the molecular weight of the polymers, the ionic composition, and the pH of the buffer

Table 1 Commonly Used Two-Phase Solvent Systems

Dextran polyethylene glycol water
Dextran Ucon water
Dextran Pluronic water
Dextran Tergitol water
Dextran Ficoll water
Ficoll polyethylene glycol water
Dextran hydroxypropyldextran water
Hydroxypropyldextran hydroxypropyldextran water
Hydroxypropyldextran polyethylene glycol water
Dextran methylcellulose water
Dextran polyvinyl alcohol water
DEAE dextran polyethylene glycol lithium sulfate water
Na dextran sulfate polyethylene glycol sodium chloride water
Na dextran sulfate methylcellulose sodium chloride water
Na dextran sulfate polyvinyl alcohol sodium chloride water
Na carboxymethyldextran polyethylene glycol sodium chloride water
Polyvinyl alcohol polyethylene glycol water
Polyvinyl pyrrolidone polyethylene glycol water
Potassium phosphate polyethylene glycol water
Potassium phosphate methoxypolyethylene glycol water
Potassium phosphate polypropylene glycol water
Ammonium sulfate polyethylene glycol water
Magnesium sulfate polyethylene glycol water

Table 2 Partition Coefficient (K) Studies of Pure UrdPase

	Aqueous two-phase solvent systems[a]				
	A	B	C	D	E
pH values	5.8	6.3	6.8	7.8	9.2
K	0.26	0.66	1.57	6.71	0.24

[a] A: PEG-1000 (24%), KH_2PO_4 (8%), KH_2PO_4 (2%); B: PEG-1000 (18%), KH_2PO_4 (10.4%), KH_2PO_4 (4.1%); C: PEG-1000 (16%), KH_2PO_4 (6.25%), KH_2PO_4 (6.25%); D: PEG-1000 (16%), KH_2PO_4 (2.5%), KH_2PO_4 (10.0%); E: PEG-1000 (16%), KH_2PO_4 (0%), KH_2PO_4 (12.5%).

solution. It is sometimes desirable to alter the pH of phase system while keeping other phase system components constant. For instance, as shown in Table 2, the partition coefficient *K*(U/L) of uridine phosphorylase (UrdPase) could range from 0.24 to 6.71, depending on the polymer composition and the pH of the phosphate buffer. Based on the *K* obtained, one can systematically select the phase composition to optimize the separation. General outlines for phase system manipulation are given in Chapter 3 of the book *Partition in Aqueous Two-Phase Systems* edited by Walter, Brooks, and Fisher [7].

III. CROSS-AXIS PLANETARY CENTRIFUGE

As shown in Figure 1, the cross-axis apparatus (model X-CPC-828) was designed to purify proteins by an aqueous two-phase solvent system. The apparatus has a pair of multilayer coiled columns made by wrapping 2.6-mm-i.d. polytetrafluoroethylene (PTFE) tubing onto holders that are mounted on two rotary shafts. Each rotary shaft is held 7.6 cm (R) from the central axis, and each column is mounted on the shaft 15 cm (L) from the center of rotation. This provides an *L*/*R* ratio of 1.5:2.5, which permits an adequate amount of polymeric phase to

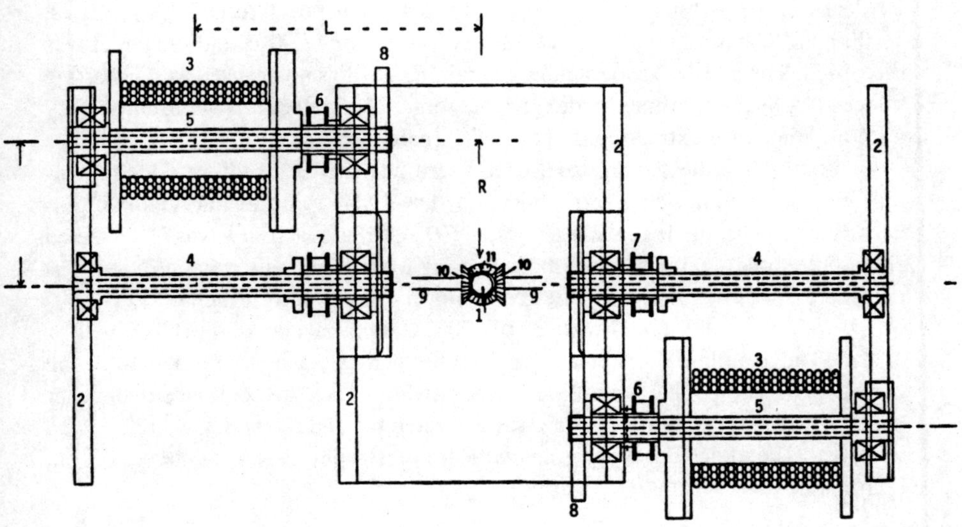

Figure 1 Cross-sectioned view of the cross-axis centrifuge (Model X-CPC-828). 1, central shaft; 2, side plates; 3, multilayer coil columns; 4, rotary shafts; 5, column holders; 6 and 7, toothed pulleys; 8, sun gears; 9, countershafts; 10, planetary miter gears; 11, stationary miter gears.

be retained in the coiled column. This ratio also provides high resolution at a lower centrifugal speed.

The two coiled columns connected in series are filled with lower (or upper) phase and the sample is injected through the sample port. Centrifugation is initiated (750 rpm) and the corresponding mobile phase is pumped through the coiled column at a flow rate of 0.5–1.0 mL/min. For most experiments, the effluent from the column outlet can be monitored by an LKB Uvicord S at 280 nm and collected in 3.0-mL fractions. Separations are usually conducted at room temperature. The X-CPC 828 is marketed by Countercurrent Technologies, Inc. (Research Triangle Park, North Carolina).

IV. APPLICATIONS

A. Purification of Recombinant Uridine Phosphorylase

1. Materials and Methods

Escherichia coli SO744 containing the plasmid pVMK27 (the plasmid containing the UrdPase gene) was grown overnight in a 16-L fermenter as described previously [18]. After centrifugation, the cells were suspended in a minimal volume of 20 mM Tris-HCl (pH 8.0) containing 5 nM $MgCl_2$, lysozyme, DNAse, and RNAse (all 20 μg/mL), and disrupted by a French pressure cell (Aminco) at 10,000 psi. The total crude extract was centrifuged at 47,000*g* and the pellet was discarded. The clarified supernatant, containing UrdPase, was used as the starting material. A later experiment indicated that removal of cell debris was unnecessary and the crude cell extract could be used directly.

Fractions collected from HSCCC were analyzed by sodium dodecyl sulfate–polyacrylamide gel electrophoresis (SDS-PAGE) on 12% acrylamide gels according to the methods of Laemmli [19]. UrdPase activity was determined spectrophotometrically as described by Magni [20]. Protein concentration was measured spectrophotometrically at 595 nm by the Bradford method [21]. Aliquots of 50 μL were mixed with 25 μL 20% trichloroacetic acid (TCA) for 1 hr at 4°C. The samples were centrifuged and the pellets washed with 5% TCA and centrifuged. The pellets were then washed twice with 1 mL cold acetone. After the last centrifugation, the acetone was aspirated off and the pellets were allowed to air dry. The molecular weight standards for electrophoresis were obtained from Pharmacia.

2. Results and Discussion

UrdPase, which catalyzes reversible phosphorolysis, is used in the pharmaceutical industry to synthesize a number of important pyrimidine nucleosides. Currently, UrdPase is produced on a large scale by recombinant DNA technology. Many chromatographic procedures, including radial flow (Q Sepharose) ion exchange chromatography and dye ligand chromatography [22], have been applied to its purification. However, HSCCC demonstrated its advantages in resolution and cost of operation.

UrdPase was selected as a model enzyme primarily because it is thermally stable and available in large quantities in both pure and crude form. Table 3 summarizes the degree of purification and recovery of UrdPase from multistep conventional purification [23]. Selection of an appropriate two-phase solvent system plays a key role for successful separation. A series of partition coefficient (K) studies using pure UrdPase provided crucial information on appropriate aqueous two-phase solvent systems. For instance, the partition coefficient K(U/L) of UrdPase in an aqueous two-phase solvent system composed of PEG-1000 and potassium phosphate buffer could range from 0.24 to 6.71 depending on the polymer composition and the pH of the phosphate buffer. As summarized in Table 2, UrdPase has K(U/L) values of 0.66 at pH 6.3, 1.57 at pH 6.8, and 6.71 at pH 7.8. Apparently, pH, surface charge, intramolecular hydrogen bonding, and hydrophilicity are factors that determine the K value of a specific protein. The partition of crude UrdPase was also measured at pH 6.8. Comparison of the UV absorption values of crude UrdPase, partitioned into the upper and lower phase, with those of pure UrdPase suggested that selection of an aqueous two-phase solvent system composed of PEG-1000 (16%) and potassium phosphate (12.5%) would have the advantage of removing impurities effectively. In Figure 2, the HSCCC elution profile of crude UrdPase (about 150 mg/6 mL), as exemplified by SDS-PAGE, revealed that the major UrdPase peak was well separated from other proteins and eluted from fraction 105 to 120, and that the majority of the impurities were eluted in the early fractions. The specific activity of the UrdPase in the pooled peak fractions was significantly improved (Table 4). The purity of the enzyme was examined on SDS polyacrylamide gel; a remarkable 95% purification was accomplished in a single step of HSCCC. The total yield of UrdPase was calculated to be approximately 70%. The enzyme activity shown in Table 4 indicates excellent recovery (more than 91%). Figure 3 shows that a supposedly pure fraction of UrdPase (175 mg/3.5 mL; lane L) derived from ion exchange chromatography was further resolved by HSCCC and a minor impurity band (fraction 50) was effectively removed from the major UrdPase (fraction 68). The same experimental conditions were applied to the crude extract of a wild type of *E. coli* and produced a 76% purification. In addition, the HSCCC system was capable of providing satisfactory separation of 4.0 g crude UrdPase.

Removal of PEG can be accomplished by absorbing the UrdPase onto a solid support such as Q or a hydrophobic column after appropriate adjustment of ionic strength and pH and washing the PEG off the column. The protein could be further purified by using selective conditions for desorption.

Table 3 Purification of UrdPase from *E. coli* B-96

Step	Volume (mL)	Total units	U/mg of protein	x-fold Purification	% Recovery
Cell extract	1180	6460	0.16	1	100
Calcium phosphate gel eluate	2000	4510	1.24	7.8	70
DEAE-cellulose chromatography	252	3320	4.17	26	51
ECTEOLA-cellulose chromatography	67	2190	10.9	68	34

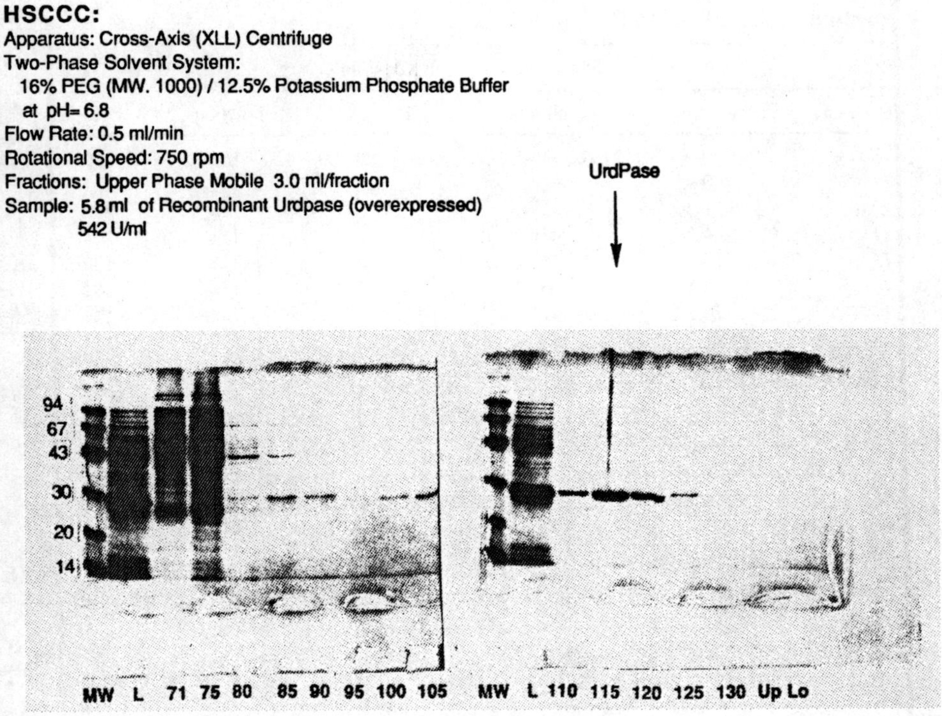

Figure 2 SDS gel profile of fractions collected from HSCCC separation of recombinant UrdPase.

B. Purification of Recombinant Purine-Nucleoside Phosphorylase

1. Material and Methods

Purine-nucleoside phosphorylase (PNPase) catalyzes the reversible phosphorolysis of purine ribonucleosides and 2-deoxyribonucleosides. The enzyme has been purified from a variety of sources, including bacteria, human erythrocytes, bovine spleen, chicken liver, rabbit liver, and, recently, *E. coli* through recombinant technology. PNPase derived from *E. coli* consists of six subunits with a molecular weight of 23,000–24,000. Since each subunit binds one molecule of each substrate, these subunits are identical. Genetically, the PNPase was produced from *E. coli* as previously described [24–26]. The crude PNP was obtained from the cell paste at a concentration of 75 mg/mL by a procedure similar to that described by Krenitsky et al. [27].

Table 4 Recovery of UrdPase in HSCCC

Load		542 U/mL	5.5–6 mL		2981-3252U
Fractions	U/3 mL	Fractions	U/3 mL	Fractions	U/3 mL
71	0	91	2.7	111	222
72	0.9	92	4.2	112	255
73	1.2	93	2.7	113	222
74	1.2	94	6.3	114	204
75	15	95	4.8	115	180
76	5.1	96	7.5	116	162
77	10.2	97	9	117	144
78	11.7	98	13.2	118	123
79	8.1	99	15.3	119	93
80	0	100	22.8	120	70
81	8.4	101	27.9	121	54
82	10.2	102	30.9	122	39
83	4.5	103	46.5	123	22.5
84	4.5	104	60	124	15
85	1.2	105	78	125	13.2
86	10	106	101	126	12
87	2.1	107	120	127	11
88	0	108	131	128	3
89	1	109	162	129	3
90	4.8	110	180	130	0
Total					2960 U
Upper phase		0 U/mL	230 mL		0 U/230 mL
Lower phase		0 U/mL	55 mL		0 U/55 mL
Recovery[a]		2960 U/3252 U	2960 U/2981 U		91–99%

[a] Recovery is calculated by dividing the total recovered enzyme (units) by initial load.

2. Results and Discussion

The isolation of electrophoretic homogeneous purine nucleoside phosphorylases from *E. coli* has been reported [24]. PNPase from *E. coli* and *S. typhimurium* have been shown to be hexamers of subunits of equal molecular weight (23,700). Conventional purification of PNPase is summarized in Table 5 [23].

The final supernatant fluid after the calcium phosphate gel treatment contained the bulk of the PNPase. DEAE-cellulose was added with stirring at 3°C to the calcium phosphate gel supernatant fluid. This resin was then packed into a column (2.5 × 5 cm). The column was washed consecutively with 100 mL of 50 mM buffer A, 75 mL of 75 mM buffer A, 25 mL of 100 mM buffer A, 265 mL of 100 mM buffer A, and, finally, 200 mL of 150 mM buffer A. The last two washes were combined and dialyzed against 24 L of deionized water at 3°C for 17 hr. The dialyzate was applied to an ECTEOLA-cellulose column (5 × 21

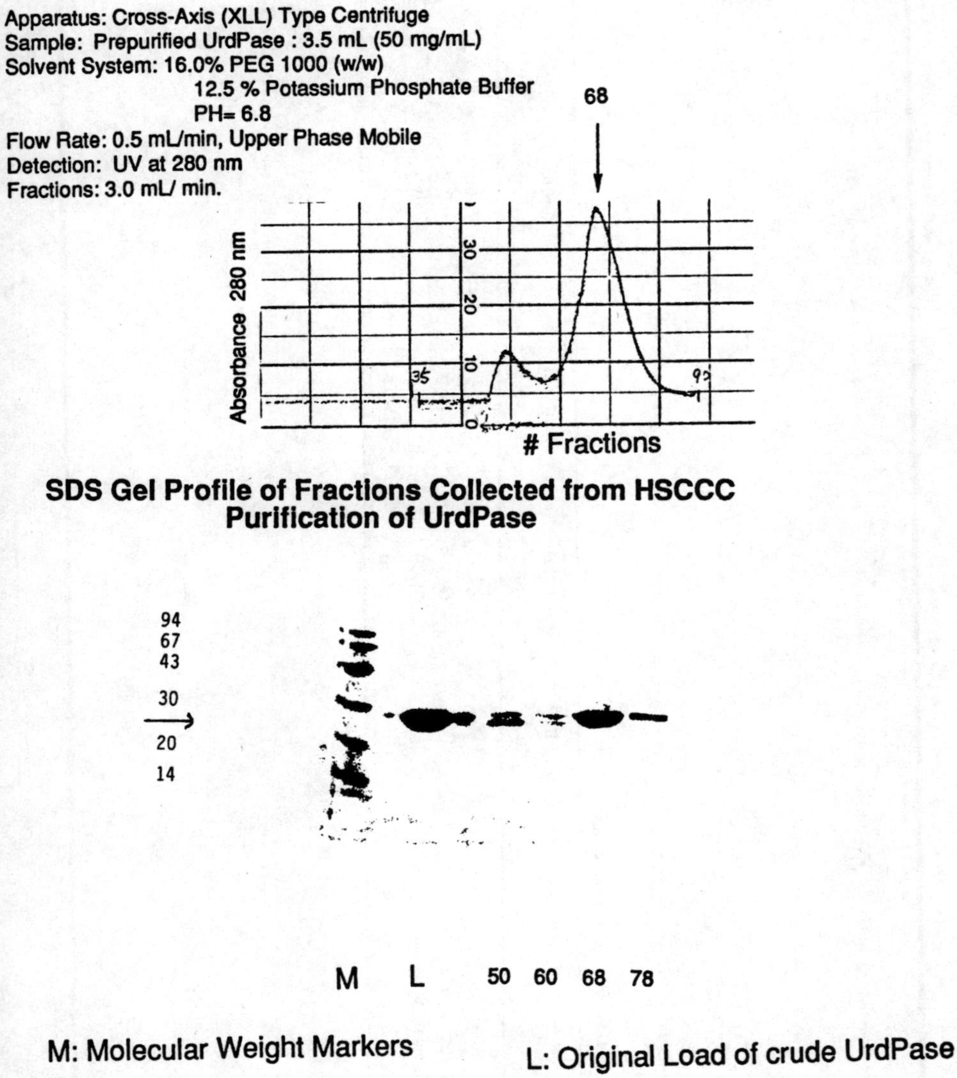

Figure 3 HSCCC separation of UrdPase.

Table 5 Purification of PNPase from *E. coli* B-96

Step	Volume (mL)	Total units	Units/mg of protein	x-fold purification	% Recovery
Cell extract	1180	7780	0.19	1	100
Calcium phosphate gel eluate	1700	6510	1.36	7.2	84
DEAE-cellulose chromatography	465	3890	7.46	39.3	50
ECTEOLA-cellulose chromatography	70	2650	27.7	146	34

cm), equilibrated with 10 mM Tris-HCl pH 8.0 (buffer C) at 25°C. After the column was washed with 500 mL of 50 mM KCl in buffer C, the enzyme was eluted with a linear gradient of 50–500 mM KCl in buffer C. The fractions having the highest specific activity were combined and evaluated. A total of 34% recovery was obtained.

Figure 4 shows the HSCCC purification of crude PNPase. PEG-1000 (16%) and potassium phosphate buffer (12.5%) at pH 6.8 was selected as the aqueous two-phase solvent system. In this case, the lower phase was used as the mobile phase. A quantitative (98%) recovery was obtained based on the calculation of enzyme activity shown in Table 6. The protein weight recovery was also in agreement with enzyme activity. Fraction 62 contains the highest amount of PNPase. The purification completed within 8 hr. Figure 5 shows the SDS gel

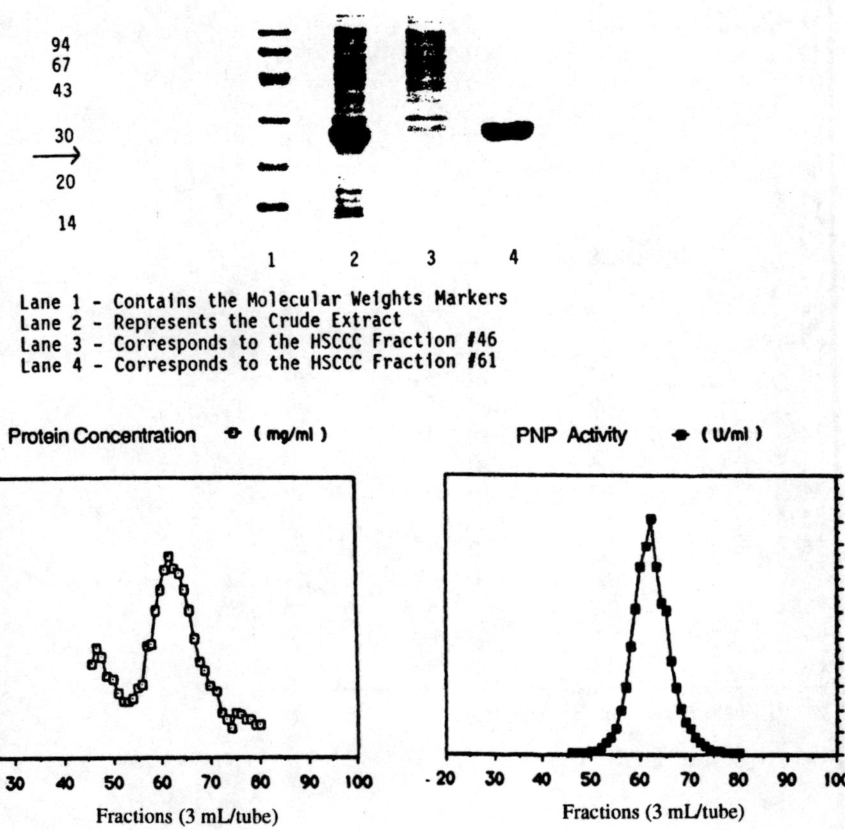

Figure 4 Analysis of PNPase fractions collected from HSCCC.

Table 6 Recovery of PNPase from HSCCC

Fractions	(U/mL)	(mg/mL)	(U/mg)
46	0.6	0.345	1.7
47	0.6	0.4	1.5
48	0.6	0.372	1.6
49	0.9	0.3	3.0
50	1.2	0.29	4.1
51	1.7	0.236	7.2
52	2.9	0.209	14
53	5.0	0.209	24
54	7.5	0.218	34
55	11	0.255	43
56	19.8	0.273	72
57	29.5	0.409	72
58	46.7	0.418	111
59	61.9	0.527	117
60	80	0.6	133
61	88.8	0.664	133
62	101.1	0.718	140
63	80.2	0.673	119
64	64.4	0.655	104
65	61	0.6	101
66	40.8	0.527	77
67	29.8	0.436	68
68	20.1	0.355	57
69	14.4	0.318	45
70	11.1	0.264	42
71	7.6	0.245	31
72	5.2	0.163	31
73	3.4	0.136	25
74	2.3	0.109	21
75	1.7	0.163	10
76	1.5	0.155	9.6
77	0.9	0.136	7.3
78	0.7	0.136	5.1
79	0.5	0.118	—
80	0.4	0.118	3.3

2417U/2460U (load) total recovery: 98%.

Apparatus : Cross-Axis (XLL)-CPC

Two-phase Solvent System : PEG (1000) 16.0% ; 12.5% Potassium Phosphate Buffer ($6.25\%\ K_2HPO_4$ + $6.25\%\ KH_2PO_4$) (pH = 6.85)

Flow rate: 0.5 ml/min, Lower phase mobile

Centrifugal Speed : 750 rpm

Fractions: 3.0 ml/ fraction, F-55 to F-140

Sample : Crude PNPase (75 mg / ml) + 7.0 ml upper phase + 7.0 ml lower phase

Retention of Stationary phase : 45 %

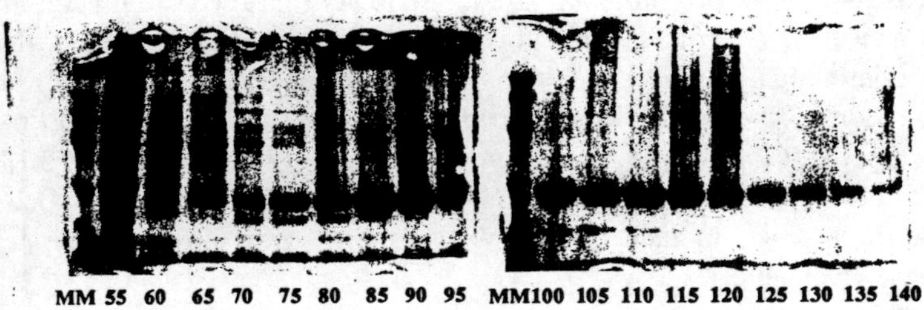

Figure 5 X-CPC: purification of recombinant PNP.

results of preparative HSCCC purification of crude PNPase (75 mg). Because of the large sample size the PNPase was distributed among a wider range of fractions (nos. 75–135). The final product is pure as shown by SDS-PAGE analysis.

C. Purification of Serum Lipoprotein

1. Materials and Methods

The aqueous two-phase system was prepared by dissolving 192 g of PEG-1000 (Sigma) and 150 g of anhydrous dibasic potassium phosphate (J. T. Baker) in 858 g of distilled water. The solvent mixture was thoroughly equilibrated in a separatory funnel at room temperature and the two phases were separated shortly before use.

Human high-density lipoprotein (HDL) and low-density lipoprotein (LDL) fractions were isolated by ultracentrifugation [28,29]. The partition coefficients of the HDL and LDL fractions and of human serum albumin and α- and γ-globulin (Sigma) were determined in the aqueous two-phase solvent system. A 0.4-

mL lipoprotein suspension, containing 5.29 mg/mL of HDL fraction and 0.76 mg/mL of LDL fraction, was partitioned in 3.0 mL of the aqueous polymer two-phase system (1.5 mL each of the upper and lower phase). A 0.8-mL aliquot of each phase was diluted with 2.0 mL of distilled water and the absorbance was measured at 280 nm with a Zeiss PM6 spectrophotometer. The partition coefficient value *K* was calculated by dividing the absorbance in the lower phase with that in the upper phase.

The sample solution was prepared by adding 1.0 g of PEG-1000 and 0.9 g of anhydrous dibasic potassium phosphate to the mixture of 3.0 mL HDL fraction (44.2 mg/mL protein) and 2.0 mL LDL fraction (11.4 mg/mL protein). Each experiment was initiated by filling the entire column with the stationary upper phase. This was followed by sample injection through the sample port. Then the apparatus was rotated at 750 rpm, and the mobile lower phase was pumped into the column at a flow rate of 0.5 mL/min. The effluent from the outlet of the column was continuously monitored with an LKB Uvicord S detector at 280 nm. The peak fractions (20–50 mL) were placed into dialysis tubing (Spectro/Pro, molecular weight cutoff 6000–8000) which was immersed into an aqueous 30% PEG-8000 solution. After 5–6 hr dialysis, the fraction was concentrated to 0.2–0.3 mL. The lipoproteins in each fraction were confirmed by 0.6% agarose gel electrophoresis (Helena Labs).

2. Results and Discussion

To achieve efficient separation of lipoproteins from other serum proteins, it is essential to optimize the partition coefficient of each component by selecting a suitable composition of the polymer phase system. It was found that an aqueous two-phase solvent system composed of 16% PEG-1000 and 12.5% dibasic potassium phosphate provides desired *K* values for the HDL (3.8) and the LDL (1.8), respectively, which are substantially different from those of other serum proteins such as albumin (0.12), α-globulin (0.05), and γ-globulin (0.02).

A mixture of LDL and HDL fractions from human plasma was eluted by using the established polymer phase system described above. As shown in Figure 6, the HDL and LDL fractions were eluted from the column in the order of their partition coefficients values and separated from other plasma proteins. The separation was completed within 12 hr and the volume of the upper stationary phase retained in the column was 45% of the total column capacity (250 mL). Fractions 43–50 and 60–70, corresponding to the first (HDL) and second peaks (LDL) in the chromatography, were examined by agarose gel electrophoresis. The third peak in the chromatogram showed no lipid staining. The latest results (Shibuzawa et al. [30]) showed that under the same experimental conditions, HDL and LDL could be separated directly from the human serum in 3 hr. In conclusion, the X-CPC/aqueous two-phase system provides a satisfactory retention of the polymeric phase at relatively high flow rates of 0.5–1.0 mL/min.

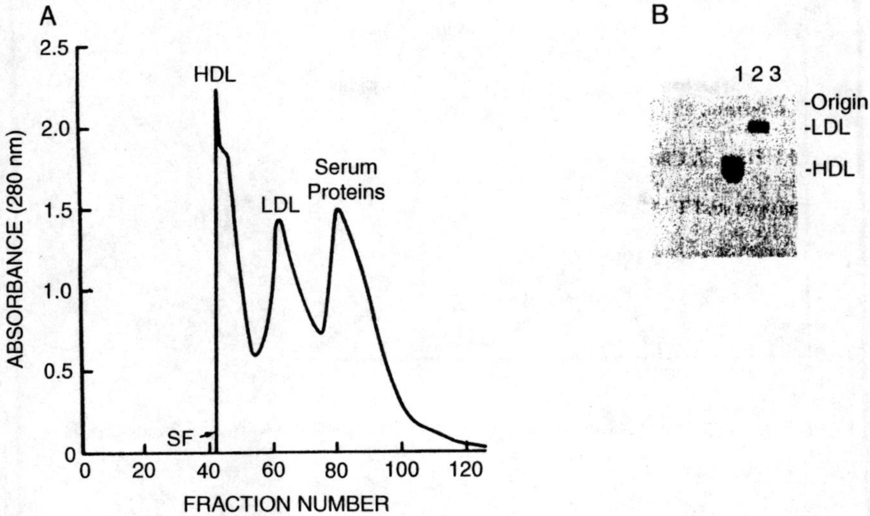

Figure 6 (A and B) Purification of serum lipoprotein.

D. Separation of Cytochrome *c*, Myoglobin, Ovalbumin, and Hemoglobin

1. Materials and Methods

PEG-1000, cytochrome *c* (horse heart), myoglobin (horse heart), ovalbumin (chicken egg), hemoglobin (bovine), and serum albumin (human and bovine) were purchased from Sigma. Protein samples were prepared by dissolving a mixture of cytochrome *c*, myoglobin, ovalbumin, and hemoglobin in about equal volumes of the upper and lower phases of the aqueous two-phase solvent system composed of 12.5% (w/w) PEG-1000 and 12.5% dibasic potassium phosphate.

In a standard separation, the coiled columns with a total capacity of 575 mL were filled with the PEG-rich upper phase, and a sample solution dissolved in 7.0 mL each of the upper phase and the lower phase was charged into the system. Then the phosphate-rich lower phase served as the mobile phase and was pumped into the column at the optimum flow rate of 1.0 mL/min.

2. Results and Discussion

Figure 7 shows the separation of the protein mixture. Cytochrome *c* was well separated whereas myoglobin and ovalbumin were only partially resolved. The hemoglobin was readily eluted when the upper phase was switched to the mobile

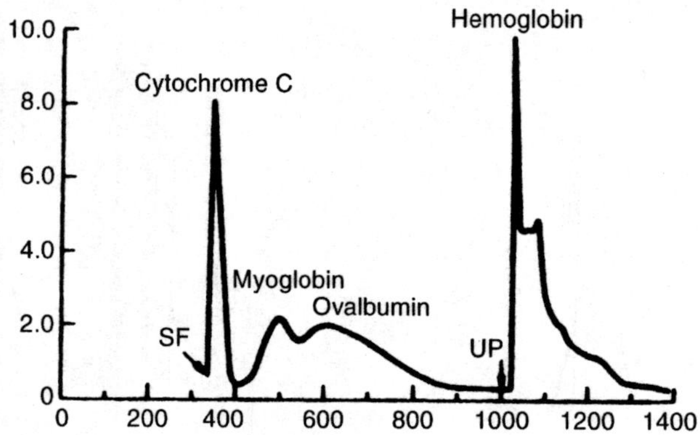

Figure 7 Separation of cytochrome *c*, myoglobin, ovalbumin, and hemoglobin.

phase. The results indicate that the X-CPC system can be used for preparative separation of a protein mixture with minimal loss of resolution. The separation was performed at 800 rpm and at a flow rate of 1.0 mL/min. The partition coefficient, calculated from the myoglobin peak, was 202 theoretical plates. The PEG–potassium phosphate two-phase solvent system was most effective for separation of proteins. The low molecular mass compounds are generally partitioned unilaterally in the upper phase regardless of the pH and the polymer composition, whereas the partition coefficients of the proteins are broadly adjusted by pH and the polymer composition.

V. CONCLUSION

Enzymes and proteins produced by recombinant DNA technology are increasingly being developed as therapeutic agents in chemotherapy and as biocatalysts in industrial processes. These rapid advances in molecular genetics have resulted in an increased awareness of the importance of protein recovery and purification, which is playing a crucial role in the isolation of a high-valued recombinant DNA derived protein in large amounts, with contaminants from the bacterial or mammalian host cell line at the ppm level.

Currently the most popular techniques for protein purification are ion exchange chromatography, hydrophobic interaction, and size exclusion chromatography. However, the resolution of these methods is limited especially in preparative scale separation. Reverse phase HPLC has gained increasing popularity. The problem with preparative HPLC is that the material tends to bind to the solid support or be denatured by the finely divided solid adsorbent, and the cost of a preparative HPLC separation is high.

During the past few years, protein isolation and purification based on the aqueous two-phase system has been studied extensively, particularly in combination with the newly developed X-CPC system. This combination system is extremely efficient for isolation of proteins and enzymes. It offers significant advantages over liquid–solid extractions because loss of valuable sample from irreversible adsorption, degradation, and denaturation is eliminated. In addition, the amount of stationary phase accessible to the solute is substantially higher than in an HPLC column of the same size, so the method is ideal for preparative separation with high sample load. The intermolecular hydrogen bonding between the protein and a polymer such as PEG or dextran stabilizes the proteins and offers very mild conditions for protein purification. In fact, some labile enzymes could be purified at room temperature by this method. In general, it appears that protein sample concentration, polymer–protein interaction, pH, and salt concentration all have significant effects on the outcome of separations. The specificity of the aqueous two-phase polymer system can be further increased by coupling affinity ligands to the dextran or PEG for continuous affinity countercurrent chromatography.

As demonstrated in this chapter, a number of recombinant proteins and serum proteins have been purified in a cost-effective manner with the X-CPC/aqueous two-phase system. In addition, we found that cell debris partitioned selectively into the lower phase of the two-phase system. Thus, it was not necessary to remove the cell debris when the crude recombinant protein was separated. The potential in process separation is quite promising because 4.0 g of crude UrdPase was successfully purified with a 500-mL X-CPC column.

Currently, a small group of scientists in the United States, Europe, and Japan are focusing on the development of X-CPC as a cost-effective method for potential downstream processing of biotech products. Existing data suggest that a preparative X-CPC apparatus with a 5-L column would be feasible.

REFERENCES

1. T. C. Ransohoff, M. K. Murphy, and H. L. Levine, *BioPharm. 3*(3): 20–26 (1990).
2. Y. Ito, *CRC Crit. Rev. Anal. Chem. 17*:65 (1986).
3. D. Conway, *Countercurrent Chromatography: Apparatus. Theory and Applications*, VCH, New York, 1990.
4. Y. Ito and N. B. Mandava, *Principle and Instrumentation of Countercurrent Chromatography: Theory and Practice*, Marcel Dekker, New York, 1988, p. 79.
5. M. Knight, Y. Ito and J. L. Sandlin, and A. M. Kask. *J. Liq. Chromatogr. 9*:791 (1986).

6. A. Albertsson, *Partition of Cell Particles and Macromolecules*, 3rd ed. John Wiley and Sons, New York, 1986.
7. H. Walter, D. E. Brooks, and D. Fisher., *Partitioning in Aqueous Two Phase System: Theory, Methods, Uses, and Applications to Biotechnology*. Academic Press, Orlando. 1985.
8. R. Kula, K. H. Kroner, and H. Hustedt, Purification of enzymes by liquid–liquid extraction. *Adv. Biochem. Eng. 24*:73–118 (1982).
9. F. Tjerneld and G. Johansson., Aqueous Two-Phase Systems for Biotechnical Use., Bioseparation, 1:255–263 (1990).
10. G. Birkenmeier, G. Kopperschlager, P. A. Albertsson, G. Johansson., F. Tjerneld, H. E. Akerlund, S. Berner, and H. Wickstroem, Fractionation of proteins from human serum by countercurrent distribution, *J. Biotechnol. 5*:115–129 (1987).
11. A. Sutherland, D. Heywood-Waddington, and T. J. Peters, Torodial coil countercurrent chromatography: a fast simple alternative to countercurrent distribution using aqueous two phase partition. *J. Liq. Chromatogr. 7*: 363–384 (1984).
12. A. Sutherland, D. Heywood-Waddington, and T. J. Peters, Countercurrent chromatography using a toroidal coil planet centrifuge: a comparative study of the separation of organelles using aqueous two-phase partition. *J. Liq. Chromatogr. 8*(12):2315–2335 (1985).
13. Y. Ito, E. Kitazume, M. Bhatnagar, and F. Trimble. Cross-axis synchronous flow-through coil planet centrifuge (type XLL). I. Design of the apparatus and studies on the retention of the stationary phase. *J. Liq. Chromatogr. 538*:59–66 (1991).
14. Y. Ito, Cross-axis synchronous flow through coil planet centrifuge (type XLL). II. Speculation on the hydrodynamic mechanism in stationary phase retention. *J. Liq. Chromatogr. 538*:67–80 (1991).
15. K. Shinomiya, J. M. Menet, H. M. Fales, and Y. Ito, Studies on a new cross-axis coil planet centrifuge for performing countercurrent chromatography: I. Design of the apparatus, retention of the stationary phase, and efficiency in the separation of proteins with polymer phase systems, *J. Chromatogr. 644*:215–229 (1993).
16. Y. Shibusawa and Y. Ito, Countercurrent chromatography of lipoproteins with a polymer phase system using the cross-axis synchronous coil planet centrifuge, *J. Chromatogr. 596*:118–122 (1992).
17. W. Lee, Y. Shibusawa, F. T. Chen, J. Meyers, J. M. Schooler, and Y. Ito, Purification of uridine phosphorylase from crude extracts of *E. coli* employing high speed countercurrent chromatography with an aqueous two-phase solvent system, *J. Liq. Chromatogr. 15*:2831–2841 (1992).
18. K. Weaver, D. Chen, L. Walton, L. Elwell, and P. Ray., *Biopharm*. July/Aug. 25–28, 1990.
19. U. K. Laemmli, *Nature 227*:680–685 (1970).
20. A. Magni, *Methods in Enzymology*, Vol. 51 (P. A. Hoffee and M. E. Jones, eds.), Academic Press, New York, 1978, pp. 290–296.
21. M. Bradford., *Anal. Biochem. 72*:248–254 (1976).
22. Vita and G. Magni, A one-step procedure for the purification of UrdPase from *E. coli, Anal. Biochem. 133*:153–156(1983).
23. T. A. Krenitsky, G. W. Koszalka, and J. V. Tuttle, *Biochemistry 20*:3615–3621 (1981).

24. F. Jensen and P. Nygaard, *Eur. J. Biochem. 51*:253–265 (1975).
25. C. Roberson and P. A. Hoffee, *J. Biol. Chem. 248*:2040–2043 (1973).
26. K. F. Jensen, Purine-nucleoside phosphorylase from *Salmonella typhimurium* and *E. coli, Eur. J. Biochem. 61*:377–386 (1976).
27. T. A. Krenitsky, *J. Biol. Chem. 251*:4405 (1976).
28. J. Havel, H. A. Eder, and J. H. Bragdon, *J. Clin. Invest. 34*:1345 (1955).
29. M. Sclavons, C. M. Cordonnier, P. M. Mailleux, F. R. Heller, J. P. Desager, and C. M. Harvengt. *Clin. Chim. Acta 153*:125 (1985).
30. Y. Shibusawa, T. Chiba, U. Matsumoto, and Y. Ito, ALS Symposium Series No. 593 ''Modern Countercurrent Chromatography.'' Edited by W. D. Conway and R. J. Petroski, chapter 11, p. 119–128, 1995.

6
Application of Countercurrent Chromatography in Inorganic Analysis

Tatiyana A. Maryutina, Piotr S. Fedotov, and Boris Ya. Spivakov
Vernadsky Institute of Geochemistry and Analytical Chemistry, Russian Academy of Sciences, Moscow, Russia

I. INTRODUCTION

Countercurrent chromatography (CCC), a support-free partition chromatography, is based on retention of either phase (stationary) of a two-phase liquid system in a rotating column under the action of centrifugal forces while the other liquid phase (mobile) is being continuously pumped through [1]. So far, the technique has been mainly studied and used for preparative and analytical separation of organic and bioorganic substances. The studies of the last several years have shown that the technique can be applied to analytical and radiochemical separation, preconcentration, and purification of inorganic substances in solutions on a laboratory scale by use of various two-phase liquid systems. A feature of CCC as chromatographic method is the absence of sorbent or solid support for retaining the stationary phase. This feature defines the main advantages of the method, such as the absence of the substances being separated due to interaction with the sorbent matrix; variety of two-phase liquid systems that may be used; easy change from one partition system to another; absence of the problem of column regeneration, high preparation capacity and possibility to change a volume of the sample solution from 0.1 to 1000 mL and more.

A few devices providing the retention of the stationary phase in the field of mass forces in the absence of solid support have been suggested [1]. Among the various possible designs, planetary centrifuge better retains the liquid stationary phase and makes possible the fastest and most efficient separation [1]. A column (or a column unit) of a certain configuration rotates around its axis and simultaneously revolves around the central axis of the device with the aid of a planetary gear. The design of the countercurrent chromatographic equipment has been described in detail [1] and therefore only a brief description of the devices for separation of inorganic elements is given in Section I.

For the separation and preconcentration of organic and inorganic substances by CCC, different two-phase liquid systems are used. The number of potentially suitable CCC solvent systems is so great that it may be difficult to select the best one. The studies of the last several years have made it possible to classify water–organic solvent systems in CCC for separation of organic compounds on the basis of the liquid phase density difference, the solvent polarity, and other parameters from the point of view of stationary phase retention in CCC columns [1–4]. Ito [1] classified some liquid systems as hydrophobic, intermediate, and hydrophilic.

The distribution of inorganic compounds as well as organic compounds is dependent on the properties of the system used, partition coefficients of substances to be separated, and parameters of the planetary centrifuge operation such as rotation and revolution speeds, direction and speed of the mobile phase pumping, internal diameter of the column, and sample volume. Such dependencies for organic compounds have been described [1–10]. However, the systems for inorganic separations are very different from those for organic separations [2,11,12] as in most cases they contain a complexing (extracting) reagent (ligand). The complexation process, its rate, and the mass transfer rate are the main factors that determine the separation efficiency. The kinetics is of particular importance because CCC separation can be a nonequilibrium process. Thus, it has been shown for several systems that the batch partition coefficients for inorganic compounds are different from dynamic ones [13–15]. Thus, the principles of the choice of two-phase liquid systems for inorganic substances separation in rotating coil columns need to be studied. The influence of mobile phase composition, reagent concentration in the solvent, type of organic solvent, and retention volume of stationary phase in the column on the chromatographic behavior of chemical elements will be described in Section II. Section II also describes the influence of the kinetic properties of several extraction systems on the separation of some elements by CCC.

Section III comprises various applications of CCC to the preconcentration and separation of some elements in inorganic analysis.

II. APPARATUS

Different types of countercurrent chromatographic apparatus with one-layer and multilayer coil separation columns have been used for inorganic analysis [13,15–

28]. Centrifugal partition chromatography, another countercurrent liquid–liquid distribution technique suitable for such separations [29–33], is discussed in Chapter 5 of this book by A. Berthod. CCC and centrifugal partition chromatography apparatus for separation of elements and chromatographic procedures are briefly described below.

A. Apparatus for CCC Separations and Chromatographic Procedure

Our chromatographic investigations [13,15,18–28] were made on a self-designed planetary centrifuge with a vertical one-layer coil column drum (scheme IV according to Ito's classification of planetary motion [1]). The rotation and revolution speeds (ω) are 350–500 rpm. The planetary centrifuge model has the following design parameters: revolution radius R = 140 mm, rotation radius r = 50 mm. The column was made of a Teflon tube with an inner diameter of 1.5 mm. The total inner capacity of the column was 20 mL.

We have also developed a new design of CCC apparatus used in our studies on inorganic separations. The centrifuge consists of six spiral columns winded onto drums of different diameters (7, 10, 15 cm) and mounted in two assemblies bearing three columns each. The tubing diameter ranges from 0.5 to 1.5 mm. The β values value ($\beta = r/R$, where r is the rotation radius and R is the revolution radius) range from 0.35 to 0.75. β is an important parameter used to determine the hydrodynamic distribution of the two solvent phases in the rotating coil. The columns can be connected in series, or they can be operated independently (important for the performance of several simultaneous experiments). In addition, the design enables the rotation axis to be changed from vertical to horizontal. The apparatus can be operated at speeds up to 1000 rpm.

For separation of inorganic substances we used the following chromatographic procedure. Before the CCC separation experiment was begun, the components of the two-phase liquid system were stirred and brought into equilibrium for mutual saturation of the phases, after which the aqueous phase was used as the mobile phase and the organic one as the stationary phase. The spiral column in the stationary mode was filled with the organic phase. After that, while the column was rotated, the aqueous phase was fed to its inlet. The mass force field that arose during rotation made it possible to retain the stationary phase in the column (V_s) while the mobile phase was continuously pumped through. The amount of the stationary phase in the column is defined by the retention factor S_f (ratio of the stationary phase volume to the total column volume). After volumetric equilibrium between the mobile and stationary phase had been established, a sample was introduced into the column. Beginning from the moment of sample introduction, fractions of the mobile phase (eluate), which passed through the column, were taken to determine the elements of interest by different methods. The pumping rate (F) is changed from 0.5 to 2.5 mL/min.

The volume of the stationary phase retained in the column depends on the extraction system nature, the rotation speed of the column, the pumping speed of the mobile phase, and the directions of rotation and pumping. When the influence of the quantity of the stationary phase in the column on the chromatographic peak shape was being investigated, a fixed volume of the stationary phase V_s^f (less than V_s allowed by experimental conditions) was introduced with the flow of the mobile phase into the column in the stationary mode. After that, while the column was rotated, the aqueous phase was continuously fed to its inlet.

A Shimadzu countercurrent chromatograph (HSCCC-1A prototype) [16], which was used for the separation of some heavy metals, holds a multilayer coil separation column on the rotary frame at a distance of 10.0 cm from the central axis of the centrifuge. The column was prepared from a single piece of approximately 150-m-long, 1.60-mm-i.d. polytetrafluoroethylene (PTFE) tubing by winding it directly onto the holder hub (10 cm in diameter). The β value ranged from 0.5 at the internal terminal to 0.6 at the external terminal. The coil capacity is approximately 300 mL. The HSCCC can be operated at a speed of 800 rpm.

The CCC apparatus described in [17] and used for the separation of rare earth elements (REEs) holds a set of three identical columns symmetrically distributed on the rotary frame at a distance of 7.6 cm from the central axis of the centrifuge. Each column holder is equipped with two planetary gears, one of which is engaged with an identical stationary sun gear mounted around the central stationary axis of the centrifuge. This gear arrangement produces a planetary motion of each column holder, i.e., one rotation around its own axis per revolution around the central axis of the centrifuge in the same direction. The other gear on the column holder is engaged with an identical gear on the rotary tube support mounted between the column holders. This gear arrangement produces counterrotation of the tube support to prevent twisting of the flow tube on the rotary frame. All column holders can be removed from the rotary frame by loosening a pair of screws on each bearing block, facilitating the mounting of the coiled column on the holder. Each multilayer coil was prepared from a single piece of approximately 100-m-long, 1.07-mm-i.d. PTFE tubing, by winding it directly onto the holder hub (15 cm in diameter), making 13 layers of the coil between a pair of flanges spaced 5 cm apart. The β values range from 0.5 at the internal terminal to 0.75 at the external terminal. Each multilayer coil consists of about 400 helical turns with approximately 90 mL capacity. The apparatus can be operated at speeds up to 1200 rpm.

B. Apparatus for Centrifugal Partition Chromatography Separations and Chromatographic Procedure

The centrifugal partition chromatography separation technique is detailed in Chapter 4 of this book and only very briefly discussed here. The technique was mainly developed on the basis of the classical countercurrent distribution method. The countercurrent distribution method was improved by the use of a totally seal joint device, which makes it possible to pump the mobile phase through the system while under centrifugation. The conventional countercurrent partition cells are replaced by microcells connected in series to form a cell cartridge. Centrifugal partition chromatographic separations of REEs [29,30] were performed with a model centrifuge containing six analytical/semipreparative cartridges each having 400 channels (2400 total channels, rotation 800 rpm). Centrifugal partition chromatographs models NMF and LLN (Sanki Engineering), containing three or six high-resolution-type cartridges with 400 microcells per cartridge equipping in series (total 1200 or 2400 microcells) at speed under centrifugation, 400–1200 rpm, were also used for the separation of REEs [31–33].

Separation of elements with centrifugal partition chromatography were carried out by the following procedures [33]. Previously the organic (stationary) phase was charged into the partition cell, and then the aqueous phase was supplied as the mobile phase by a pressurized pump until the volume balance between the two phases was equilibrated. A sample solution was injected through a sample loop to the partition cells. Each solute was eluted with the mobile phase.

III. SOLVENT SYSTEMS AND EXTRACTING AGENTS: THEORETICAL AND PRACTICAL CONSIDERATIONS

A. Two-Phase Liquid Systems and CCC Equations

At least the following three groups of requirements should be met for proper selection of a two-phase liquid system in CCC [5,18]. First, the system must be composed of two immiscible phases. In the case of organic-aqueous two-phase systems, the organic phase consists of one solvent or of a mixture of solvents. For preconcentration and separation of inorganic species, a stationary phase containing an extracting reagent in an organic solvent is applied in most cases. Of importance is also the selection of aqueous (mobile) phase as its composition affects the separation efficiency and the rate of chromatographic band movement. The mobile phase components should not interfere with subsequent analyses. Solutions of inorganic acids and their salts are most often used, the mobile phase may also contain specific complexing agents that can bind one or several elements under separation. Two-phase aqueous systems on the basis of water-soluble polymers may be used too.

Second, one of the phases (stationary one) must be retained in rotating the column to a required extent. The amount of stationary phase in the rotating column depends on the parameters of the planetary centrifuge, its operating conditions, and the properties of the two-phase systems.

Third, the stationary phase should permit separate elution of the elements into the mobile phase. In analytical applications, the stationary phase should also provide preconcentration of the elements to be determined, if necessary.

The selectivity of solvent systems can be estimated by determination of the partition coefficients for each element. The batch partition coefficients D^{bat} are calculated as the ratio of the component concentration in the organic phase to that in the aqueous phase. The dynamic partition coefficients of compounds D^{dyn} are determined from an experimental elution curve by Eq. (1):

$$D^{\mathrm{dyn}} = \frac{V_r - V_m}{V_s} \tag{1}$$

Here V_r is the retention volume, V_m is the volume of the mobile phase inside the column, $V_m = V_c - V_s$, where V_c is the total column volume, V_s is the volume of the stationary phase retained in the column.

Equations for the calculations of D^{dyn}, number of theoretical plates (N), separation factors (α), and peak resolution (R_s) have been set up on the basis of classical expressions used in extraction chromatography [4,5,10].

The chromatographic efficiency of the separation is measured by N and is connected with the peak sharpness.

$$N = (4t_r/W)^2 \tag{2}$$

where t_r is the retention time of the solute and W is the peak base width expressed in time units as t_r.

$$R_s = \frac{2(t_{r,2} - t_{r,1})}{W_1 + W_2} \tag{3}$$

$$\alpha = D_2^{\mathrm{dyn}}/D_1^{\mathrm{dyn}} \tag{4}$$

The retention of a substance depends on its partition coefficient in the biphasic liquid system used. The partition coefficients of organic substances are substantially related to the phase polarity difference [2]. The partition coefficients of ions depend dramatically on the nature and concentration of extracting reagent (extractant) in the solvent, type of organic solvent, composition of the mobile phase, and kinetic parameters of the system. The batch partition coefficients for inorganic compounds may be different from the dynamic ones.

Stationary phase on the basis of extracting reagents of different types (cation exchange, anion exchange, and neutral) was used for preconcentration and separation of various trace elements.

B. Influence of the Composition of the Mobile Phase and Kinetic Parameters of the System on the Chromatographic Behavior of Elements

The composition of the mobile phase affects the partition coefficients of inorganic substances and the separation efficiency. The chromatographic behavior of same elements in systems on the basis of cation exchange and neutral extracting reagents may serve as example.

1. Systems with Cation Exchange Reagents

Di-(2-ethylhexyl)phosphoric acid (D2EHPA) is a cation exchange (acidic) metal extractant. The chemism of metal cation extraction from a diluted aqueous solution can be described by Eq. (5) [34]:

$$M^{z+}{}_{(aq)} + zHB_{(org)} = MB_{z(org)} + zH^{+}{}_{(aq)} \tag{5}$$

where HB is the acidic extractant, MB_z is the adduct species whose formation is responsible for retention, and M^{z+} is the cation involved.

As is seen from this equilibrium, an increase in the hydrogen ion concentration in the aqueous phase leads to a decrease in the partition coefficient for the metal. The equilibrium describes correctly the extraction process when trace amounts of metals are extracted, i.e., when the amount of reagent needed for complex formation is much less than its total concentration in the extraction system [34]. The number of extractant molecules interacting with the metal ion depends on the charge and coordination number of the metal cation extracted from the aqueous solution.

Araki et al. [31,32] investigated the separation of lighter rare earth metal ions with D2EHPA by centrifugal partition chromatography. It has been shown that the retention volume of elements strongly depends on the concentration of HCl in the mobile phase, which is one of the key factors responsible for effective separations.

The chromatographic behavior of trivalent Eu in the system with D2EHPA at different HCl concentrations in aqueous phase in the use of the planetary centrifuge with one-layer coil column is shown in Figure 1 [13]. The elution of Eu(III) begins in all cases immediately after a volume of the mobile phase equal to V_m has passed through the column. The peak width increases with the decrease in HCl concentration. The values of D^{bat} and D^{dyn}, calculated from the elution curves for Eu(III) by Eq. (1), are given in Table 1. The difference in the shapes of elution curves 3 and 4 is due to the significantly different partition coefficients for Eu(III) at different HCl concentrations (from 0.25 to 0.2 mol/L). It is of interest that at a constant V_s value (12 mL) the shapes of the elution curves for Eu(III) are not

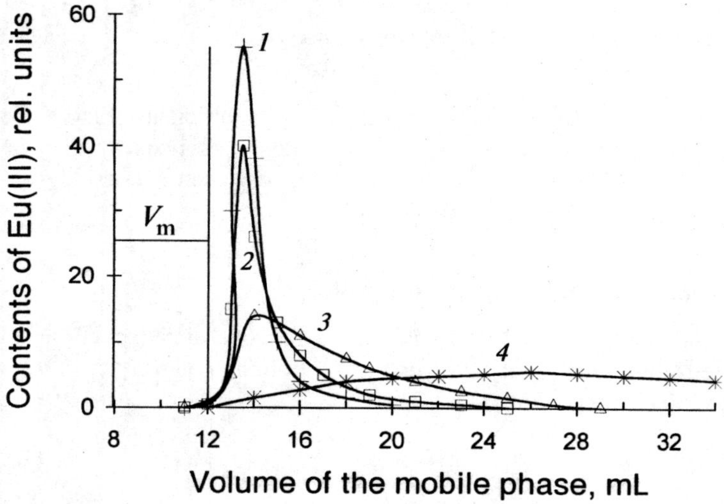

Figure 1 Chromatographic behavior of Eu(III) in 0.05 mol/L D2EHPA/*n*-decane/HCl systems. $S_f = 0.5$. HCl concentration in the mobile phase, mol/L: 1, 0.70; 2, 0.40; 3, 0.25; 4, 0.20. $F = 1$ mL/min.

influenced by the rotation speed (from 350 to 500 rpm) and pumping speed of the mobile phase (from 0.4 to 1.5 mL/min).

Kinetic peculiarities of the interaction between D2EHPA and various metal ions have been relatively well studied. Eu(III) is known to be extracted much more quickly than Fe(III) in systems with D2EHPA [35].

The equilibrium for Fe(III) extraction in D2EHPA*n*-decane HCl systems is attained within several hours under batch extraction conditions. When the chromatographic behavior of Fe(III) in these systems is examined, it would not be correct to discuss dynamic partition coefficients since under the same conditions

Table 1 Batch and Dynamic Partition Coefficients for Eu(III) in the 0.05 mol/L D2EHPA/*n*-Decane/HCl System

C_{HCl} (mol/L)	D^{bat}	D^{dyn}
0.70	0.06	0.12
0.40	0.12	0.12
0.25	0.20	0.18
0.20	0.80	1.20

a part of Fe(III) is eluted from the column, whereas the other part is extracted into the stationary phase and is not eluted even after the volume of the mobile phase equal to 10 V_m has been passed through the column. Consequently, the first part of Fe(III) should have a partition coefficient of about 0.1 and the second part of Fe(III) about 10 or higher. About 50% of Fe(III) is eluted in the case of 0.05 mol/L and 5% in the case of 0.5 mol/L D2EHPA (Figure 2).

Therefore, for the chromatographic data interpretation additional studies of kinetic properties of the systems used and determination of Eu(III) and Fe(III) mass transfer coefficients (*k*) are required. These studies were made by use of a stirred diffusion cell [13]. The volumes of organic and aqueous phases in the cell were 10 and 30 mL, respectively; the initial concentration of the element in the aqueous phase was 5 ppm; the total volume of the aqueous phase fractions taken for concentration measurements were less than 3 mL.

The mass transfer coefficients were determined with the use of *t* dependence of $-\ln(1 - E_t)$.

$$E_t = (C_0 - C_t)/(C_0 - C^*) \tag{6}$$

where C_0 is initial, C_t current, and C^* equilibrium concentration of the element in the aqueous phase, mol/L. It is known that

$$-\ln(1 - E_t) = kat \tag{7}$$

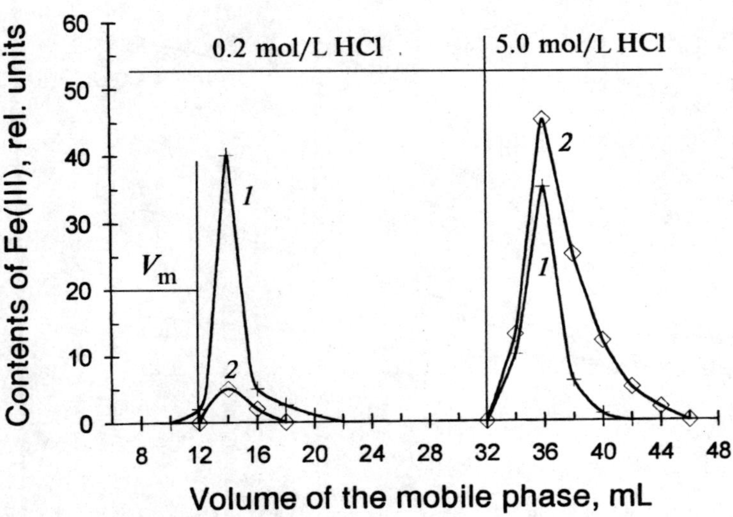

Figure 2 Chromatographic behavior of Fe(III) in D2EHPA/*n*-decane/HCl systems. $S_f = 0.5$. Stationary phase: 1, 0.05; 2, 0.5 mol/L D2EHPA. Step elution. $F = 1$ mL/min.

where $a = S/V$ (S, phase interface area, cm^2; V, aqueous phase volume, mL; k, mass transfer coefficient, cm/sec). The tangent of an angle of t dependence of $-\ln(1 - E_t)$ is equal to ka.

Figure 3 illustrates t dependencies of $-\ln(1 - E_t)$ for Eu(III) and Fe(III) extraction in the systems on the basis of D2EHPA. It can be seen that dependencies 1 and 2 for Fe(III) extraction (systems containing 0.05 and 0.5 mol/L D2EHPA, respectively) and dependence 3 for Eu(III) extraction (system containing 0.5 mol/L D2EHPA) are straight lines, whereas the dependence 4 for Eu(III) extraction (system containing 0.5 mol/L D2EHPA) is more complicated. The formation of thin condensed films at the interface in Eu-containing systems with high D2EHPA concentration [35] may explain why the Eu(III) extraction retards with time.

The t dependencies of $-\ln(1 - E_t)$ for Fe(III) extraction in D2EHPA-based systems are straight lines (Figure 3) but the mass transfer coefficients are much lower than in the case of Eu(III) (Table 2). Besides, the extraction of Fe(III) in the systems with D2EHPA is practically irreversible. For instance, the equilibrium partition coefficient (D^{bat}) of Fe(III) in the 0.05 mol/L D2EHPA/*n*-decane/2.0 mol/L HCl system is about 1 in the extraction, whereas in the back-extraction this value is about 10 (Table 2). The mechanisms of the extraction and back-

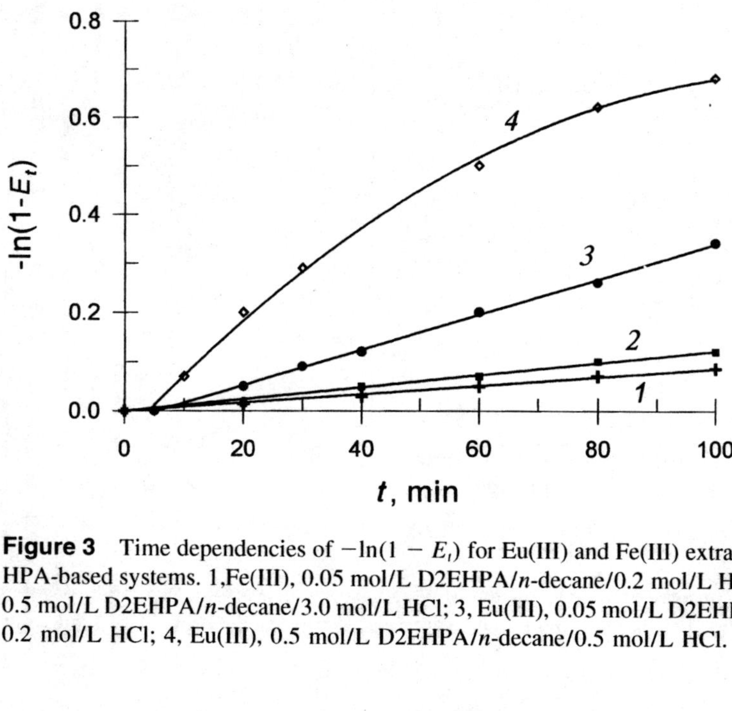

Figure 3 Time dependencies of $-\ln(1 - E_t)$ for Eu(III) and Fe(III) extraction in D2E-HPA-based systems. 1,Fe(III), 0.05 mol/L D2EHPA/*n*-decane/0.2 mol/L HCl; 2, Fe(III), 0.5 mol/L D2EHPA/*n*-decane/3.0 mol/L HCl; 3, Eu(III), 0.05 mol/L D2EHPA/*n*-decane/ 0.2 mol/L HCl; 4, Eu(III), 0.5 mol/L D2EHPA/*n*-decane/0.5 mol/L HCl.

Table 2 Mass Transfer Coefficients (k) and Batch Partition Coefficients (D^{bat}) for Eu(III), Fe(III), Ta(V), Nb(V), Hf(IV), and Zr(IV) in D2EHPA-Based Extraction Systems

Element	Extraction system	k(cm/sec)	D^{bat}
Eu(III)	0.05 mol/L D2EHPA/n-decane/0.2 mol/L HCl	2.0×10^{-4}	0.8
	0.5 mol/L D2EHPA/n-decane/0.5 mol/L HCl	6.7×10^{-4}[a]	5.2
Fe(III)	0.05 mol/L D2EHPA/n-decane/0.2 mol/L HCl	4.0×10^{-5}	8.6
	0.05 mol/L D2EHPA/n-decane/2.0 mol/L HCl	1.2×10^{-4}[b]	10.9
	0.5 mol/L D2EHPA/n-decane/3.0 mol/L HCl	9.3×10^{-5}	3.1
Ta(V)	0.5 mol/L D2EHPA/n-decane/2.0 mol/L HNO_3	5.7×10^{-5}	12.1
Nb(V)		1.3×10^{-4}	33.8
Hf(V)		6.5×10^{-4}	>100
Zr(V)		7.1×10^{-4}	>100

[a] $t < 30$ min (see Figure 3, curve 4).
[b] Back-extraction.

extraction of Fe(III) in the system under investigation are apparently different. It should be noted that the mass transfer coefficients for Eu(III) and Fe(III) extraction are not influenced by changing HCl concentration (from 0.2 to 3.0 mol/L), all other factors being the same.

The corresponding values of mass transfer coefficients of Fe(III) for two different D2EHPA concentrations (Figure 2) are 4.0×10^{-5} and 9.3×10^{-5} cm/s. Hence, the greater the magnitude of the mass transfer coefficient, the less the quantity of Fe(III) eluted immediately after the V_m volume has been passed through the column. This can be explained by the increase in the batch partition coefficient for Fe(III) ($D^{bat} = 18.6$ if 0.05 mol/L D2EHPA is used and $D^{bat} > 50$ if 0.5 mol/L D2EHPA is used) also contributes to the degree of Fe(III) extraction. In both cases Fe(III) retained by the stationary phase can be eluted only by a highly concentrated HCl solution, since the mechanism of back-extraction of Fe(III) is different from that of extraction.

Similar curves circumscribe the chromatographic behavior of Ta(V) and Nb(V) (elements kinetically more inert than Fe) in the 0.5 mol/L D2EHPA/n-decane/2.0 mol/L HNO_3 system (Figure 4). As is seen, a part of Ta(V) (40%) and Nb(V) (35%) is not extracted and eluted after the V_m volume has passed through the column. Zr(IV) and Hf(IV) are extracted quantitatively. The results of the study on Ta(V), Nb(V), Hf(IV), Zr(IV) extraction kinetics in the 0.5 mol/L/D2EHPA/n-decane/2.0 mol/L HNO_3 system are presented in Figure 5 and Table 2. The t dependencies of $-\ln(1 - E_t)$ are straight lines for all elements, but the mass transfer coefficients for Ta(V) and Nb(V) are appreciably lower than for Hf(IV) and Zr(IV).

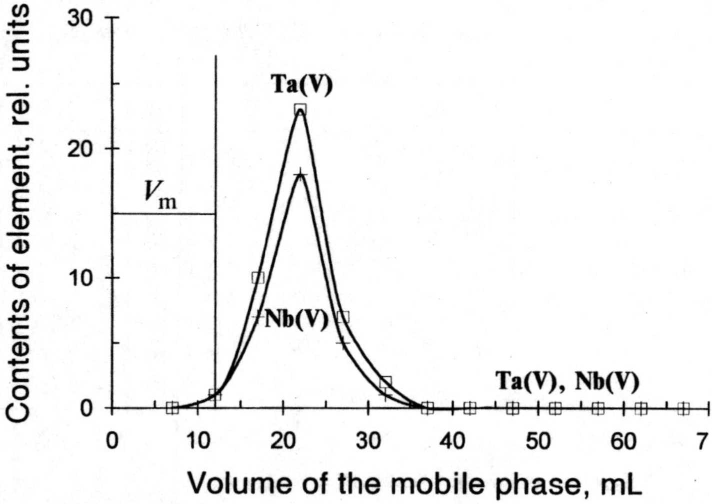

Figure 4 Chromatographic behavior of Ta(V), Nb(V), Hf(IV), and Zr(IV) in 0.5 mol/L D2EHPA/*n*-decane/2.0 mol/L HNO_3 system. $S_f = 0.5$. $F = 1$ mL/min.

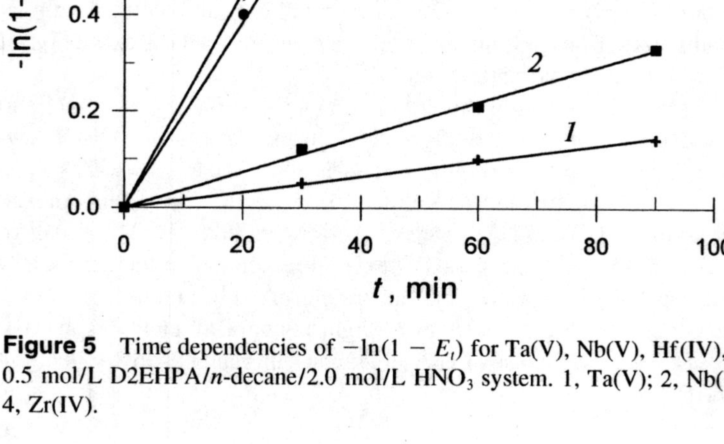

Figure 5 Time dependencies of $-\ln(1 - E_t)$ for Ta(V), Nb(V), Hf(IV), and Zr(IV) in 0.5 mol/L D2EHPA/*n*-decane/2.0 mol/L HNO_3 system. 1, Ta(V); 2, Nb(V); 3, Hf(IV); 4, Zr(IV).

The extraction system used provides relatively high values of the equilibrium partition coefficients for Ta(V) and Nb(V). However, the quantitative extraction is not attained because of a low rate of the mass transfer process. Comparison of the chromatographic behavior of the elements under investigation and their mass transfer coefficients allows us to assume the following. To concentrate species with mass transfer coefficients $k < 10^{-4}$ cm/sec under the experimental conditions used ($V_c = 24$ mL), their partition coefficients must be 100 or higher. Otherwise, the species will be partly eluted after the V_m volume is passed through the column.

Therefore, when kinetically inert compounds are concentrated in the stationary phase under CCC conditions, a column with a larger total volume should be used. However, in the separation of any species, a larger column volume results in a longer separation time and broader chromatographic peaks [17].

2. Systems with Neutral Extractants

Tetraphenylmethylenediphosphine dioxide (TPMDPD) is a neutral extractant. The extraction of metals by neutral organophosphorous reagents occurs with substitution of water molecules and solvation of neutral compounds, which are initially present in the aqueous phase. A neutral compound of metal cation with anion or anionic ligand is extracted into the organic phase. One of two equilibria may be used to describe the extraction process:

$$M^{z+}{}_{(aq)} + zA^{-}{}_{(aq)} + yB_{(org)} = MA_z(B)_{y(org)} \tag{8}$$

$$MA_{z(aq)} + yB_{(org)} = MA_z(B)_{y(org)} \tag{9}$$

where A^- is anionic ligand present in the aqueous phase and B is neutral organophosphorous reagent.

Neutral organophosphorous compounds are known to be advantageously distinguished from acidic organophosphorous compounds by their kinetic properties. Thus, the equilibrium attainment in an extraction system on the basis of neutral organophosphorous reagents is faster and the mass transfer more efficient.

The chromatographic behavior of Eu(III) in the 0.005 mol/L TPMDPD in chloroform/0.5 mol/L NH_4SCN/HCl system at different HCl concentrations is shown in Figure 6. The width of the chromatographic peaks increases with a decrease in HCl concentration. However, in contrast to systems with D2EHPA, Eu(III) was retained in the separation column after a volume of V_m had passed off. The values of mass transfer coefficients of Eu(III) in 0.005 mol/L TPMDPD/chloroform/0.5 mol/L NH_4SCH–0.5 mol/L HCl extraction system is 1.8×10^{-3} cm/sec (t dependence is a straight line). The values of D^{dyn} and D^{bat} for Eu(III) in the systems under investigation are given in Table 3. The values of D^{dyn} are lower than those of D^{bat}. This fact testifies that the elution of Eu(III) is not an

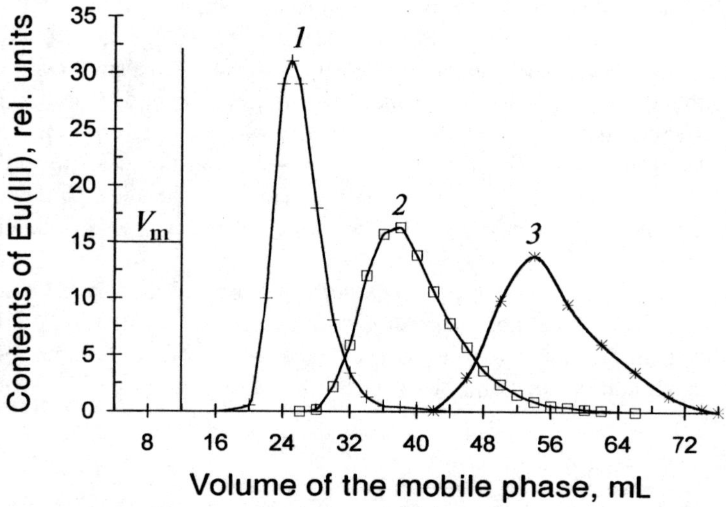

Figure 6 Chromatographic behavior of Eu(III) in 0.005 mol/L TPMDPD chloroform/ 0.5 mol/L NH_4SCN/HCl system. $S_f = 0.5$. HCl concentration in the mobile phase, mol/ L: 1, 1.0; 2, 0.6; 3, 0.4. $F = 1$ mL/min.

equilibrium process. Asymmetry of the chromatographic peaks also shows that the equilibrium has not been attained under dynamic conditions.

The nature of the extraction system and its kinetic properties may have a greater effect on the chromatographic process than the parameters of the planet centrifuge operation. As we have shown [13,15], the two extraction systems on the basis of D2EHPA and of TPMDPD are characterized by different kinetic properties and different values of mass transfer coefficients for Eu(III). Figure 7 illustrates once more the importance of the contribution from chemical kinetics to the separation process. Two peaks of Eu(III) are quite different due to the

Table 3 Batch and Dynamic Partition Coefficients for Eu(III) in the 0.005 mol/L TPMDPD/Chloroform/0.5 mol/L NH_4SCN/HCl System

C_{HCl}(mol/L)	D^{bat}	D^{dyn}
1.0	1.8	0.9
0.6	2.7	1.8
0.4	3.8	3.0

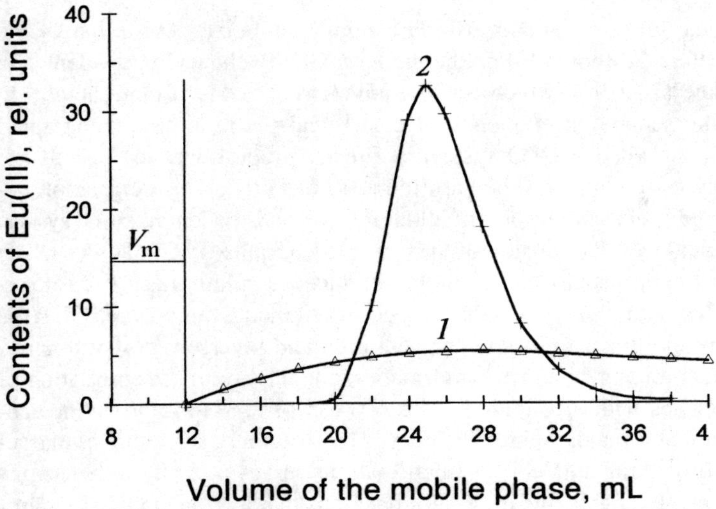

Figure 7 Influence of mass transfer coefficient on the Eu(III) peak shape. 1, 0.05 mol/L D2EHPA/*n*-decane/0.20 mol/L HCl system, = $S_f = 0.5$. $D^{dyn} = 1.2$. $k = 2.0 \times 10^{-4}$ cm/sec. 2, 0.005 mol/L TPMDPD/chloroform/1.0 mol/L HCl/0.5 mol/L NH_4SCN system. $S_f = 0.5$. $D^{dyn} = 0.9$. $k = 1.8 \times 10^{-3}$ cm/sec.

kinetic peculiarities of the extraction systems used though the corresponding values of dynamic partition coefficients are close (1.2 and 0.9, respectively). Hence, the mass transfer coefficients determine the type of elution, which is necessary for the element separation under given experimental conditions. The systems providing high values of k (about 10^{-3} cm/sec) are suitable for isocratic separation as well as for separation by step elution, whereas kinetically more inert systems (k is about 10^{-4} cm/sec) require only step elution. If the elements under investigation are kinetically very inert (k is about 10^{-5} cm/sec), some difficulties arise even when these elements are being concentrated. High partition coefficients (about 10^2) are necessary for the quantitative extraction of such elements into the stationary phase.

Freiser et al. [29,30] investigated the influence of kinetic factors on the separation efficiency taking as example the behavior of platinum group metals by a trioctylphosphine oxide (TOPO)/heptane/HCl system in a centrifugal partition chromatographic apparatus. It was found that increasing chloride concentration and decreasing TOPO concentration provided better centrifugal partition chromatographic efficiencies for Pd(II) [30]. It has been noted [30] that the chromatographic efficiency in separations involving distribution of metal complexes reveals an influence of the kinetics of complex formation and dissociation on the chromatographic efficiencies and affords a semiquantitative description of the kinetic parameters. Authors [30] define the term CETP, channel equivalent of a theoretical plate (2400/*N*), which is the equivalent of reduced plate height, to characterize the separation efficiency of centrifugal partition chromatography. The CETP for the Pd(II)-TOPO system is directly proportional to the half-life $t_{1/2}$ for the back-extraction and the partition ratio of Pd(II). This correlation between CETP and $t_{1/2}$ has allowed attribution of the band broadening in the system to the slow kinetics of dissociation of the extracted complex $PdCl_2(TOPO)_2$. The dissociation rate constant for this complex in aqueous solution has been determined to be 168 $mol^{-1}\ sec^{-1}$ by the stopped-flow kinetics procedure. CETP is also directly proportional to TOPO concentration and inversely proportional to chloride concentration. The kinetic information obtained from the correlation of column efficiencies with distribution ratios (*D*) can be used to improve the centrifugal partition chromatographic efficiency. The customary screening of extraction reactions involving single-stage equilibrations serves not only to determine *D* values but also to enable one to discard reactions that are too slow to be chromatographically useful. Even when these experiments indicate that the extraction reactions are too fast to be measured by ordinary means, the sensitivity of countercurrent and centrifugal partition chromatographic techniques to reaction rates are dramatically greater. Slow reactions that have half-lives of even 1 sec may affect the column efficiency.

C. Influence of Reagent Concentration in the Solvent and Type of Organic Solvent on the Chromatographic Behavior of Elements

Reagent concentration in the organic solvent and type of organic solvent also affect the elution curve shape and, therefore, the dynamic coefficients.

Let us discuss the effect of cation exchange reagent (D2EHPA) concentration on the separation of REEs taking the Nd–Sm pair as example. A twofold increase in the extractant concentration results in a better selectivity of Sm extraction (Table 4): the batch partition coefficient for Sm increases by a factor 10,

Table 4 Batch and Dynamic Partition Coefficients for Sm(III) and Nd(III) in D2EHPA-Based Extraction Systems

	D^{bat}		D^{dyn}	
Extraction system	Nd	Sm	Nd	Sm
0,5 mol/L D2EHPA/*n*-decane/0.50 mol/L HCl	0.5	1.3	0.7	1.5
1,0 mol/L D2EHPA/*n*-decane/0.50 mol/L HCl	0.7	>10	0.8	>10

whereas the recovery of Nd remains practically the same. Thus, favorable conditions are fulfilled for the separation of Nd and Sm ($\alpha > 10$).

Figure 8 illustrates the chromatographic behavior of Nd and Sm in D2EHPA/*n*-decane/HCl systems at different reagent concentrations. The both elements are eluted simultaneously in the 0.5 mol/L D2EHPA/*n*-decane/0.5 mol/L HCl system, although the batch partition coefficients for Nd and Sm (Table 4) make their isocratic separation formally possible. The elution of the batch elements begins just after the volume V_m has been passed off and the resultant peaks are somewhat dispersed because of insufficient efficiency of the mass transfer between the phases. In the use of 1 mol/L D2EHPA/*n*-decane/0.5 mol/L HCl

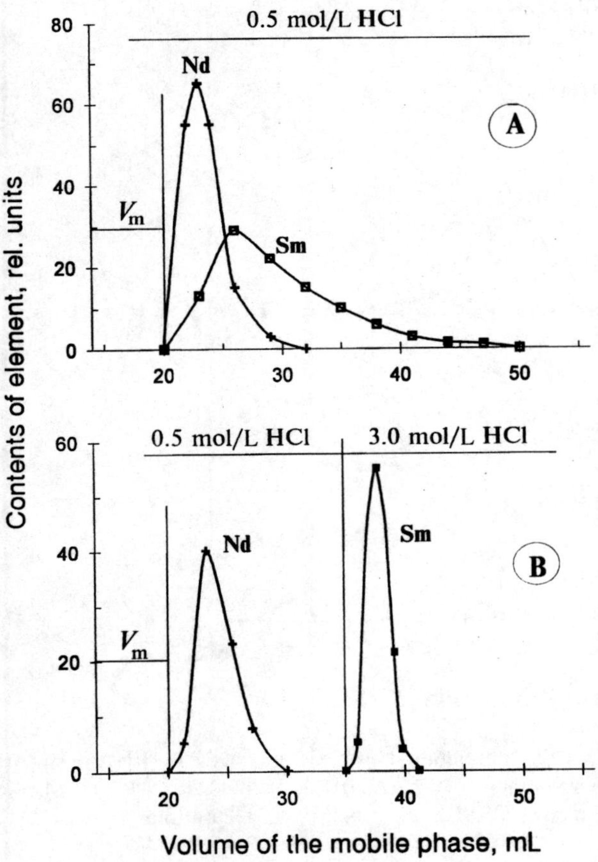

Figure 8 Influence of D2EHPA concentration in *n*-decane on Nd(III) and Sm(III) peak shapes. D2EHPA concentration, mol/L: A, 0.5; B, 1.0.

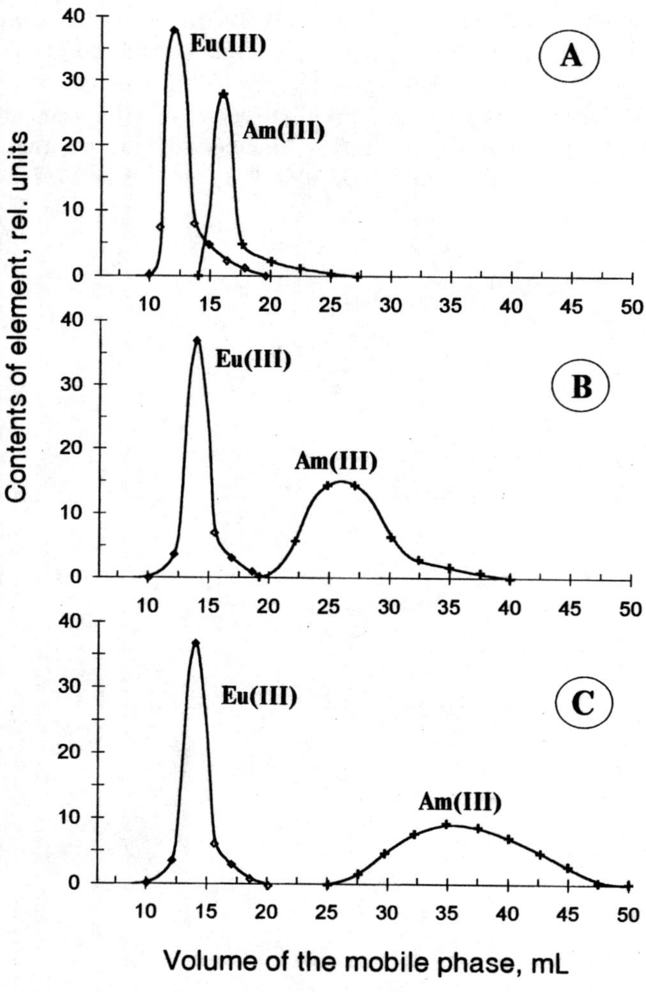

Figure 9 Influence of TPMDPD concentration in chloroform on the Eu(III) and Am(III) peak shapes. TPMDPD concentration, mol/L: A, 0.002; B, 0.0025; C, 0.003. Mobile phase: 1 mol/L HCl + 0.5 mol/L NH_4SCN. $S_f = 0.75$. $F = 1.1$ ml/min.

system, neodymium is eluted just after the volume V_m has passed through the column (Figure 8B), samarium being concentrated in the stationary phase. With an increase in HCl concentration in the mobile phase up to 3 mol/L, the D values for REE drop dramatically, and Sm is eluted into a small (6 mL) eluate fraction. The mass transfer coefficient values for the systems under investigation are approximately equal to 5×10^{-4} cm/sec. This enables the elements having $D^{bat} > 10$ to be retained in the stationary phase after the V_m volume has passed off.

Similar effects of the reagent concentration on the elution curve shapes are observed in the use of reagents of other classes. Figure 9 illustrates the chromatographic behavior of Eu(III) and Am(III) at different TPMDPD concentrations in chloroform [25]. It is seen that an increase in the reagent concentration in the organic phase leads to better separation. Table 5 gives the D^{dyn} values for Eu(III) and Am(III) and the separation factors $\alpha = D_{Am}/D_{Eu}$, calculated from the elution curves. However, a rather large volume of the mobile phase (15–20 mL) is required for the elution of americium from the column.

The type of organic solvent may very often have a great effect on the chromatographic process, the reagent concentration in the solvent and the mobile phase composition being the same. Figure 10 shows the elution curve shapes for orthophosphate ions back-extracted from a stationary phase on the basis of 5% dinonyltin dichloride (DNTDC) (anion exchange extractant) in methyl isobutyl ketone (MIBK) (curve A) and in *n*-decanol (curve B). The elution of orthophosphate was carried out after its preconcentration into the stationary phase from 1 mol/L HNO_3 solution. In the case of MIBK as organic solvent, orthophosphate was quantitative eluted from the stationary phase with 0.5 mol/L NaCl in 0.5 mol/L HCl. Using *n*-decanol instead of MIBK leads to smearing of elution curve.

Araki et al. reported [31,32] the influence of the solvent nature on the centrifugal partition chromatographic separation of REE in the use of D2EHPA as extractant and aqueous HCl solution as mobile phase. The authors concluded

Table 5 Dynamic Partition Coefficients for Fe(III) and Am(III) and Their Separation Factor (α) in the TPMDPD/Chloroform/1 mol/L HCl/0.5 mol/L NH_4SCN Systems

	D^{dyn}		
C_{TPMDPD} (mol/L)	Am(III)	Eu(III)	α
0.0020	0.8	0.5	1.6
0.0025	1.4	0.6	2.3
0.0030	2.0	0.7	2.8

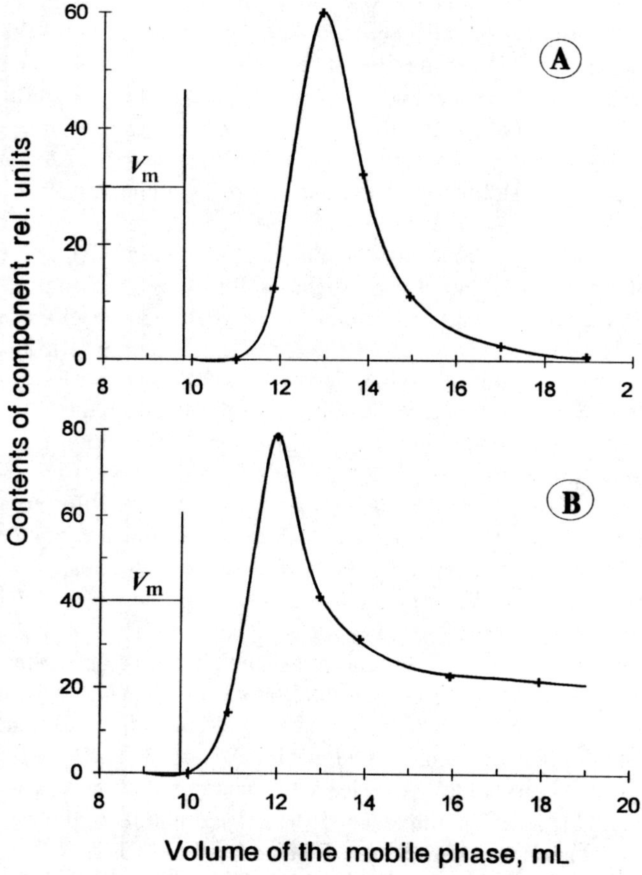

Figure 10 Chromatographic behavior of orthophosphate in 5% DNTDC/organic solvent/0.5 mol/L NaCl/0.5 mol/L HCl system. Solvent: A, MIBK; B, *n*-decanol. S_f = 0.5. F = 1 mL/min.

that more polar solvents are effective for both shortening the retention time and narrowing the peaks of REEs having higher atomic numbers, i.e., *n*-heptane for La(III), Ce(III), Pr(III), and Nd(III); toluene for Sm(III), Eu(III), and REE ions slightly heavier; and chloroform for much heavier ions around Yb(III). It should be noted, however, that the three solvents studied are different not only in their polarity. Further investigations of the effect of solvent nature on the countercurrent and centrifugal partition chromatographic separations seem to be necessary.

D. Influence of the Stationary Phase Retention Volume on the Chromatographic Behaviour of Elements

The element elution depends not only on the mobile and stationary phase compositions but also on the operation conditions of the planetary centrifuge, which influence the quantity of the stationary phase in the column. It has been shown [5,10] that the retention of the stationary phase is dependent on the physicochemical properties of the solvent system, the internal diameter of the column, the filling procedure, the rotational speed of the CCC apparatus, and the flow rate. The retention of liquid stationary phase increases with the centrifugal rotation speed and decreases with the flow rate [5,10]. The physicochemical properties of the pure liquids are important. Low viscosity, high interfacial tension, and high density differences are desirable [10].

Figure 11 (the origin of the coordinates corresponds to the volume of the mobile phase in the column) shows the curve shape for Eu(III) and its position as a function of the stationary phase retention volume in the column containing TPMDPD in chloroform as extractant. The chromatographic peak shifts to the left and narrows if the extractant quantity is twice lower (V_s^f = 6 mL, peak 1), all other factors being the same. The values of D^{dyn} calculated from the chromatograms are the same and equal to 1.8.

Berthod [10] observed a similar effect in the CCC separation of anthranoid pigments.

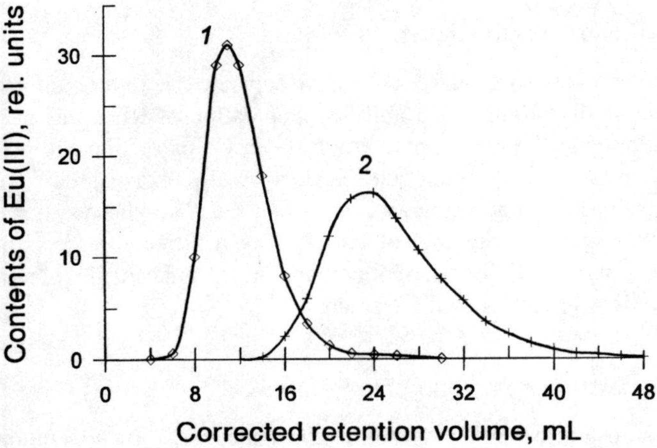

Figure 11 Influence of the stationary phase retention volume on the shape of the elution curves of Eu(III) in the 0.005 mol/L TPMDPD/chloroform/0.6 mol/L HCl/0.5 mol/L NH_4SCN system. 1, S_f = 0.25 (V_s = 6 mL); 2, S_f = 0.5 (V_s = 12 mL).

IV. PRECONCENTRATION, SEPARATION, AND DETERMINATION OF ELEMENTS

A. Procedures for Separation and Preconcentration

The following procedures may be used for separation of inorganic substances.

The first one is a standard procedure (isocratic or gradient separation). The substances are separated due to the difference in their distribution coefficients on sample injection into the column with the mobile phase flow.

The second procedure is separation of elements after their preliminary preconcentration. To reduce the elution time and eluate volume and to achieve maximum efficiency of separation, we used the following variant of the procedure. The sample is introduced into the column with a mobile phase that provides high distribution coefficients ($D > 10$) for the elements of interest; the elements are preconcentrated at this stage. Then by step changing the eluents, successive elution of the elements is performed using small volumes of eluents providing $D^{dyn} < 0.5$. Isocratic separation can be also applied after a preconcentration step. As we know, the described procedures for the separation after preconcentration were used in chromatographic techniques other than CCC. The mass transfer coefficients (k) should also be taken into account in the development of element separation methods. As we have shown [13,15], high values of k ($>10^{-3}$ cm/sec) ensure the possibility of isocratic separation. Step elution should be used if k values are relatively low ($<10^{-4}$ cm/sec).

B. Separation of Rare Earth Elements

Trivalent cations of REEs are particularly difficult to separate due to their almost equal diameters and similar complexing abilities. Separation of REE was performed by high-performance liquid chromatography (HPLC), extraction chromatography, and solvent extraction. Extraction reagents of different classes were applied to extraction separation and concentration of REEs. Two-phase systems on the basis of such reagents in organic solvents that meet the requirements of partition chromatography can also be used for countercurrent and centrifugal partition chromatographic separations of the elements.

1. Centrifugal Partition Chromatographic Separation

The first work on the use of centrifugal partition chromatography for separation of inorganic species was published in 1988 by Araki [31] and devoted to the separation of lighter REEs. The separations were carried out on the apparatus described above by use of a D2EHPA-based extraction system. The possibility was shown of the separation of La, Ce, Pr; La, Pr, Nd; Pr, Nd; and Nd, Sm by use of variable HCl concentration (from 0.04 to 0.15 mol/L), and of Sm, Eu using toluene instead of *n*-heptane as organic solvent [31]. A separation run took about 5 hr. The peak resolution can be improved by increasing the number of microcelles (from 1200 to 2400) [32]. For the binary mixture of Pr and Nd, whose separation is important for REE technologies, an increase in the number of microcells for obtaining a higher number of theoretical plates was shown to be effective. Later Araki et al. [36] published their studies on the centrifugal partition chromatographic behavior of the heavier PEE (Sm, Eu, Gd, Tb, Dy, Ho, Er, Tm, Yb, Lu) in the use of 0.01 mol/L D2EHPA in chloroform at different HCl concentrations in the mobile phase and different temperatures. The effect of temperature was examined for a binary mixture of Dy and Er. It was shown that the retention times were not significantly changed with temperature, but the number of theoretical plates was considerably increased with temperature. The separation for the REEs ranged from 2 to 4 hr.

Centrifugal partition chromatographic separation of light lanthanoids in D2EHPA-based systems using a CCl_4–*n*-paraffin mixture as diluent and HNO_3 as mobile phase was investigated in [37]. It was shown that the separation efficiency could be improved not only by optimization of the operation conditions but also by the selection of diluent. The effect of the *n*-paraffin chain length (number of carbon atoms: 6–15) was examined; the best separation efficiency was achieved with the longest, *n*-pentadecane. On the basis of the results obtained, mutual separation of Pr, Sm, Eu, and Gd was successfully performed in the 30% D2EHPA 15% CCl_4 55% *n*-pentadecane–0.5 mol/L HNO_3 extraction system (Figure 12).

The application of 0.02 mol/L 2-ethylhexylphosphinic acid mono-2-ethylhexyl ester solution in kerosene also helped to achieve the separation of some heavy REEs by centrifugal partition chromatography [37]. The required pH value of the mobile phase was adjusted by (H, Na)Cl_2CHCOO. The peak resolution R_s for the Gd,Tb and Tb,Dy pairs are 1.33 and 1.17, with a separation factor α of 4.71 and 2.70, respectively. The number of theoretical plates N for the bands of Gd, Tb, and Dy are 108.2, 51.4, and 41.6. The volumes of the separated element fractions are as large as 100–500 mL.

Separation of light trivalent REEs (Ce, Pr, Eu) by centrifugal partition chromatography was studied also in a 30% tributyl phosphate (TBP)*n*-dodecane extraction system by the aid of salting-out effect of $LiNO_3$ [30]. By substituting *n*-dodecane with CCl_4 as diluent of TBP, use of $LiNO_3$ up to 5 mol/L was realized. Effective separation of Pr and Eu was achieved for 90 min although the concentration effect of the lanthanoids showed the difference of the apparent D values between the centrifugal partition chromatography and the batch experiments. The separation methods described above may be more useful for preparative than analytical purposes because of rather long separation times and large fraction volumes.

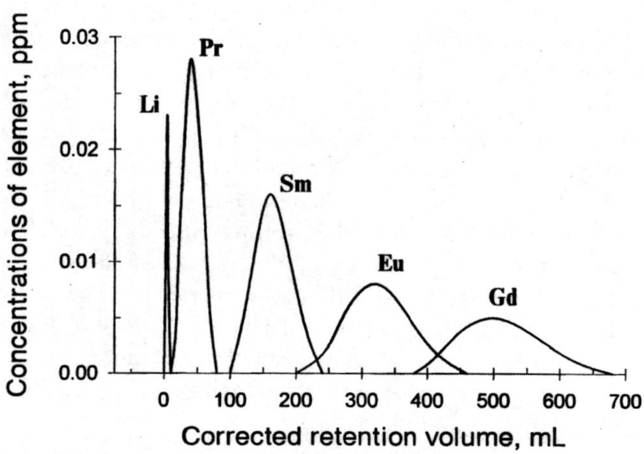

Figure 12 Separation of Li, Pr, Sm, Eu, and Gd by CPC. Stationary phase: 30% D2EHPA/15% CCL_4/55% *n*-pentadecane. Mobile phase: 0.5 mol/L HNO_3. Rotational speed: 800 rpm. (From Ref. 37.)

Centrifugal partition chromatography was used for analytical scale separations of trivalent REEs by 0.1 mol/L Cyanex 272 [bis(2,4,4-trimethylpentyl)phosphoric acid] in heptane as stationary phase and water at the appropriate pH as mobile phase [14]. It was demonstrated for the first time that a mixture of light and heavy REEs can be efficiently separated in a single run by centrifugal partition chromatography using gradient pH elution.

2. CCC Separation

We first used the CCC technique for the separation of the REE pairs La,Ce; Ce,Eu; and Sm,Eu [18,22]. A 0.5 mol/L D2EHPA in *n*-decane was applied as the stationary phase. The quantitative extraction and full separation of the elements is achieved with the use of step elution. The separation of La and Ce was performed for 100 min; of Sm and Eu for 45 min.

Kitazume [17] separated REE on a high-speed centrifuge equipped with three multilayer coils connected in series by use of a two-phase system on the basis of D2EHPA. La, Pr, Nd were well resolved in 2.5 hr, while the peak resolution between La and Pr is 6.95 and that between Pr and Nd is 1.74.

To further demonstrate the capabilities of the method, a one-step separation of all 14 REEs was performed by applying an exponential gradient of HCl concentration in the mobile phase [17]. Figure 13 shows the chromatogram of all 14 REEs resolved in less than 5 hr [17]. This chromatogram looks impressive

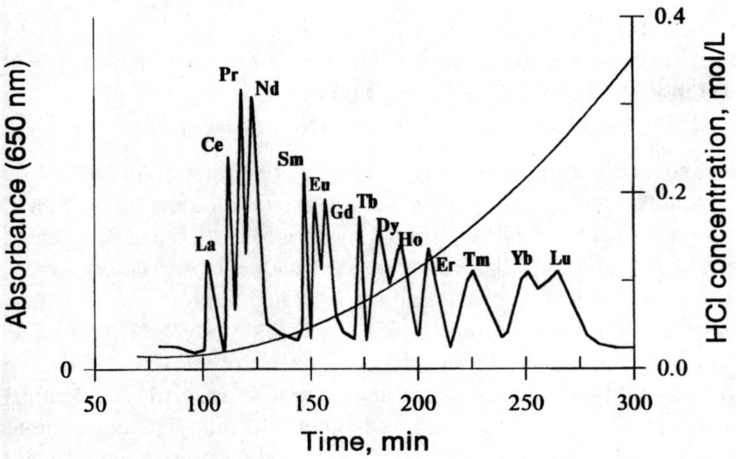

Figure 13 Gradient separation of 14 REEs obtained by HSCCC (planetary centrifuge with 7.6-cm revolution radius and three multilayer coiled columns connected in series; total capacity, 270 mL). Stationary phase: 0.003 mol/L D2EHPA in *n*-heptane. Mobile phase: exponential gradient of hydrochloric acid concentration from 0 to 0.3 mol/L. Sample: 14 REE chlorides each 0.001 mol/L in 100 μL water. Revolution speed: 900 rpm; $F = 5$ mL/min. (From ref. 17.)

and the procedure promising for separation on preparative scale. For application in analysis, the eluate volume (1500 mL) is too large.

3. Group Preseparation and Determination of Rare Earth Elements in Geological Samples

Due to high preparation capacity and the possibility of using most liquid–liquid extraction systems, CCC can be applied as a preconcentration and preseparation technique for various trace elements before their instrumental determination [18–22]. The technique (described in Section I) was utilized for preconcentration and separation of REEs from major constituents of various geological samples and after their decomposition for subsequent determination by inductively coupled plasma atomic emission spectrometry (ICP-AES) and mass spectrometry (MS) [19,20].

Three extraction systems on the basis of D2EHPA, TOPO, and diphenyl(dibutylcarbamoylmethylphosphine) oxide (Ph_2Bu_2) were shown to be applicable to the group separation of REEs from dissolved samples of rocks, ores, and minerals (basalts, granites, dolomite, fluorite-barite-hydropatite ore, syenite, etc.):

0.5 mol/L D2EHPA *n*-decane HCl (system 1)
0.3 mol/L D2EHPA + 0.02 mol/L TOPO/*n*-decane + MIBK (v/v = 3:1)/1.0 mol/L NH_4NO_3/6 mol/L HCl (system 2)
1.0 mol/L Ph_2Bu_2/chloroform/3.0 mol/L HNO_3 (system 3)

Some practical capabilities of CCC for the group separation of REEs from matrix components of geological samples are illustrated in Figure 14. An extraction system on the basis of D2EHPA (system 1) was used in this case. The separation is attained by step elution. Alkali, alkaline earth elements, and other elements with distribution coefficients less than 0.5 are separated from REE at the stage of their preconcentration from 0.1 mol/L HCl. By using 3 mol/L HCl REEs are selectively eluted from the stationary phase. To remove Fe(III), U(VI), Th, Mo(IV), and Ti from the stationary phase, the column is washed with 5 mol/L HCl. The total time of the separation cycle is about 40 min. The concentrates obtained are aqueous solutions of REE ready immediately for instrumental analysis. Table 6 illustrates the separation of some matrix components of dolomite and fluorite-barite-hydropatite ore by use of system 1 and subsequent atomic emission analysis. The high capacity of the separation column facilitates one-stage separation of REEs from most of matrix components. ICP-AES determination of REEs in their concentrates obtained from these two samples as well as from a basalt sample (Table 7, system 1) has shown that only iron, present in the concentrates in essential quantities, may interfere with REE determination.

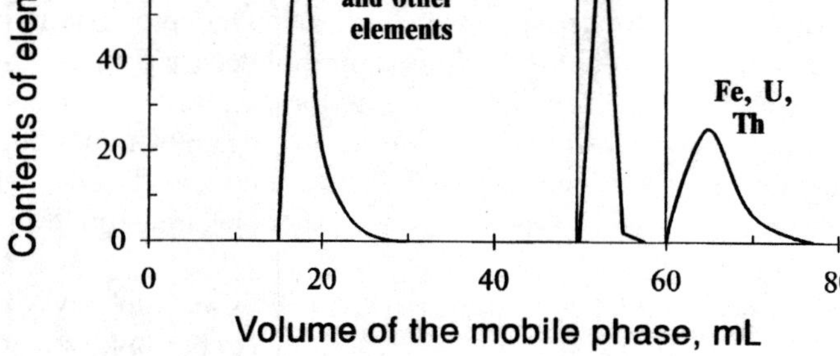

Figure 14 CCC separation of REEs from some macrocomponents present in geological samples. Stationary phase: 0.5 mol/L D2EHPA in *n*-decane. $S_f = 0.5$, $F = 1.2$ mL/min.

Table 6 Separation of Some Matrix Components by CCC Recovery of REEs from Aqueous Solutions of Geological Samples in 0.5 mol/L D2EHPA/*n*-decane/0.1 mol/L HCl System (μg)

	Dolomite		Fluorite-barite-hydropatite ore	
Element	Content in the sample solution	Content in the REE concentrate	Content in the sample solution	Content in the REE concentrate
Cr	5.5	≤0.05	5.20	≤0.05
Mn	17.6	≤0.01	3.80	≤0.05
Fe	1700.0	330.00	624.00	72.00
Co	1.4	≤0.03	0.66	≤0.03
Ni	2.5	≤0.10	0.80	≤0.10
Cu	1.9	≤0.02	0.80	≤0.02
Al	3800.0	≤0.20	54.20	2.50
Sr	4.3	≤0.50	38.00	≤0.50
Mo	5.0	≤0.05	1.10	≤0.05
Cd	0.5	≤0.02	0.26	≤0.02
Sn	30.0	≤0.20	1.80	≤0.20
Sb	20.0	≤1.00	3.30	≤1.00
Ba	32.0	≤0.01	40.00	≤0.01
Pb	12.0	≤0.20	1.80	≤0.20
Ti	2.0	≤0.10	2.70	≤0.20
Zr	6.0	≤0.03	1.10	≤0.03
B	32.0	≤0.05	1060.00	2.00
V	5.0	≤0.02	0.47	≤0.02

However, the residual amounts of Fe as well as Ti and Al do not interfere if the interelement corrections described in Ref. 19 are used. A good agreement with the certified values for light REEs and Y was obtained with system 1 (Table 7). Tm, Yb, and Lu are partially retained in the stationary phase during elution with 3 mol/L HCl and can be removed only by 5 mol/L HCl at the stage of the column regeneration. Other two-phase systems are required for quantitative and selective simultaneous separation of both light and heavy REEs.

Figure 15 and Table 8 exemplify the separation of the sum of REEs by use of system 2. Alkali, alkaline earth elements, Cu, Pb, Cr, and many other elements remain in the aqueous phase (1 mol/L NH_4NO_3) at the REE concentration stage (sample solution pH 2.0–2.3, ascorbic acid as masking agent). The change of eluent for 6.0 mol/L HCl makes possible the elution of REE. This system was used in the analysis of granite [20] and basalt (Table 7, system 2). The obtained concentrates contain some amounts of Ti, Al, and Fe; their interfer-

Table 7 ICP-AES Determination of REEs (ppm) in a Reference Sample of Basalt BM-1 After Their Group Separation in Three Systems.[a]

	Systems			
Element	1	2	3	C.V.[b] [38,39]
La	10.7 ± 0.3	10.0 ± 1.0	9.1 ± 0.3	9.0 ± 1.2
Ce	18.0 ± 3.0	19.0 ± 2.0	17.0 ± 3.0	22.0 ± 2.0
Pr	5.0 ± 1.0	4.2 ± 0.3	—	—
Nd	15.0 ± 2.0	14.0 ± 1.0	16.0 ± 3.0	16.0 ± 3.0
Sm	3.6 ± 0.1	3.0 ± 0.5	3.2 ± 0.2	3.6 ± 0.3
Eu	1.0 ± 0.1	1.0 ± 0.1	1.1 ± 0.1	1.12 ± 0.07
Cd	1.8 ± 0.1	1.9 ± 0.3	2.3 ± 0.3	—
Tb	1.3 ± 0.3	1.5 ± 0.2	1.4 ± 0.2	0.9 ± 0.3
Ho	—	1.4 ± 0.3	1.1 ± 0.2	—
Er	—	—	—	—
Tm	<0.02	0.50 ± 0.05	0.4 ± 0.1	—
Yb	<0.03	1.7 ± 0.2	2.0 ± 0.5	3.0 ± 0.5
Lu	<0.01	0.34 ± 0.03	0.32 ± 0.03	0.41 ± 0.07
Y	22.0 ± 1.0	19.0 ± 3.0	24.0 ± 2.0	27.0 ± 3.0

[a] 0.5 mol/L D2EHPA/*n*-decane/HCl (1)
0.3 mol/L D2EHPA/0.02 mol/L TOPO/*n*-decane–MIBK (v/v = 3:1)–1.0 mol/L NH_4NO_3/6 mol/L HCl (2)
1.0 mol/L Ph_2Bu_2/chloroform/3.0 mol/L HNO_3 (3)
[b] C.V., certified values.

ences can be avoided [19]. Thus, system 2 allows us to determine 13 rare earth elements and Y. One separation run takes 1 hr.

System 3 makes it possible to separate all REEs by the following procedure. A geological sample solution in 3.0 mol/L HNO_3 is introduced into the column; then an additional 20 mL of 3.0 mol/L HNO_3 is pumped through to remove matrix elements from the organic phase. The next 5-mL portion of the same acid solution elutes the total REE contents. Then the column is washed with doubly distilled water.

This system can be used only for the analysis of samples containing relatively high quantities of REEs as the preconcentration is not achieved in this case. The ultimate concentrates contain some amounts of matrix components (Table 9), which do not interfere with the ICP-AES determination (Table 7, system 3).

It should be stressed that all of the three systems provide the recovery of REEs into a small eluate volume using one chromatographic run.

The proposed methods of group separation of REEs by CCC have advantage due to their simplicity, versatility, and relatively short separation times, and

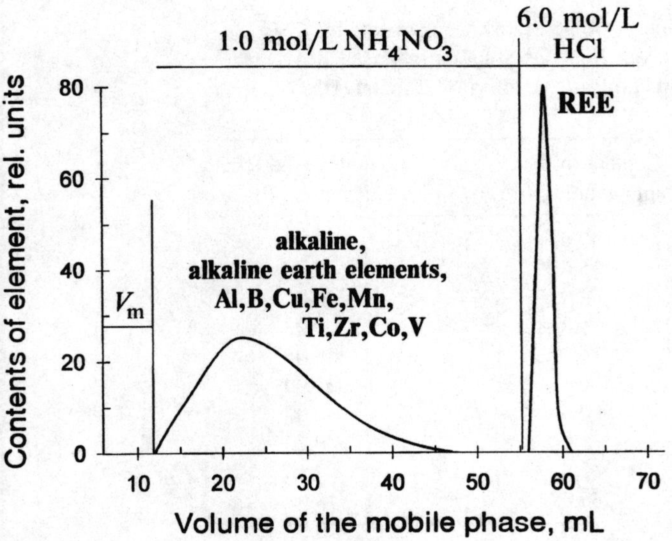

Figure 15 CCC separation of REE from some macrocomponents present in geological samples. Stationary phase: 0.3 mol/L D2EHPA/0.02 mol/L TOPO/*n*-decane + MIBK (3:1). $S_f = 0.7$, $F = 0.9$ mL/min.

Table 8 Separation of Some Matrix Components by CCC Recovery of REEs from Aqueous Solution of Granite GM Sample in 0.3 mol/L D2EHPA/0.2 mol/L TOPO/*n*-decane + MIBK (v/v = 3:1)–1.0 mol/L NH_4NO_3/6 mol/L HCl System

Element	Content in the sample solution (μg)	Content in the REE concentrate (μg)
Al	3600.0	920.00
Ba	17.0	<0.01
Co	0.2	<0.03
Cu	0.7	<0.02
Cr	0.6	<0.05
Fe	685.0	132.00
Ti	61.0	30.00
Mn	17.0	<0.01
Sr	6.7	<0.50
Zr	7.5	<0.03
V	0.6	0.02

Table 9 Separation of Some Matrix Components by CCC Recovery of REEs from Aqueous Solution of Basalt BM-1 Sample in 1.0 mol/L Ph_2Bu_2/chloroform/3.0 mol/L HNO_3 System

Element	Content in the sample solution (μg)	Content in the REE concentrate (μg)
Al	4290.0	480.00
Ba	125.0	2.00
Cu	2.0	<0.02
Co	1.8	<0.03
Cr	6.0	<0.05
Fe	3350.0	512.90
Mn	560.0	32.00
Pb	1.0	<0.20
Ni	2.8	<0.10
Ti	330.0	115.00
Sr	10.0	<0.50
Zr	5.0	1.50
V	9.5	3.00
W	4.4	2.00

can compete with the precipitation method [39] and other chromatographic methods [26,39–42].

The determination of Nd and Sm isotopes concentrations in rock samples is another important problem that can be solved with the use of CCC. The ratio of these isotopes allows us to judge the age of rocks. Mass spectrometry is one of the most successful techniques for the determination of Nd and Sm isotopes. MS measurements make it possible to estimate the isotopic composition of these and other elements with the required accuracy (with errors not greater than 0.1%) in very small samples containing several nanograms of test components. The quality of MS analysis depends on the purity of element fractions to be analyzed. As Nd and Sm have isotopes of equal mass, Nd must be completely separated from Sm. Besides, the isolated Nd and Sm fractions should be free of some concomitant elements (Ba, Ca, Fe, and others) that might interfere with the ion emission in the mass spectrometer.

The 1.0 mol/L D2EHPA/*n*-decane/HCl system allowed us to separate Nd and Sm as well as to separate both elements from the majority of main constituents of geological materials. Figure 16 illustrates the separation of Nd and Sm in a step elution mode. At the first stage (sample introduction), Nd and Sm are concentrated in the stationary phase, whereas Rb, Sr, and most matrix elements (Al, Ca, Mg, Na, K, Mn, Ba, etc.) are mainly eluted. Then an additional 5 mL of

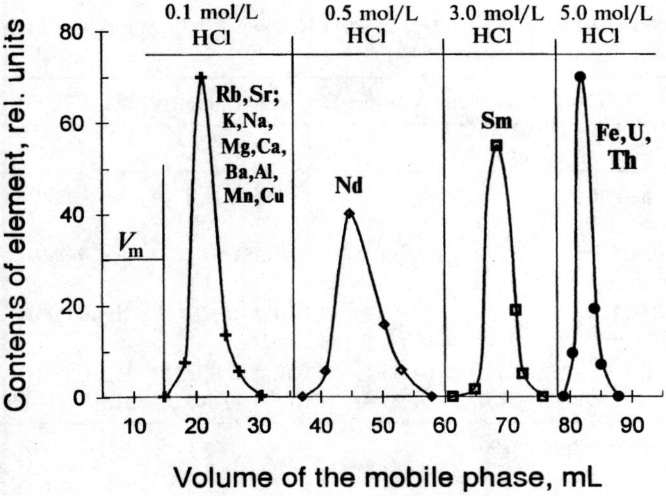

Figure 16 CCC separation of Nd and Sm from some macrocomponents of geological samples in 1.0 mol/L DEHPA/*n*-decane/HCl system. $S_f = 0.42$ ($V_s^f = 10$ mL). $F = 1$ mL/min.

0.1 mol/L HCl are pumped through for ultimate removal of matrix components. Afterward Nd and Sm are isolated into two small fractions by use of different eluents. (Other REEs are distributed somehow between the two fractions, but they do not interfere with MS determination of Nd and Sm.) At the column regeneration stage (passing of 5.0 mol/L HCl), Fe, Th, and U are eluted, while Zr, Hf, and Nb remain in the stationary phase as they are irreversibly extracted by D2EHPA. However, Zr, Hf, and Nb content in geological samples is usually very low, and their accumulation at a high extractant concentration does not affect the analytical results, at least after several dozens of separation cycles. Depending on the purpose, the whole Nd and Sm fractions can be taken (if quantitative isolation is necessary), or the fractions are used partly (if very pure Nd and Sm fractions are required for estimating the isotopic ratios). The results of Nd and Sm separation from a geological sample solution are presented in Table 10.

C. Other Elements

1. Zirconium, Hafnium, Niobium, Tantalum

Group preseparation of Zr(IV), Hf(IV), Ta(V), and Nb(V) is desirable in the analysis of geological samples of different composition containing low amounts of these elements.

Table 10 MS Determination of Rb, Sr, Nd, and Sm Recovered by use of CCC from Aqueous Solution of a Geological Sample CV-5/H[a]

	Contents of element in the fractions separated (μg)			
Fractions	Rb	Sr	Nd	Sm
Rb	0.512 ± 0.064 (RSD = 0.10)	<0.01	<0.0001	<0.0001
Sr	<0.001 (RSD = 0.12)	26.970 ± 3.940	<0.0001	<0.0001
Nd	<0.001	<0.05 (RSD = 0.07)	1.610 ± 0.163	<0.0001
Sm	0.018 ± 0.003 (RSD = 0.10)	0.110 ± 0.033 (RSD = 0.18)	0.052 ± 0.033 (RSD = 0.40)	0.229 ± 0.037 (RSD = 0.09)

[a] Weighed sample 30 mg; $n = 3$; $p = 0.90$.

Preconcentration of Zr and Hf and separation from Fe (macrocomponent that is most often a source of spectral interferences) makes it possible to decrease the interelement influence and to improve the sensitivity of their determination by ICP-AES. Preconcentration and separation of Zr and Hf on a planetary centrifuge with one-layer coil column in a system on the basis of a mixture of two extracting reagents, D2EHPA and *N*-benzoyl-*N*-phenylhydroxylamine (BPHA), was achieved (Figure 17). The separation was performed by step elution. In the first stages (mobile phases 1 and 2), Zr and Hf were concentrated in the stationary phase, while Fe and other elements were eluted. Then Zr and Hf were eluted by use of oxalic acid solution.

A system with tetraoctylethylendiamine (TOEDA) in chloroform can be applied to preconcentration of Zr, Hf, Nb, and Ta for their determination in geological samples (Figure 18). At the concentration stages (mobile phases 1 and 2), alkali, alkaline-earth and rare earth elements, Fe, Al, and Cu were removed from the column (D^{bat} for Zr, Hf, Nb, Ta are about 100). Then Zr, Hf, Nb, and Ta were separated into 10 mL of 2 mol/L HCl. Zn and Cd were eluted with 1 mol/L HNO_3 at the stationary phase regeneration stage.

Using 0.001 mol/L 1-phenyl-3-methyl-4-benzoylpyrazolone-5 in MIBK as stationary phase, separation of Zr and Hf can be achieved [43]. The separation was performed by step elution: Hf was eluted by 3 mol/L HNO_3 and Zr by 0.1 mol/L Na_2SO_4 in 3 mol/L HNO_3.

2. Strontium, Calcium, Barium, Rubidium

Extraction systems on the basis of crown ethers are known to be suitable for the separation of Rb, Ca, Sr, and Ba [43,44]. An example of the separation of stron-

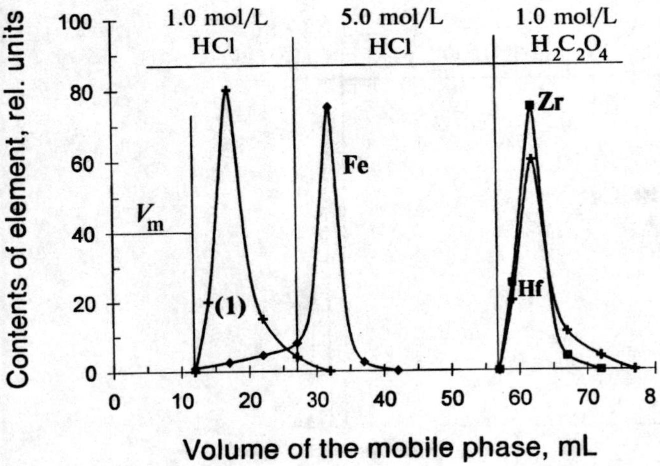

Figure 17 Preconcentration and separation of Zr and Hf from some elements. Stationary phase: 0.25 mol/L D2EHPA/0.05 mol/L BPHA/chloroform. $S_f = 0.18$, $F = 1.0$ mL/min. (1) Cu, Al, Co, Ni, Zn, Cd, alkali, alkaline earth, rare earth, and other elements.

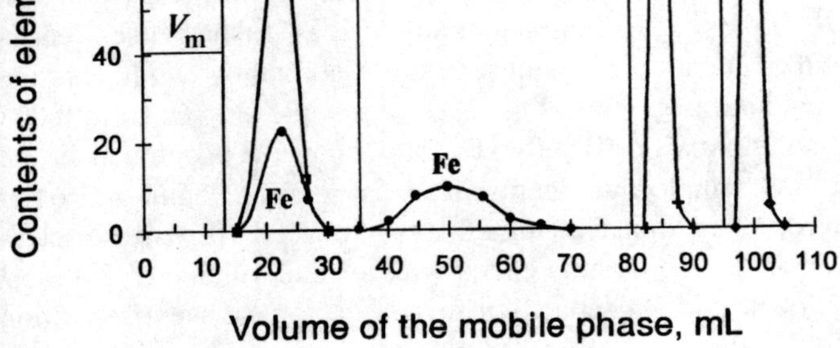

Figure 18 Preconcentration and separation of Zr, Hf, Nb, Ta from matrix components. Stationary phase: 0.1 mol/L TOEDA/chloroform. Mobile phase: 1, 0.1 mol/L HCl + 0.01 mol/L $H_2C_2O_4$; 2, 0.1 mol/L HCl + 0.5 mol/L $H_2C_4H_4O_6$; 3, 2.0 mol/L HCl; 4, 1.0 mol/L HNO_3. $S_f = 0.50$. $F = 1.0$ mL/min.

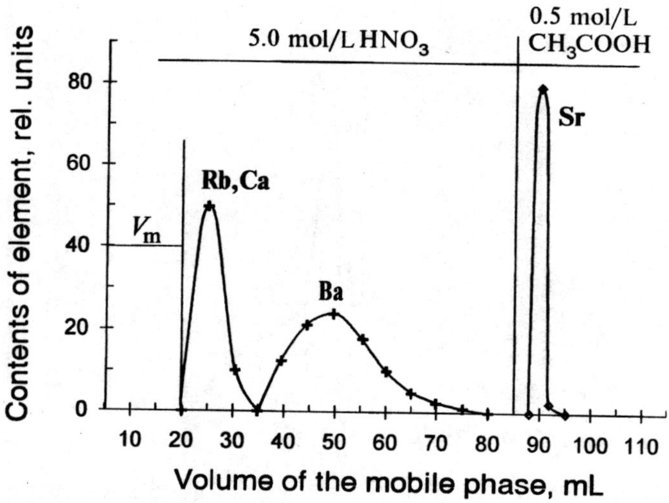

Figure 19 Separation of Sr, Rb, Ca, Ba from other elements. Stationary phase: 0.1 mol/L DCH18C6 in chloroform. $S_f = 0.25$ ($V_s^f = 6$ mL). $F = 1$ mL/min.

tium from the other elements on a planetary centrifuge with one-layer coil column in the 0.1 mol/L DCH18C6/chloroform/5.0 mol/L HNO_3 system is given in Figure 19. A sample is introduced to the flow of 5.0 mol/L HNO_3. At the concentration step, Rb, Ca, and Ba are removed with the acid flow. Rb and Ca are eluted together with Al, Mg, Na, K, and Mn. The change of the eluent for 0.5 mol/L CH_3COOH solution results in the elution of strontium in a small-volume fraction.

Selective separation of Rb can also be achieved in a DCH18C6-based system but with the use of picrate (Pi^-) instead of nitrate as Pi^- is the better counterion for metal–crown ether cationic complexes [44]. A chromatogram illustrating the separation of rubidium is given in Figure 20. A sample is injected to 0.005 mol/L picric acid solution with pH 4.0–6.0. Rubidium is concentrated in the stationary phase at the sample introduction stage, whereas calcium and other elements are removed. An additional 20 mL of 0.005 mol/L HPi is passed through for complete elution. Then rubidium is eluted with a small volume of 2.0 mol/L HCl. The results of Sr and Rb separation from a geological sample solution are presented in Table 10.

3. Cesium and Strontium

CCC can be used for solving radioanalytical problems, such as for the separation of radionuclides of Cs and Sr on a planetary centrifuge with a one-layer column

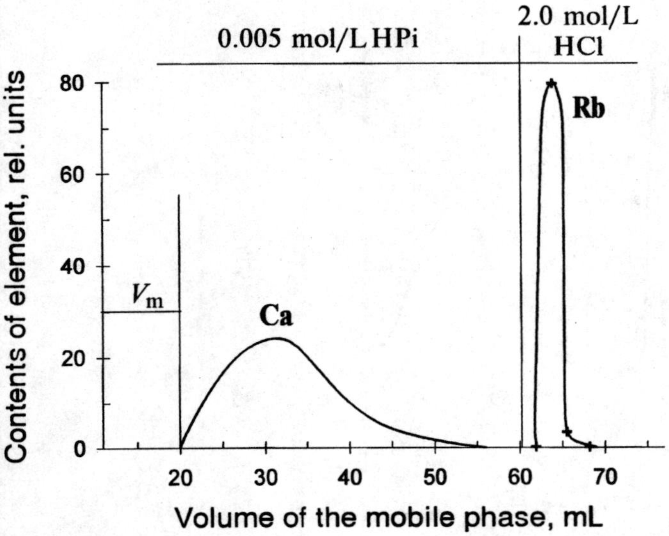

Figure 20 Preconcentration and separation of Rb. Stationary phase: 0,03 mol/L DCH18C6 in chloroform. S_f = 025 (V_s^f = 6 mL). F = 1 mL/min.

described in Section I [18,22,23]. Systems of 0.01 mol/L cobalt dicarbollide (CD)/nitrobenzene/HNO_3 solution and 0.05 mol/L DCH18C6/3% D2EHPA/chloroform/HNO_3 solution were used for this purpose. Figure 21 illustrates the separation of Cs and Sr using two different stationary phases and the same mobile phase. The use of the stationary phase on the basis of CD enables the concentration of Cs and Sr to be achieved. Elution of Cs and Sr from the stationary phase on the basis of DCH18C6 and D2EHPA into a mobile phase, containing barium nitrate, polyethylene glycol (PEG), and nitric acid, was studied. Different variants of cesium and strontium separation obtained by changing the composition of the mobile phase (at the same extractant concentration in the stationary phase) can be realized (Figure 22). The quantitative elution and complete separation of Cs and Sr by means of one eluent or by step elution is possible. The latter procedure is faster and enables stripping of the elements by smaller mobile phase volumes. Table 11 gives the number of theoretical plates, peak resolution, and separation factor values for the elements calculated from Eqs. (2)–(4) for the same F and ω values. The value of N varies from 77 to 836, which is enough for most inorganic separations. Besides, very selective two-phase systems for CCC separations are available. It should be also noted that an increase in S_f does not result in higher N values.

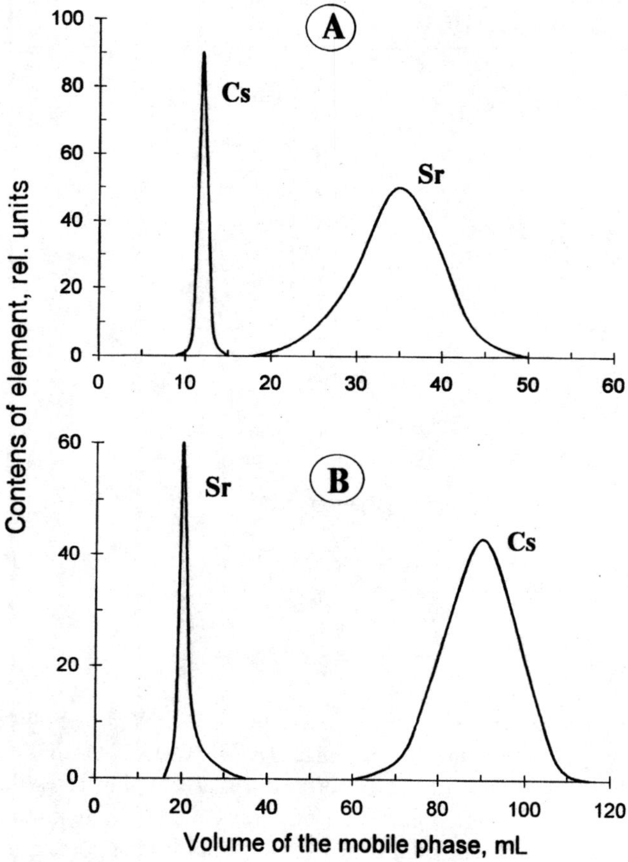

Figure 21 Separation of Cs and Sr. Mobile phase: 0.001 mol/L $Ba(NO_3)_2$ in 0.1 mol/L HNO_3. F = 1 mL/min. Stationary phase: (A) 0.05 mol/L DCH18C6/3% D2EHPA/chloroform; S_f = 0.5; (B) 0.01 mol/L cobalt dicarbollide/nitrobenzene; S_f = 0.12.

4. Transplutonium and Rare Earth Elements

Two extraction systems, based on bidentate neutral organophosphorous reagents, were shown to be applicable to the group separation of trivalent transplutonium elements (TPEs) from trace and macroamounts of REEs [25,28]:

> 0.005 mol/L tetraphenylmethylenediphosphine dioxide (TPMDPD)–chloroform–0.5 mol/L NH_4SCN–1 mol/L HCL/0.025 mol/L hydroxyethylidenediphosphonic acid (HEDPA) (system 1); 0.02 mol L 2,4,6-tris[ditolyphos-

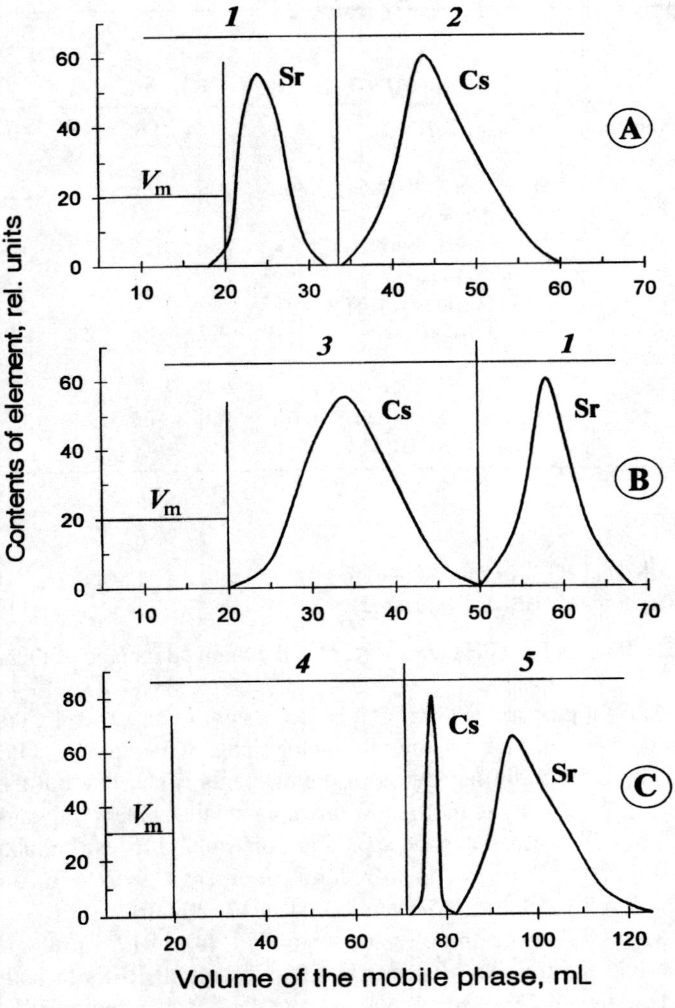

Figure 22 Separation of Cs and Sr. Stationary phase: 0.01 mol/L cobalt dicarbollide in nitrobenzene. Mobile phase: 1, 0.001 mol/L $Ba(NO_3)_2$ in 0.1 mol/L HNO_3; 2, 5×10^{-4} mol/L $Ba(NO_3)_2$ + 0.008 mol/L PEG-300 in 0.5 mol/L HNO_3; 3, 0.008 mol/L PEG-300 in 0.1 mol/L HNO_3; 4, 5×10^{-4} mol/L HNO_3; 5, 0.001 mol/L $Ba(NO_3)_2$ + 0.008 mol/L PEG-300 in 0.1 mol/L HNO_3. $S_f = 0.12$. $F = 1$ mL/min.

Table 11 Number of Effective Theoretical Plates (N), Peak Resolution R_s, and Separation Factor (α) for the Separation of Cs and Sr by CCC

Stationary phase	Mobile phase	N		R_s	α_s
		Cs	Sr		
10^{-2} mol/L cobalt dicarbollide in nitrobenzene	0.1 mol/L HNO_3	246	470	3.9	20
	10^{-3} mol/L $Ba(NO_3)_2$ in 0.1 mol/L HNO_3	159	736	3.6	23.7
	10^{-3} mol/L $Ba(NO_3)_2$ + 0.25% PEG-300 in 0.1 mol/L HNO_3	836	126	1.8	4.3
0.05 mol/L DCH18C6 + 3% D2EHPA in chloroform	0.1 mol/L HNO_3	77	106	2.4	8.0
	10^{-3} mol/L $Ba(NO_3)_2$ in 0.1 mol/L HNO_3	150	105	2.2	5.3

phoryl]-1,3,5-triazine (Tol-triazine)–chloroform–0.5 mol/L NH_4SCN–1 mol/L HCl/0.025 mol/L HEDPA (system 2)

Higher partition coefficients for TPE than for REE is the main advantage of these systems.

Taking the Am, Eu pair and a TPMDPD-based system as an example, the influence of reagent concentration on the chromatographic separation of TPE from REEs was studied. Isocratic separation of the elements is possible but the separation by use of two eluents is more convenient as smaller eluent valumes may be applied. Figure 23A illustrates the separation of Am and Eu with 0.005 mol/L TPMDPD. First, 99.5% Eu is eluted by 5 mL of eluent 1, then 98% Am is stripped out with 3 mL of eluent 2. Separation takes 17–20 min.

To develop a method for the group separation of TPE from REEs, preliminary studies of the batch extraction of Am, Cm, Bk, Cf, and all REEs in both systems were made. Am and Cm were shown to have the lowest (among TPE) and very close partition coefficients, whereas La and Ce were found to possess the highest (among REEs) and close partition coefficients. Therefore, the Am/Ce separation factor characterizes the group separation of TPE and REEs.

The effect of the total REE concentration on the separation of TPE (Am) from REE (Ce) in both systems was investigated [45]. Under the chosen conditions, when the sum of REEs is eluted with a first portion of the eluent, the Am peak position and the Am/Ce separation factor depend on the REE concentration. Increasing the REE concentration first results in better Am retention and higher

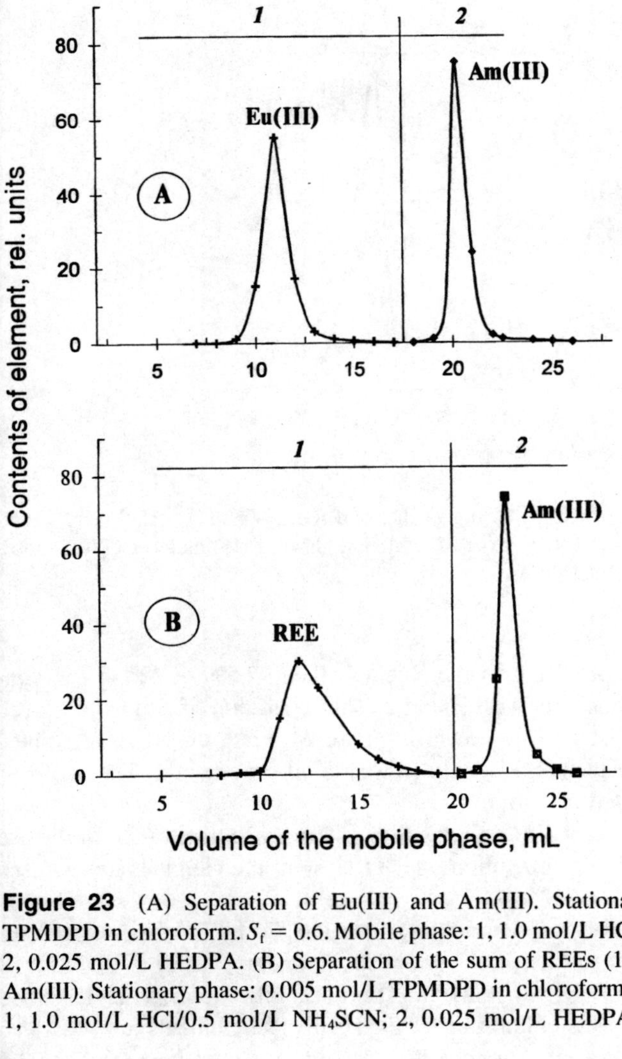

Figure 23 (A) Separation of Eu(III) and Am(III). Stationary phase: 0.005 mol/L TPMDPD in chloroform. $S_f = 0.6$. Mobile phase: 1, 1.0 mol/L HCl + 0.5 mol/L NH_4SCN; 2, 0.025 mol/L HEDPA. (B) Separation of the sum of REEs (10 mg of total REEs) and Am(III). Stationary phase: 0.005 mol/L TPMDPD in chloroform. $S_f = 0.6$. Mobile phase: 1, 1.0 mol/L HCl/0.5 mol/L NH_4SCN; 2, 0.025 mol/L HEDPA.

separation factor, probably due to a salting-out effect of the REE. Further increasing the REE concentration leads to a reverse effect: the higher the REE concentration, the lower Am retention (and the separation factor). This can be explained by a decrease in the free reagent concentration when the extraction of REEs is noticeable.

Figure 23B illustrates the separation of Am (TPE) from Ce (10 mg of total REE) in the system on the basis of 0.005 mol/L TPMDPD. First, 95.4% of REE

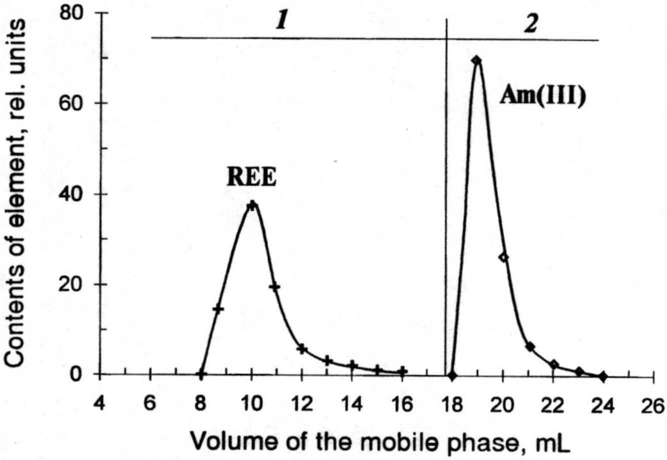

Figure 24 Separation of REEs (20 mg of the total REEs) from Am stationary phase: 0.02 mol/L Tol-triazine/chloroform. $S_f = 0.6$. Mobile phase: 1, 1.0 mol/L HCl/0.5 mol/L NH_4SCN; 2, 0.025 mol/L HEDPA.

without Am is eluted into 8 mL of the eluent 1, then 97.5% of Am (TPE) containing no REE is stripped out with eluent 2. The separation of Am from Ce (20 mg of total REE) in system 2 is shown in Figure 24. First, the REEs are eluted with 8 mL of eluent 1, then Am is isolated with 6 mL of eluent 2. The elements are completely separated in 20 min.

The Am, Eu pair was also separated in systems with nitric acid. In the use of the Tol-triazine/chloroform/3 mol/L HNO_3 system the elements are isolated isocratically.

Figure 25 illustrates the separation of Am and Eu by bidentate extracting reagent [the 0.035 mol/L tris(diphenylphosphinylmethyl)benzene (DPPMB)/chloroform/3.0 mol/L HNO_3 system]. In difference with the systems described above Eu is retained in the column better than Am. The elements are separated isocratically. First, 97.5% of Am containing no Eu is eluted into 4 mL of 3.0 mol/L HNO_3. Then 99.65% of Eu containing no Am is isolated.

5. Some Heavy Metals

Separation of Co(II), Cu(II), Fe(II), Fe(III), Mg, and Ni was performed with a high-speed CCC coil planet centrifuge equipped with a multilayer coil [16]. The separation by use of D2EHPA in *n*-heptane (stationary phase) and diluted citric acid (mobile phase) can be optimized by selecting a proper acid concentration. Continuous detection of the elements was performed by direct current plasma

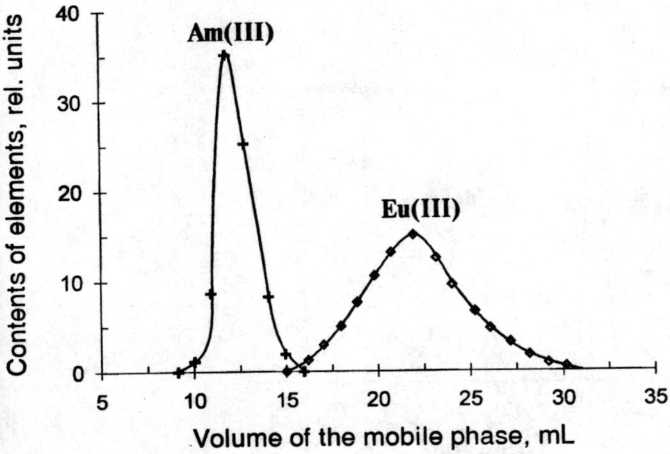

Figure 25 Separation of Am and Eu in the 0.035 mol/L DPPMB/chloroform/3 mol/L HNO_3 system.

atomic emission spectrometry. Each element peak was well resolved at high partition efficiencies ranging from 200 to 3600 theoretical plates. An example of isocratic separation of Cu(II), Cd(II), and Mn(II) is given in Figure 26 [16]. The partition efficiencies N ranged from 630 (Cu) to 270 (Mn).

Isocratic separation of Fe(II) and Fe(III) was achieved at 0.002 mol/L D2EHPA concentration in the stationary phase and at 0.14 mol/L citric acid as the mobile phase [16]. The partition efficiency of the separation was 3700 for Fe(II) and 410 for Fe(III), while the peak resolution between the peaks was 2.46.

For step separation of Fe(II) and Fe(III) we used a stationary phase of 0.005 mol/L TPMDPD in *n*-decane and a planetary centrifuge with a one-layer coil column [15]. Trace amounts of Cd were separated from a large excess of Zn by step elution using 30% TBP in heptane as stationary phase on the same device [22].

6. Platinum Group Metals

The separation of Pd(II) from Pt(II), Ir(III), and Rh(III) with TOPO in heptane using centrifugal partition chromatography was investigated [29]. The separation of Pd from Pt is achieved in the 0.5 mol/L TOPO in heptane/0.1 mol/L HCl system with a resolution of 1.54 or <0.3% peak overlap. The retention volumes of Pt and Pd are 110 mL and 135 mL, respectively ($D_{Pt} = 0.11$, $D_{Pd} = 1.4$, and $N = 360$). The same system was used for the separation of Pd from the other

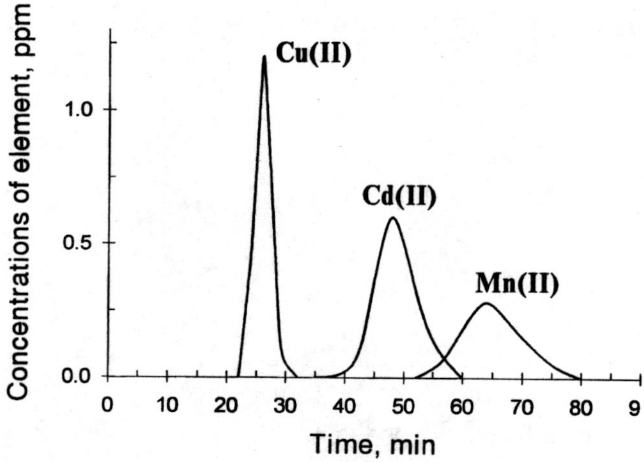

Figure 26 Isocratic separation of Cu, Cd, and Mn by HSCCC (planetary centrifuge with 10.0-cm revolution radius and one multilayer coiled column, total capacity = 300 mL). Stationary phase: 0.2 mol/L D2EHPA in *n*-heptane. Mobile phase: 0.05 mol/L citric acid. Sample: each element 14.1 μg. Revolution speed: 800 rpm. $F = 5.0$ mL/min. (From Ref. 16).

platinum group metals. Partial resolution of Pt(II) from Ir(III) and Rh(III) was obtained. However, complete separation of the four platinum group metals was not possible using TOPO [29].

7. Orthoposphate and Pyrophosphate

CCC can be used for anion separation. We investigated the separation of ortho- and pyrophosphate ions on a planetary centrifuge with a one-layer column [18,22,24]. A 5% solution of dinonyltin dichloride in MIBK was selected as the stationary phase for concentration and separation of ortho- and pyrophosphates. This reagent extracts phosphate ions with high partition coefficients from nitric acid solutions. For back-extraction of ortho- and pyrophosphates, acidic chloride solutions should be used. Quantitative concentration and complete separation of ortho- and pyrophosphates are attained by use of step elution (Figure 27). Concentration of the phosphorous forms was carried out from 1 mol/L HNO_3 solution (eluent 1). After that, orthophosphate was eluted from the stationary phase with 0.5 mol/L NaCl in 0.5 mol/L HCl (eluent 2), and pyrophosphate with 3 mol/L HCl (eluent 3). Phosphate ions were determined in the eluates on a flow injection analyzer in accordance with the procedure described [18,24].

The threefold excess of arsenate interferes with the determination. The

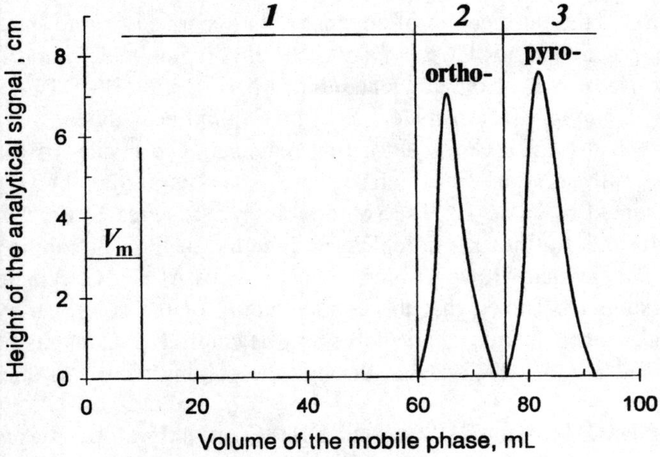

Figure 27 Concentration and separation of ortho- and pyrophosphate ions. Stationary phase: 5% dinonyltin dichloride in MIBK. Mobile phase: 1, 1 mol/L HNO_3; 2, 0.5 mol/L NaCl in 0.5 mol/L HCl; 3, 3 mol/L HCl. $S_f = 0.5$. $F = 1.5$ mL/min.

range of determinable ortho- and pyrophosphate concentrations in these mixtures is from 0.01 to 5 μg/mL when the sample volume changes from 100 to 0.5 mL [24].

8. Purification of Salt Solutions

Together with Dr. P. Tschoepel (Max Plank Institute for Metal Research, Dortmund, Germany), the applicability of CCC on a planet centrifuge to purification of salt solution (to gain high-purity reagents, which, after evaporation, can be used for fusion decomposition purposes in trace analysis of high-tech ceramic materials and other materials) has been investigated.

There is a difference in the aims of analytical preconcentration and purification procedures: in case of purification, the purified constituent of aqueous solution is the purposeful component, whereas the microelements are considered as impurities to be separated and discarded. In this case, the maximum possible number of microelements should be separated in one chromatographic run. Other requirements to be met in the use of any purification method are connected with the purity of all chemicals applied and with the necessary minimum decontamination of the solutions involved, which may be due to their contact with the device materials. In the case of a planetary centrifuge, the solutions are in contact only with Teflon, which is one of most inert materials known.

The possibility of the purification of aqueous solutions of inorganic salts, such as $(NH_4)_2SO_4$, NH_4F, and NH_4Cl, from a number of common metal impurities was shown by use of *N,N*-hexamethylenedithiocarbamic acid (HMDTCA), diethylammonium diethyldithiocarbamate, 8-hydroxyquiniline, dibenzo-18-crown-6 (DB18C6), and DCH18C6 as extracting reagents. The results of the purification of the salt solution from microelements as impurities by 1% HMDTCA are presented in Table 12. The solution analyzed is free of Fe, Cu, Zn, Co, Cd, Ni, Mn; this testifies the complete purification of the salt solution from the elements enumerated. However, certain amounts of Al and Cr remain in the solution. It should be noted that the contamination of the solutions by the column material, reagents, and organic solvents was controlled. Contents of elements in water and 2 mol/L HCl passed through the column were less than the detection limits.

A mixture of HMDTCA, DCH18C6, and DB18C6 in chloroform makes possible the purification of 1% $(NH_4)_2SO_4$ solution (pH 5.5) from K, Fe, Cu, Zn, Co, Cd, Ni, Al, and Mn (their concentrations in the purified solution are below detection limits for such a sensitive determination technique as electrothermal atomic absorption spectrometry). Remaining in the purified salt were 5% Ca, 8% Mg, and 55% Cr.

Therefore, complete purification of the salt solution from contaminant microelements can be attained by varying the organic phase composition.

Table 12 Purification of 1% Solution of $(NH_4)_2SO_4$ from Fe, Cu, Zn, Co, Cd, Ni, Al, Mn, and Cr by 1% HMDTCA in Chloroform (V_s = 8.5 mL)

Element	Concentration in salt solution (ng/mL)	
	Before purification	After purification
Fe	15.0	<0.1
Zn	2.5	<0.1
Cd	1.6	<0.1
Co	15.0	<0.1
Cu	15.0	<0.2
Ni	15.0	<0.1
Mn	15.0	<0.2
Al	10.0	1.5
Cr	5.0	2.5

Table 13 Application of Two-Phase Systems for Preconcentration and Separation of Elements by CCC

Elements	Stationary phase	Mobile phase		Ref.
		Preconcentration	Separation	
Ce, Eu	0.5 mol/L D2EHPA[1] in *n*-decane	0.1 mol/L HCl	Ce: 0.5 mol/L HCl Eu: 2.0 mol/L HCl	18
Sm, Eu	0.5 mol/L D2EHPA[1] in *n*-decane	0.1 mol/L HCl	Sm: 0.5 mol/L HCl Eu: 1 mol/L HCl	22
Nd, Sm	1 mol/L D2EHPA[1] in *n*-decane	0.1 mol/L HCl	Nd: 0.5 mol/L HCl Sm: 3.0 mol/L HCl	21
Ce, Pr, Eu, Ba	30% TBP[2] 35% *n*-decane, 35% CCl_4		Ba, Ce, Pr, Eu: 5.0 mol/L $LiNO_3$ + 1.0 mol/L HNO_3	
La, Ce, Pr, Nd, Sm, Eu*	0.1 mol/L D2EHPA[1] in *n*-heptane		La, Ce, Pr: 0.05 mol/L HCl La, Ce: 0.04 mol/L HCl La, Pr, Nd: 0.07 mol/L HCl Nd, Sm: 0.14 mol/L HCl	31, 32
Er, Yb*	0.1 mol/L D2EHPA in chloroform		Er, Yb: 0.1 mol/L HCl	32
Sm, Eu*	0.1 mol/L D2EHPA[1] in toluene		Sm, Eu: 0.03 mol/L HCl	31
La, Pr, Nd, Sm, Eu, Gd, Dy, Tb, Ho, Er, Tm, Yb*	0.1 mol/L Cyanex 272[3] in heptane		Aqueous solutions of different pH (1.36–2.8)	14
La, Ce, Pr, Nd, Sm, Eu, Gd, Tb, Dy, Ho, Er, Tm, Yb, Lu	0.003 mol/L D2EHPA[1] in *n*-heptane		Exponential gradient of HCl (0–0.3 mol/L)	17
La, Ce, Pr, Nd	0.02 mol/L D2EHPA[1] in *n*-heptane		La, Ce, Pr, Nd: 0.02 mol/L HCl	17
Eu, Fe(II), Fe(III)	0.005 mol/L TPMDPD[4] in chloroform		Eu, Fe(II):0.9 mol/L HCl + 0.5 mol/L NH_4SCN Fe(III): 2.0 mol/L HCl	15

Table 13 Continued

Elements	Stationary phase	Mobile phase: Preconcentration	Mobile phase: Separation	Ref.
Separation[a] of REE from matrix components	0.5 mol/L D2EHPA[1] in *n*-decane	REE: 0.1 mol/l HCl	Group separation Cr, Mn, Co, Ni, Cu, Al, Sr, Mo, Cd, Sn, Sb, Ba, Pb, Ti, Zr, B, V: 0.1 mol/L HCl; REE: 3.0 mol/L HCl; Fe, U, Th: 5.0 mol/L HCl	19
Sm, Eu, Gd	0.1 mol/L D2EHPA[1] in chloroform		Sm, Eu, Gd: 0.02 mol/L HCl	36
Gd, Tb, Dy	0.1 mol/L D2EHPA[1] in chloroform		Gd,Tb,Dy: 0.04 mol/L HCl	36
Eu, Gd, Tb, Dy	0.1 mol/L D2EHPA[1] in chloroform		Eu, Gd, Tb, Dy: 0.03 mol/L HCl	36
Dy, Ho, Er	0.1 mol/L D2EHPA[1] in chloroform		Dy, Ho, Er: 0.03 mol/L HCl (t = 45°C)	36
Ho, Er, Tm	0.1 mol/L D2EHPA[1] in chloroform		Ho, Er, Tm: 0.08 mol/L HCl in 20% v/v ethylene glycol–water	36
Er, Tm, Yb	0.1 mol/L D2EHPA[1] in chloroform		Er, Tm, Yb: 0.1 mol/L HCl in 20% v/v ethylene glycol–water	36
Er, Tm, Yb	0.1 mol/L D2EHPA[1] in chloroform		Er, Tm, Yb: 0.1 mol/L HCl in 20% v/v ethylene glycol–water (t = 45°C)	36
Tm, Yb, Lu	0.1 mol/L D2EHPA[1] in chloroform		Tm, Yb, Lu: 0.12 mol/L HCl; Tm, Yb, Lu: 0.14 mol/L HCl	36

La, Pr, Nd	0.02 mol/L $(EHPA)_2$[5] in kerosene		La, Pr, Nd: 0.1 mol/L $(H, Na)Cl_2CHCOO$ in 20% v/v ethylene glycol–water, pH 2.50	37
Sm, Eu, Gd	0.02 mol/L $(EHPA)_2$[5] in kerosene		Sm, Eu, Gd: 0.1 mol/L $(H,Na)Cl_2CHCOO$ in 20% v/v ethylene glycol–water, pH 1.91	37
Gd, Tb, Dy	0.02 mol/L $(EHPA)_2$[5] in kerosene		Gd, Tb, Dy: 0.1 mol/L $(H, Na)Cl_2CHCOO$ in 20% v/v ethylene glycol–water, pH 1.48	37
Separation[b] of REE from matrix components	0.1 mol/L TOPO[6] in MIBK	REE: 1.0 mol/L NH_4NO_3	Group separation Al, Ba, Cu, Co, Fe, Sr, Pb, Mn, Ti, Zr, Cr, Ni, V: 1.0 mol/L NH_4NO_3; REE: 6.0 mol/L HCl	20
	0.3 mol/L D2EHPA[1] + 0.02 mol/L TOPO[6] in *n*-decane + MIBK (3:1)	REE: 1.0 mol/L NH_4NO_3 0.01 mol/L	Group separation Al, Ba, Cu, Co, Fe, Sr, Pb, Mn, Ti, Zr, Cr, Ni, V: 1 mol/L NH_4NO_3; REE: 6 mol/L HCl	
	1 mol/L Ph_2Bu_2[7] in chloroform		Group separation Al, Ba, Cu, Co, Fe, Sr, Pb, Mn, Ti, Zr, Cr, Ni, V; REE: 3 mol/L HNO_3	
Ortho- and pyrophosphate ions	5% DNTDC[8] in MIBK	1 mol/L HNO_3	Ortho-: 0.5 mol/L NaCl in 0.5 mol/L HCl Pyro-: 3 mol/L HCl	18, 22, 24

Table 13 Continued

Elements	Stationary phase	Mobile phase		Ref.
		Preconcentration	Separation	
Eu(III), AM(III)	0.005 mol/L TPMDPD in chloroform		Am(III): 1.0 mol/L HCl + 0.5 mol/L NH_4SCN Eu(III): 0.025 mol/L HEDPA[9]	25
	0.035 mol/L TDPPMB[10] in chloroform		Eu(III), Am(III): 3 mol/L HNO_3	
	0.005–0.03 mol/L TPMDPD in chloroform		Eu(III), Am(III): 1 mol/L HCl + 0.5 mol/L NH_4SCN	
	0.07 mol/L Tol-triazine[11] in chloroform		Eu(III), Am(III): 3.0 mol/L HNO_3	28
Sum of REE (10 mg), Am(III)	0.005 mol/L TPMDPD in chloroform		REE: 1 mol/L HCl + 0.5 mol/L NH_4SCN Am(III): 0.025 mol/L HEDPA	28
Sum of REE (20 mg), Am(III)	0.02 mol/L Tol-triazine[11] in chloroform		REE: 1.0 mol/L HCl + 0.5 mol/L NH_4SCN Am(III): 0.025 mol/L HEDPA	28
Rb, Ca, Ba, Sr	0.1 mol/L DCH18C6[12] in chloroform	Sr: 5 mol/L HNO_3	Rb, Ca; Ba: 5.0 mol/L HNO_3; Sr: 0.5 mol/L CH_3COOH	
Ca, Rb	0.03 mol/L DCH18C6[12] in chloroform	Rb: 0.005 mol/L picric acid	Ca: 0.005 mol/L picric acid; Rb: 2.0 mol/L HCl	21
Cs, Sr	0.01 mol/L CD[13] in nitrobenzene	5×10^{-4} mol/L HNO_3	Cs, Sr: 5×10^{-3} mol/L $Ba(NO_3)_2$ + 0.25% PEG 300 in 0.1 mol/L HNO_3	22
Sr, Cs	3% D2EHPA[1] + 0.05 mol/L DCH18C6[12] in chloroform		Sr, Cs: 10^{-3} mol/L $Ba(NO_3)_2$ in 0.1 mol/L HNO_3	22
Pt(II), Ir(III), Pd(II), Rh(III)*	0.5 mol/L TOPO[6] in heptane		Pd, Pt, Ir, Rh: 0.1 mol/L HCl	29

Separation of Zr, Hf from matrix components	0.025 mol/L BPHA[14] + 0.25 mol/L D2EHPA[1] in chloroform	Zr, Hf: 1.0 mol/L HCl 5.0 mol/L HCl	Group separation Cu,Al, Ni, Co, Zn, Cd, alkali, alkaline earth, rare earth elements: 1.0 mol/L HCl; Fe: 5.0 mol/L HCl; Zr, Hf: 1.0 mol/L $H_2C_2O_4$	
Separation of Zr, Hf, Nb, Ta from matrix components	0.1 mol/L TOEDA[15] in chloroform	Zr, Hf, Nb, Ta, Zn, Cd: 0.1 mol/L HCl + 0.01 mol/L $H_2C_2O_4$; 0.1 mol/L HCl + 5% ascorbic acid	Cu, Al, Co, Ni, Fe, alkali, alkaline earth, rare earth elements: 0.1 mol/L HCl + 0.01 mol/L $H_2C_2O_4$; Fe: 0.1 mol/L HCl + 5% ascorbic acid; Zr, Hf, Nb,Ta: 2.0 mol/L HCl; Zn, Cd: 1.0 mol/L HNO_3	
Zr, Hf	10^{-3} mol/L PMBP[16] in MIBK		Zr: 3.0 mol/L HNO_3; Hf: 0.1 mol/L Na_2SO_4 in 3.0 mol/L HNO_3	27
Ni, Co, Mg, Cu	0.2 mol/L D2EHPA[1] in heptane		Ni,Co, Mg,Cu: 7×10^{-3} mol/L citric acid	16
Cu, Cd, Mn	0.2 mol/L D2EHPA[1] in heptane		Cu, Cd, Mn: 5.9×10^{-2} mol/L citric acid	16
Fe(II), Fe(III)	0.002 mol/L D2EHPA[1] in heptane		0.14 mol/L citric acid	16

[1] D2EHPA, di-(2-ethylhexyl)phosphoric acid.
[2] TBP, tri-*n*-butyl phosphate.
[3] Cyanex 272, bis(2,4,4-trimethylpentyl)phosphoric acid.
[4] TPMDPD, tetraphenylmethylenediphosphine dioxide.
[5] $(EHPA)_2$, 2-ethylhexylphosphonic acid mono-2-ethylhexyl ester.
[6] TOPO, trioctylphosphine oxide.
[7] Ph_2Bu_2, diphenyl(dibutylcarbamoylmethylphosphine)oxide.
[8] DNTDC, dinonyltin dichloride.
[9] HEDPA, hydroxyethylidenediphosphonic acid.
[10] DPPMB, 1,2,4-tris(diphenylphosphinylmethyl)benzene.
[11] Tol-triazine, 2,4,6-tris[ditolylphosphoryl]-1,3,5-triazine.
[12] DCH18C6, dicyclohexano-18-crown-6.
[13] CD, cobalt dicarbollide.
[14] BPHA, *N*-benzoyl-*N*-phenylhydroxylamine.
[15] TOEDA, tetraoctylethylenediamine.
[16] PMBP, 1-phenyl-3-methyl-4-benzoylpyrazolone-5.
* Separation by centrifugal partition chromatography.
[a] Masking agent: citric acid.
[b] Masking agent: ascorbic acid.

V. CONCLUSIONS

The application of CCC in inorganic analysis looks promising because various two-phase liquid systems, providing the separation of a variety of inorganic species, may be used for separation of trace elements. The characteristics of the two-phase extraction systems, applied in the separation and preconcentration of inorganic species, are given in Table 13. Some experiments are currently in progress to investigate other possibilities for the application of CCC.

REFERENCES

1. Y. Ito, in *Countercurrent Chromatography. Theory and Practice* (N.B. Mandava and Y. Ito, eds.), Marcel Dekker, New York, 1988.
2. A. Berthod and N. Schmitt. *Talanta 40*:1489 (1993).
3. J.-M. Menet, D. Thiebaut, R. Rosset, J. E. Wesfreid, and M. Martin. *Anal. Chem. 66*:168 (1994).
4. W. D. Conway, *Countercurrent Chromatography: Apparatus, Theory and Applications*, VCH, New York, 1990.
5. J.-M. Menet, M.-C. Rolet, D. Thiebaut, R. Rosset, and Y.Ito, *J. Liq. Chromatogr. 15*:2883 (1992).
6. A. P. Foucault and F. Le Goffic, *Analusis 19*:227 (1991).
7. O. Bousquet, A. P. Foucault and F. Le Goffic, *J. Liq. Chromatogr. 14*:3343 (1991).
8. A. P. Foucault, O. Bousquet, F. Le Goffic, *J. Liq. Chromatogr. 15*:2691 (1992).
9. A. P. Foucault, O. Bousquet, F. Le Goffic, and J. Cazes, *J. Liq. Chromatogr. 15*: 2721 (1992).
10. A. Berthod, *J. Chromatogr. 550*:677 (1991).
11. S. Drogue, M.-C. Rolet, D. Thiebaut, and R. Rosset. *J. Chromatogr. 593*:363 (1992).
12. W. D. Conway, *J. Liq. Chromatogr. 13*:2409 (1990).
13. P. S. Fedotov, T. A. Maryutina, V. M. Pukhovskaya, and B. Ya. Spivakov, *J. Liq. Chromatogr. 17*(16):3491 (1994).
14. S. Muralidharan, R. Cai, and H. Freiser, *J. Liq. Chromatogr. 13*:3651 (1990).
15. P. S. Fedotov, T. A. Maryutina, A. A. Pichugin, and B. Ya. Spivakov, *Russian J. Inorg. Chem. 38*:1878 (1993).
16. E. Kitazume, N. Sato, Y. Saito, and Y. Ito, *Anal. Chim. 65*:2225 (1993).
17. E. Kitazume, M. Bhatnagar, and Y. Ito, *J. Chromatogr. 538*:133 (1991).
18. Yu. A. Zolotov, B. Ya. Spivakov, T. A. Maryutina, V. L. Bashlov, and I. V. Pavlenko, *Fresenius Z. Anal. Chem. 35*:938 (1989).
19. V. M. Pukhovskaya, T. A. Maryutina, O. N. Grebneva, N. M. Kuz'min, and B. Ya. Spivakov. *Spectrochim. Acta 48B*:1365 (1993).
20. V. M. Pukhovskaya, O. N. Grebneva, T. A. Maryutina, N. M. Kuz'min, and B. Ya. Spivakov, *Spectrochim. Acta 50B*:5 (1995).
21. P. S. Fedotov, S. F. Karpenko, A. V. Lyalikov, V. G. Spiridonov, T. A. Maryutina, and B. Ya. Spivakov, *Zh. Anal. Khim.* (in Russian) (in press).

22. B. Ya. Spivakov, T. A. Maryutina, V. L. Bashlov, V. M. Pukhovskaya, and Yu. A. Zolotov, *Proceedings of 5th Japan–USSR Symposium on Analytical Chemistry*, Sendai and Kiryu, Japan, 1990, p. 241.
23. B. Ya. Spivakov, T. A. Maryutina, and Yu. A. Zolotov, *Proceedings of the International Solvent Extraction Conference*, Japan, July 1990, Elsevier, 1992, p. A451.
24. T. A. Maryutina, B. Ya. Spivakov, L. K. Shpigun, I. V. Pavlenko, and Yu. A. Zolotov, *Zh. Anal. Khim.* (in Russian) *45*:665 (1990).
25. M. K. Chmutova, T. A. Maryutina, B. Ya. Spivakov, and B. F. Myasoedov, *Radiokhimia* (in Russian) *6*:56 (1992).
26. N. M. Kuz'min, V. M. Pukhovskaya, G. M. Varshal, B. Ya. Spivakov, T. A. Maryutina, M. P. Volynets, V. A. Ryabukhin, N. N. Chkhetiya, O. N. Grebneva, and V. M. Pavlutskaya, *Zh. Anal. Khim.* (in Russian) *48*:898 (1993).
27. I. V. Pavlenko, V. L. Bashlov, B. Ya. Spivakov, and Yu. A. Zolotov, *Zh. Anal. Khim.* (in Russian) *44*:827 (1989).
28. M. K. Chmutova, *Proceedings of Third Finnish–Russian Symposium on Radiochemistry*, Helsinki, 1994, p. 44.
29. Y. Surakitbanharn, S. Muralidharan, and H. Freiser, *Solvent Extr. Ion Exch. 19*:45 (1991).
30. Y. Surakitbanharn, S. Muralidharan, and H. Freiser, *Anal. Chem. 63*:2642 (1991).
31. T. Araki, T. Okazawa, Y. Kubo, H. Ando, and H. Asai, *J. Liq. Chromatogr. 11*:267 (1988).
32. T. Araki, T. Okazawa, Y. Kubo, H. Asai, and H. Ando, *J. Liq. Chromatogr. 11*(12): 2473 (1988).
33. H. Abe, S. Usuda, H. Takeishi, and S. Tachimori, *J. Liq. Chromatogr. 16*:2661 (1993).
34. G. Ghersini, in *Extraction Chromatography* (T. Braun and G. Ghersini, eds.), Elsevier, Budapest, 1975, p. 68.
35. V. V. Tarasov, G. Ya. Yagodin, and A. A. Pichugin, *Itogi Nauki i Tekhn. Neorg. Khim.* (in Russian) *11*:1 (1984).
36. T. Araki, H. Asai, H. Ando, N. Tanaka, K. Kimata, K. Hosoya, and H. Narita, *J. Liq. Chromatogr. 13*:3673 (1990).
37. H. Abe, S. Usuda, H. Takeishi, and S. Tachimori. *J. Liq. Chromatogr. 17*:1821 (1994).
38. K. Govindaraja, *Geostand. Newslett. 13*:1 (1989).
39. S. Abbey and F. S. Gladney, *Geostand. Newslett. 10:1* (1986).
40. E. Bauer-Wolf, W. Wegscheider, S. Posch, and G. Knapp, *Talanta 40*:9 (1993).
41. I. W. Croudace and S. Marshall, *Geostand. Newslett. 15*:139 (1991).
42. P. S. Watkins and S. Novan, *Chem. Geol. 95*:131 (1992).
43. B. S. Mohite and S. M. Khopkar, *Analyst 112*:191 (1987).
44. B. S. Mohite and S. M. Khopkar, *Talanta 32*:565 (1985).
45. M. K. Chmutova, L. A. Ivanova, G. V. Bodrin, Yu. M. Polikarpov, and B. F. Myasoedov, *Radiokhimia* (in Russian) *36*:320 (1994).

7
Chiral Separation by High-Speed Countercurrent Chromatography

Ying Ma and Yoichiro Ito
National Heart, Lung, and Blood Institute, National Institutes of Health, Bethesda, Maryland

Alain Berthod
Centre National de la Recherche Scientifique, University of Lyon 1, Villeurbanne, France

I. INTRODUCTION

Chromatographic separations of solutes are usually based on such parameters as hydrophobicity, molecular charge, and size of the molecules. In chiral separations, where these parameters are common among the racemates, the method requires an additional parameter to obtain enantioselectivity. There are three ways to separate enantiomers: (1) The racemic pair of enantiomers can be derivatized by an optically pure reagent. The two diastereoisomers obtained can be separated by classical liquid chromatography. (2) A chiral molecule can be added to the mobile phase so that diastereoisomers are reversibly formed inside the column and are separated by the classical achiral stationary phase. (3) A chiral selector is bonded to the solid stationary phase, which becomes capable of separating some enantiomeric pairs [1,2].

The rapidly increasing demand for chiral drugs has produced a great improvement in enantioselective technologies. Resolution of optical isomers by liquid column chromatography using chiral stationary phases is becoming very popular, and currently more than 100 different kinds of chiral solid phases are available in the market [1,2]. These chiral phases are mainly divided into two classes: one is biopolymers such as proteins and cellulose, and the other, synthetic selectors such as Pirkle's chiral stationary phases [3]. Except for the cellulose chiral stationary phase, these chiral columns are produced by chemically bonding the chiral selector to a solid support that serves as a stationary phase. However, manufacturing such a chiral stationary phase often requires a series of time-consuming and complicated processes, resulting in high cost of the chiral column. Consequently, the separations are mostly limited to an analytical scale. In this connection, countercurrent chromatography (CCC) is considered to be a possible alternative for performing preparative scale separations of chiral compounds.

Countercurrent chromatography [4–6] is a generic term for support-free liquid–liquid partition chromatography that can be used for the separation of a variety of enantiomers via the addition of a chiral selector (CS) to the liquid stationary phase. The method might offer various advantages over the conventional high-performance liquid chromatographic (HPLC) technique. For example, it may eliminate the costly CS immobilization process. Also, if suitable CS molecules are available, the CCC column could be repeatedly used for the separation of various enantiomers simply by dissolving the selected CS in the liquid stationary phase.

The existing CCC systems may be classified into two categories, i.e., hydrostatic and hydrodynamic equilibrium systems [4]. The hydrostatic system uses a stable gravitational or centrifugal force field to retain the stationary phase while the mobile phase gently percolates through it. Hydrostatic machines easily retain any liquid stationary phases. However, they have a limited chromatographic efficiency. The hydrodynamic systems utilize a coiled column rotated in a force field to produce an Archimedean screw effect. This results in an efficient hydrodynamic mixing of the two phases in the coil, hence the system yields a high partition efficiency that is critical in chiral separations. However, it may be difficult to retain some liquid stationary phases.

In the past the hydrostatic CCC systems, such as droplet CCC [7], rotation locular CCC (RLCCC) [8], and centrifugal partition chromatography (centrifugal droplet CCC) [9], have been used for the separation of chiral compounds. However, none of these techniques is considered satisfactory for the preparative purposes in terms of sample size, resolution, and/or separation time. The advent of the high-speed CCC technique [10] based on the hydrodynamic system improved both the partition efficiency and separation times. In selected cases, sizable separations were completed in a couple of hours.

Recently, the high-speed CCC technique has been successfully applied to the separation of racemates by the use of a Pirkle-type CS in the organic stationary phase [11,12]. Our studies revealed that both analytical (mg) and preparative (gs) separations can be performed simply by adjusting the amount of CS in the liquid stationary phase in the standard separation column. In general, the best results were achieved by applying the high CS concentration in the stationary phase and adjusting the hydrophobicity of the solvent system. Furthermore, the method allows computation of the formation constant of the CS–enantiomer complex, one of the most important parameters for studies on the mechanism of enantioselectivity, which cannot be easily obtained by the conventional HPLC technique. In addition, a large-scale separation of enantiomers can be performed by pH-zone–refining CCC, a new preparative separation technique described in detail elsewhere [13,19].

In this chapter we review the separation of racemates by various CCC techniques with special emphasis on the results recently obtained by high-speed CCC in our laboratory.

II. SEPARATION OF RACEMATES BY HYDROSTATIC CCC SYSTEM

A. Droplet CCC

Droplet CCC (DCCC)—a classical technique—utilizes a unit gravity to form a procession of droplets of the mobile phase through the stationary phase along a vertical column [7]. Using this CCC method, complete resolution of DL-isoleucine was achieved by Takeuchi et al. [14]. With a two-phase solvent system composed of *n*-butanol/50 mM acetate buffer (pH 5.5) containing 1 mM copper(II) ion and 2 mM *N*-*n*-dodecyl-L-proline, 2.6 mg of isoleucine was completely resolved in about 60 hr (Figure 1). In this solvent system, essentially all of the copper ions are distributed into the organic phase forming a complex with a chiral selector, *N*-*n*-dodecyl-L-proline, which transfers the DL-amino acid into the organic phase. There are significantly different distribution ratios between other amino acid enantiomers in these solvent systems, and therefore it may be possible to separate those amino acid enantiomers by more efficient CCC techniques. In order to achieve good resolution of the enantiomers in this method, the concentration of DL-amino acid to be resolved must be kept substantially below the concentration of copper ions in the solvent system. This example was chosen to show an early chiral CCC separation. The 3-day duration for the separation of only 2.6 mg shows that it is impractical to use this method for preparative separations.

Another example of early chiral separation by DCCC was reported by Oya and Snyder [15]. One hundred milligrams of bicyclo[2,2,1]-hept-5-ene-2-carboxylic acid racemates was resolved in about 56 hr using the solvent system composed of chloroform/methanol/water (7:13:8) (Figure 2). In this separation the aqueous phase was used as the mobile phase where (−)-*R*-2-aminobutanol was added as a chiral resolving agent. The separation was based on the difference in partition coefficient of the ionic complexes formed between the carboxylic acid racemates and aminobutanol. Eluted fractions need further purification, because the chiral resolving agent is present in the mobile phase. The experiment duration and the latter step render the method impractical.

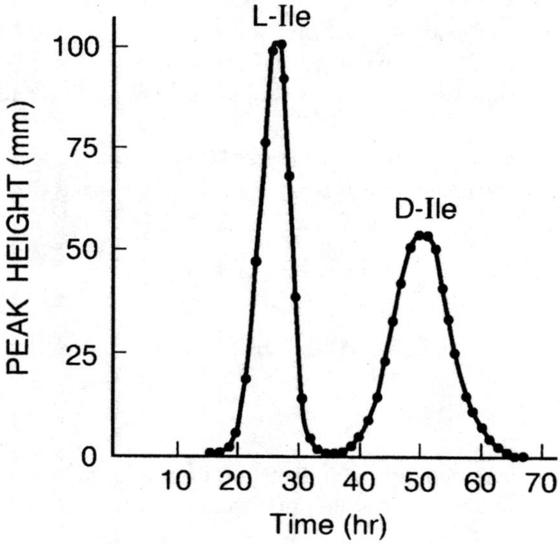

Figure 1 Chiral separation of L- and D-isoleucine by droplet CCC. The stationary and mobile phases were prepared by equilibrating equal volumes of 50 mM acetate buffer (pH 5.5) containing 1 mM copper(II) acetate and *n*-butanol containing 2 mM C_{12}-proline. The column consisted of 400 pieces of Teflon tubing of two sizes, 40 cm × 4 mm i.d. and 50 cm × 1 mm i.d., alternately in series. The separation was effected in the descending mode at a flow rate of 1.1 mL/min and 10-mL fractions were collected.

B. Rotation Locular CCC

Rotation locular CCC uses a series of locular columns each consisting of a tubular structure with multiple partition compartments called ''locules.'' The solute partitioning takes place within each locule of the tilted column where the two solvent phases are efficiently mixed by rotation of the column [8]. The separation of norephedrine racemates by RLCCC has been described by Domon et al. [16]. Using an RLCCC instrument with a 16 × 37 locules, partial resolution of 200 mg norephedrine was obtained in 4 days. The solvent system is composed of 0.5 M sodium hexafluorophosphate (pH 4) aqueous solution (stationary phase) and 0.3 M solution of (R,R)-di-5-nonyl tartrate in 1,2-dichloroethane (mobile phase). In this separation, hexafluorophosphate salts of the norephedrine recemates were separated by partition between the aqueous stationary phase and the organic mobile phase containing the tartrate ester as a chiral selector. Here again, the eluted fraction needs further purification to eliminate the chiral resolving agent. We note that the separated amount was 200 mg per run.

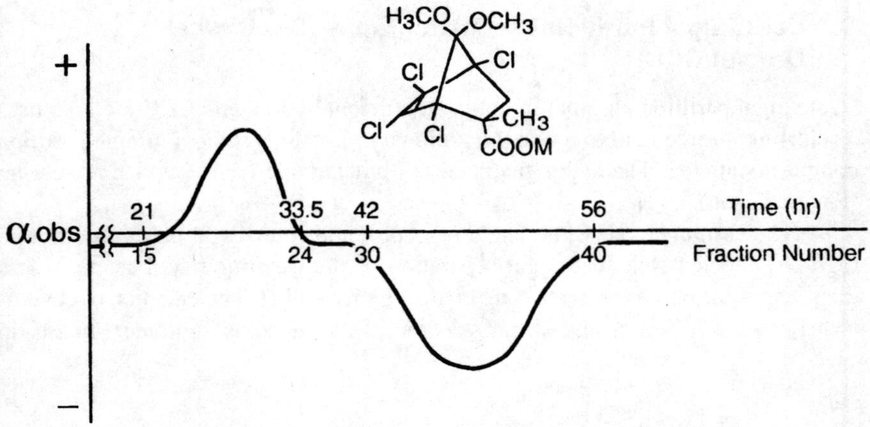

Figure 2 Observed optical rotation of DCCC fractions from chiral resolution of the carboxylic acid, after removal of (−)-(*R*)-2-aminobutanol. Experimental conditions: solvent system, chloroform/methanol/water (7:13:8) (pH 7, 0.01 M phosphate buffer); sample size, 100 mg; elution, descending mode at a flow rate of about 10 mL/hr.

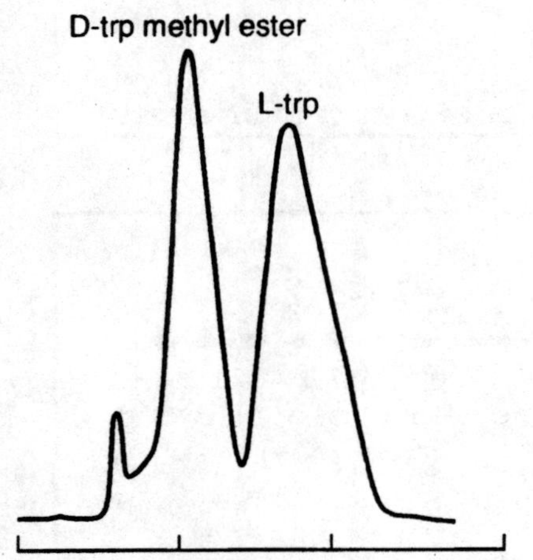

Figure 3 CCC chromatogram showing the separation of L-tryptophan from D-tryptophan methyl ester after incubation of about 90 mg of the racemic ester. Stationary phase, water; mobile phase, butanol; flow rate, 4 mL/min, ascending mode; apparatus, Sanki CPC-LLN with 12 cartridges (2400 channels); revolution, 700 rpm.

C. Centrifugal Partition Chromatography (Centrifugal Droplet CCC)

Centrifugal partition chromatography, an efficient hydrostatic CCC method, uses a centrifugal force field to retain the stationary phase in a train of minute partition compartments [9]. The apparatus is currently available from Sanki Engineering (Kyoto, Japan).

The combined use of an enzyme reactor and centrifugal partition chromatography has led to a successful separation of the DL-amino acid esters in less than 3 hr, as reported by Armstrong et al. (Figure 3) [17]. The enzyme, α-chymotrypsin, was chosen as the chiral selector in the aqueous stationary phase, in

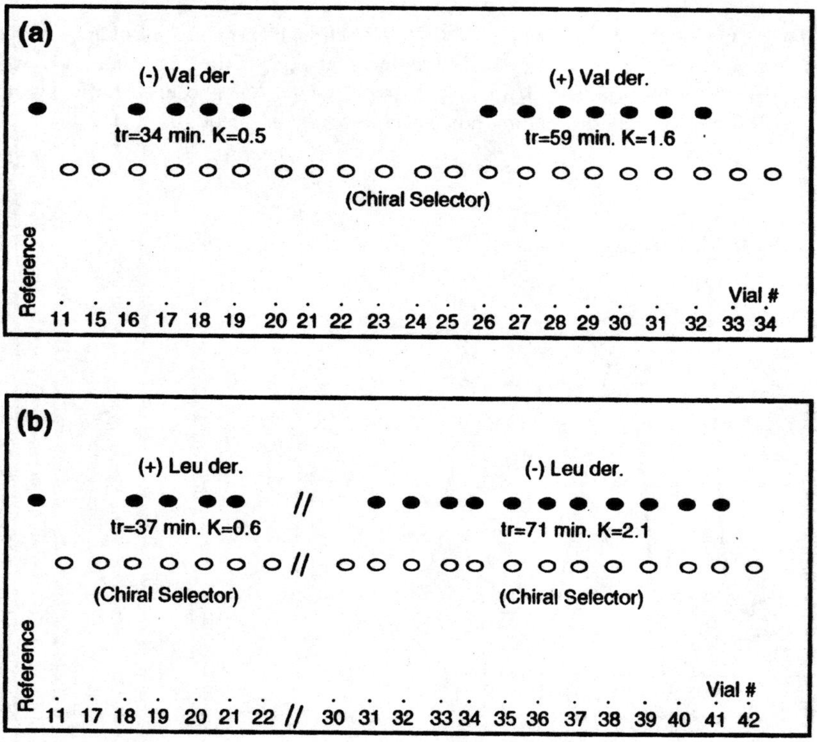

Figure 4 TLC analysis of the chiral resolution of (±)-DNB-valine (a) and (±)-DNB-leucine (b) by centrifugal partition chromatography. CCC conditions: solvent system, heptane/ethyl acetate/methanol/water (3:1:3:1), descending mode; flow rate, 5 mL/min; rotation speed, 1200 rpm; back-pressure, 3 MPa; $V_{stationary}/V_{mobile} \sim 1$; CS, 0.01 M in upper phase; sample, 2 mL of a 0.001 M amino acid derivative; fractions, every 2 min (10 mL).

which the L enantiomer would bind to the active site of the immobilized enzyme for hydrolysis while the D enantiomer would not. As much as 880 mg DL-tryptophan methyl ester could be separated in less than 3 hr into the D-tryptophan methyl ester and L-tryptophan. This work shows the drastic reduction of the experiment duration obtained with modern hydrostatic apparatuses.

Recently, Oliveros et al. successfully separated racemates of amino acid derivatives by centrifugal partition chromatography using a Pirkle's CS in its free form [18]. Using *N*-dodecanoyl-L-proline-3,5-dimethylanilide as a CS in the two-phase solvent system composed of heptane, ethyl acetate, methanol, and water at a volume ratio of 3:1:3:1, (±)-*N*-(3,5-dinitrobenzoyl)-*tert*-butylvalinamide and -leucinamide were each completely resolved in less than 2 hr. Because a large amount of CS in the mobile phase interfered with direct detection of a small amount of the analyte, the CCC fractions were further subjected to thin-layer chromatographic (TLC) analysis (Figure 4). Although the sample size was limited to a submilligram range, this work is the first example of complete resolution of nonionic racemates by CCC.

III. THEORETICAL ASPECTS OF CHIRAL CCC

CCC uses biphasic liquid systems. This is the main difference with liquid chromatography. In CCC, each time the mobile phase is modified, the stationary phase can be changed as well. The chiral selector added to the stationary phase can partition with the mobile phase. Taking into account all possible partitioning processes, a theoretical model is derived. It will be illustrated by two experimental cases exposed later in this chapter.

A. General Model

Figure 5 shows the equilibria occurring between the racemates A_+ and A_-, the chiral selector CS, the organic stationary phase, and the aqueous stationary phase. The *K* values correspond to the partition ratio of the molecules or complexes. The apparent partition ratio, the one obtained experimentally, are noted as *K*′. The *K*f values are the formation constants of the complexes in both phases.

For the (+)-enantiomer, the partition ratio is:

$$K_+ = [A_+]_{org} \, / \, [A_+]_{aq} \tag{1}$$

The apparent partition ratio is

$$K'_+ = \{[A_+]_{org} + [CSA_+]_{org}\} \, / \, \{[A_+]_{aq} + [CSA_+]_{aq}\} \tag{2}$$

Similar equations for the (−)-enantiomer are easily derived changing plus for minus in Eqs. (1) and (2). The CS partitions as well:

	Kf_{org}
Organic stationary phase	$[A_+]_{org} + [A_-]_{org} + [CS]_{org} \rightleftarrows [CSA_+]_{org} + [CSA_-]_{org}$
	$K_+ \Uparrow\Downarrow$ ----- $K_- \Uparrow\Downarrow$ ---- $K_{CS} \Uparrow\Downarrow$ -------- $K_{CSA+} \Uparrow\Downarrow$ ----- $K_{CSA-} \Uparrow\Downarrow$
Aqueous mobile phase	$[A_+]_{aq} + [A_-]_{aq} + [CS]_{aq} \rightleftarrows [CSA_+]_{aq} + [CSA_-]_{aq}$
	Kf_{aq}

Figure 5 Schematic diagram of chemodynamic equilibrium between the racemates ($A_{\pm}$) and chiral selector (CS) in the separation column.

$$K_{CS} = [CS]_{org} \;/\; [CS]_{aq} \tag{3}$$

$$K'_{CS} = \{[CS]_{org} + [CSA_+]_{org} + [CSA_-]_{org}\} \;/\; \{[A_+]_{aq} + [CSA_+]_{aq} + [CSA_-]_{aq}\} \tag{4}$$

The formation constants, *K*f, are expressed by:

$$Kf_{+org} = [CSA_+]_{org} \;/\; \{[A_+]_{org}[CS]_{org}\} \tag{5}$$

The three other constants, Kf_{-org}, Kf_{+aq}, and Kf_{-aq}, are easily derived from Eq. (5).

Using Eqs. (1)–(5) and rearranging, it is possible to express the apparent, and experimentally measured, partition ratio for the (+)-enantiomer as:

$$K'_+ = K_+ \{(1 + Kf_{+org}[CS]_{org}) \;/\; (1 + Kf_{+aq}[CS]_{aq})\} \tag{6}$$

The K'_- expression is similar to Eq. (6) with the plus signs switched to minus signs.

B. Application to Actual Experiments

The CS should be selected so that its K_{CS} value is higher than 100. It means that there will be practically no chiral selector in the aqueous phase. This allows CS-free recovery of the separated enantiomers and avoids further purification and on-line detection problems. Also, it greatly simplifies the theoretical development. All complexation reactions in the aqueous phase can be ignored, i.e., $[CS]_{aq} \approx 0$ and $[CSA]_{aq} \approx 0$. Equation (6) becomes

$$K'_+ = K_+ \{1 + Kf_{+org}[CS]_{org}\} \tag{7}$$

Here $[CS]_{org}$ is a difference between the initial concentration of the CS and the concentration of the $CSA_{\pm}$ complexes, i.e.,

$$[CS]_{org} = [CS]_{initial} - [CSA_{+}]_{org} - [CSA_{-}]_{org} \quad (8)$$

When $[A]_{org} \ll [CS]_{initial}$, $[CS]_{org}$ approaches $[CS]_{initial}$; hence Eq. (7) may be re-written:

$$K'_{+} \approx K_{+} \, (1 + Kf_{+org}[CS]_{initial}) \quad (9)$$

and the formation constant, Kf_{+org}, can be approximated by the following equation:

$$Kf_{org} \approx \{(K'_{+}/K_{+}) - 1\}/[CS]_{initial} \quad (10)$$

The partitioning of the free + and − isomers in an achiral liquid system is identical, i.e.,

$$K_{+} = K_{-} = K_{D} \quad (11)$$

The K'_{+}, K'_{-}, and K_D parameters can be respectively computed from the chromatograms obtained with and without the CS in the stationary phase, using the conventional equation:

$$K = (V_r - V_m)/(V_c - V_m) \quad (12)$$

where V_r, V_m, and V_c indicate the retention volume of the analyte, the mobile phase volume in the column, and the total column capacity, respectively.

The validity of Eq. (9) has been examined by a series of experiments where small amounts (0.1–1 mg) of enantiomers (DNB–amino acids) were separated using various concentration (0.5, 1, 2, and 4 g in 200 mL of the organic phase) of the CS (*N*-dodecanoyl-L-proline-3,5-dimethylanilide) in the stationary phase. An acidic solvent system composed of hexane/ethyl acetate /methanol/10 mM HCl (8:2:5:5, by volume) was used to protonate the analyte molecules in the aqueous mobile phase. The experimental K' values for each enantiomer were plotted vs. the initial CS concentrations in the organic stationary phase. Straight lines were obtained for both enantiomers. The intercepts were almost identical corresponding to the K_D value. The slopes of the straight lines corresponded to the product $Kf_{org}K_D$ from which the formation constant could be obtained (Table 1). For each pair of racemates examined, the average value of the intercepts (K_D in Table 1) closely matched the partition ratio (K'_D in Table 1) directly obtained from the experiment using CS-free solvent system. These results indicate that Eq. (9) is useful for computing the formation constants, Kf, of various analyte–CS pairs. As shown in Table 1, all formation constants of K_{-} enantiomers lie in a narrow range between 30 and 50, whereas those of K_{+} enantiomers varies in a broader range from 100 to 250 and correlate with the length of the hydrocarbon chain at the asymmetrical carbon in both aliphatic and aromatic groups. The +

Table 1 Parameters of the K vs $[CS]_{initial}$ Lines

Solute	Form	Slope	Intercept	r	$K_D(K'_D)$	Kf (L/mol)
DNB-phenylglycine	−	6.09	0.150	0.963	0.155 (0.146)	39
	+	17.3	0.159	0.983	0.155 (0.146)	112
DNB-phenylalanine	−	9.72	0.184	0.995	0.213 (0.190)	46
	+	28.8	0.242	0.995	0.213 (0.190)	135
DNB-valine	−	4.73	0.142	0.996	0.146 (0.146)	32
	+	18.6	0.149	0.999	0.146 (0.146)	131
DNB-leucine	−	9.92	0.284	0.999	0.285 (0.280)	35
	+	71.8	0.285	0.998	0.285 (0.280)	252

CS, Chiral selector, *N*-dodecanoyl-L-proline-3,5-dimethylanilide; DNB, 3,5-dinitrobenzoyl; K_D, racemate partition ratio by linear regression analysis; K'_D: racemate partition ratio by experiment; Kf: formation constant of the enantiomer with the CS molecule.

form of the four DNB–amino acid derivatives studied has between three and seven times more affinity for the L-proline derivative CS than the − form.

The enantioselectivity factor, α, is the ratio of the K' partition coefficients for the two enantiomers:

$$\alpha = K'_+/K'_- = (1 + Kf_{+org}[CS]_{initial})/(1 + Kf_{-org}[CS]_{initial}) \quad (13)$$

The α vs. $[CS]_{initial}$ curves are hyperboles tending toward the asymptotic value Kf_+/Kf_- for very high CS concentrations.

C. Ionizable Compounds

With an increasing pH mobile phase, pH zone–refining CCC separates ionizable compounds in the order of the increasing pK_a values [13]. Assuming that (1) the chiral selector is located in the organic phase, (2) it is not ionizable itself, (3) the complexed isomers does not partition with the aqueous phase, and (4) the ionized form of the analyte has no affinity for the organic phase (i.e., it cannot interact with the CS), Figure 6 shows the chemodynamic equilibrium in a portion of the separation column where the organic stationary phase is in the upper half and the aqueous mobile phase in the lower half.

In the aqueous phase the acidic analytes are partly dissociated to form anions that are not soluble in the hydrophobic organic phase, whereas in the organic phase these molecules are fully protonated, forming CS complexes according to their formation constants. When equilibrium is reached, the following set of equations is given for each racemate:

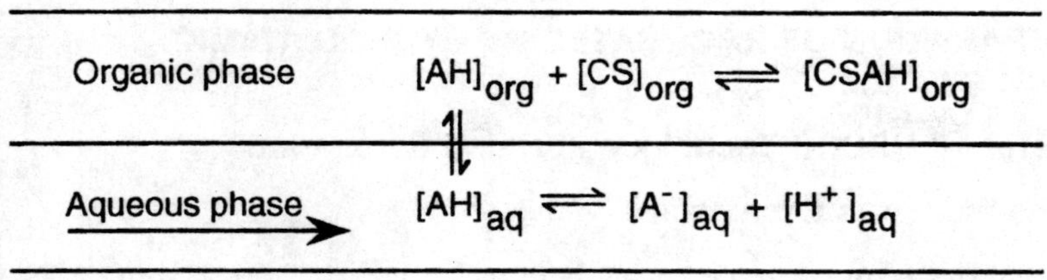

Figure 6 Schematic diagram of chemodynamic equilibrium between ionic analytes, CS, and their complexes in the separation column (pH-zone-refining).

$$K = ([\mathrm{AH}]_{\mathrm{org}} + [\mathrm{CSAH}]_{\mathrm{org}})/([\mathrm{AH}]_{\mathrm{aq}} + [\mathrm{A}^-]_{\mathrm{aq}}) \tag{14}$$

$$K_{\mathrm{D}} = [\mathrm{AH}]_{\mathrm{org}}]/[\mathrm{AH}]_{\mathrm{aq}} \tag{15}$$

$$K_{\mathrm{a}} = [\mathrm{AH}]_{\mathrm{aq}}/[\mathrm{A}^-]_{\mathrm{aq}}[\mathrm{H}^+]_{\mathrm{aq}} \tag{16}$$

$$K\mathrm{f} = [\mathrm{CSAH}]_{\mathrm{org}}/[\mathrm{CS}]_{\mathrm{org}}[\mathrm{AH}]_{\mathrm{org}} \tag{17}$$

where K, K_{D}, K_{a}, and K_{f} represent the partition coefficient, the partition ratio, the dissociation constant, and the CS complex formation constant of each racemate, respectively. From these equations, we obtain [11]:

$$\mathrm{pH} = \mathrm{p}K_{\mathrm{a}} + \log\{(K_{\mathrm{D}}/K)(1 + [\mathrm{CS}]_{\mathrm{org}}K\mathrm{f}) - 1\} \tag{18}$$

In pH-zone–refining CCC, the peak resolution is mainly determined by the difference in pH between the two zones. Since K values for all analytes in pH-zone–refining CCC are equal to that of the retainer acid (K_{r}), and pK_{a} and K_{D} are common for both enantiomers, the zone pHs values of enantiomers A_+ and A_- are given by Eq. (18) in the following equations:

$$\mathrm{pH}_{z+} = \mathrm{p}K_{\mathrm{a}} + \log\{(K_{\mathrm{D}}/K_{\mathrm{r}})(1 + [\mathrm{CS}]_{\mathrm{org}}Kf_+) - 1\} \tag{19}$$

$$\mathrm{pH}_{z-} = \mathrm{p}K_{\mathrm{a}} + \log\{(K_{\mathrm{D}}/K_{\mathrm{r}})(1 + [\mathrm{CS}]_{\mathrm{org}}K\mathrm{f}_-) - 1\} \tag{20}$$

which gives

$$\Delta\mathrm{pH}_{z+/-} = \log\{([\mathrm{CS}]_{\mathrm{org}/z+}K\mathrm{f}_+ + 1 - K_{\mathrm{r}}/K_{\mathrm{D}})/([\mathrm{CS}]_{\mathrm{org}/z-}K\mathrm{f}_- + 1 - K_{\mathrm{r}}/K_{\mathrm{D}})\} \tag{21}$$

where $[\mathrm{CS}]_{\mathrm{org}/z+}$ and $[\mathrm{CS}]_{\mathrm{org}/z-}$ are the free CS concentrations in the A_+ and A_- zones, respectively. When $K\mathrm{f}_+ > K\mathrm{f}_-$, we have the sequence $[\mathrm{CS}]_{\mathrm{org}/z+} < [\mathrm{CS}]_{\mathrm{org}/z-} < [\mathrm{CS}]_{\mathrm{initial}}$. Equation (21) indicates that the chiral resolution can be improved by increasing the $K_{\mathrm{r}}/K_{\mathrm{D}}$ ratio and/or by choosing a CS with a large $K\mathrm{f}_+/K\mathrm{f}_-$ value with the enantiomeric pair to be separated. It also implies that an increase of the CS concentration will give a higher peak resolution.

IV. SEPARATION OF RACEMATES BY HYDRODYNAMIC CCC SYSTEM

A. Standard HSCCC Technique in Chiral Separation

1. Analytical Scale Separations

The separations were performed using both our prototype fabricated in the National Institutes of Health (NIH) machine shop and a commercial high-speed CCC centrifuge purchased from P.C. Inc. (Potomac, Maryland). (A comparable instrument is also available from Pharma-Tech Research Corporation, Baltimore, Maryland). The following procedure was used: The column is first entirely filled with the stationary phase that contains a desired amount of CS. Usually we leave about 10% of CS-free stationary phase at the end of the column. This will prevent the contamination of the eluate by CS that may be carried over by the mobile phase. After sample solution is injected through the sample port, the mobile phase is pumped into the column while the column is rotated. Separations can be carried out by successive injections of samples without renewing the column phase containing the chiral selector.

Figure 7 shows the separation of four pairs of DNB–amino acid enantiomers by the standard CCC technique using a two-phase solvent system composed of hexane/ethyl acetate/methanol/10 mM HCl (8:2:5:5) and *N*-dodecanoyl-L-proline-3,5-dimethylanilide as a CS [11]. Because of its high hydrophobicity, this

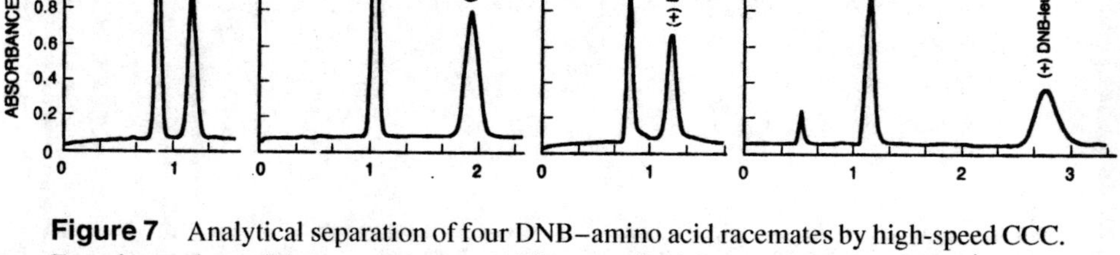

Figure 7 Analytical separation of four DNB–amino acid racemates by high-speed CCC. Experimental conditions: apparatus, multilayer coil high-speed CCC centrifuge with a semipreparative column of 1.6 mm i.d. and 330-mL capacity; solvent system, hexane/ethyl acetate/methanol/10 mM HCl (8:2:5:5), *N*-dodecanoyl-L-proline-3,5-dimethylanilide (2 g) was added to the organic stationary phase (200 mL) as a chiral selector; sample: 5–10 mg of each DNB–amino acids dissolved in 5 mL of solvent consisting of equal volumes of each phase; flow rate, 3.3 mL/min in the head-to-tail elution mode; revolution, 800 rpm; stationary phase retention, 65% of total column capacity.

CS partitions almost entirely into the organic stationary phase ($K > 100$). All analytes are well resolved in 1–3 hr.

As clearly shown from the chromatograms, the peak resolution of DNB-phenylalanine is much greater than that of DNB-phenylglycine, suggesting that the longer hydrocarbon chain attached at the asymmetrical carbon increases the retention time. Similarly, the peak resolution between DNB-leucine racemates is far greater than that between DNB-valine racemates. These results indicate that the chain length of the hydrocarbon attached to the asymmetrical carbon may produce an important effect on the chiral selection.

Two other chiral selectors were also tested by HSCCC: *N*-(2′,6′-dimethyl-piperidine)-6-methoxy-α-methyl-2-naphthalene ethanamide and *N*-dodecanoyl-6-methoxy-α-methyl-2-naphthalene ethanamide. The DNB-leucine enantiomers were resolved by both chiral selectors with the same solvent system (Figure 8A,B) [20]. An increased length of the side chain at the asymmetrical center of the CS not only alters the separation factor but increases the hydrophobicity of CS to promote its partition toward the organic stationary phase.

All chiral selectors successfully used in the above studies are similar to the chiral stationary phase that has been introduced by Pirkle et al. for the HPLC separation of racemic DNB–amino acid *t*-butylamide [3]. This may indicate that the chiral selectors used as the HPLC solid stationary phase can also be utilized effectively for HSCCC in their free form, i.e., without immobilization to the solid support.

The studies were continued in our laboratory to investigate the effects of various factors such as concentration and distribution of CS in the stationary phase and hydrophobicity of the solvent system on the chiral separation in HSCCC.

Effect of Concentration of Amount of CS in the Stationary Phase

A series of experiments was performed to investigate the effects of the CS concentration in the organic stationary phase on chiral resolution. Figure 9 shows a set of chromatograms of four DNB–amino acid racemates obtained by introducing various amounts of CS in the stationary phase [12]. In the CS-free separation (top row), all racemates formed a single peak as expected and eluted at similar retention times shortly after the mobile phase front. As the amount of CS was increased, each racemate was resolved where (−)-enantiomers with less affinity to the CS eluted earlier while (+)-enantiomers with a stronger affinity to the CS retained longer in the column.

Figure 10 illustrates the effects of the amounts of CS in the stationary phase on the separation factor (10A) and peak resolution (10B) [12]. The separation factors for all racemates become greater as the amount of CS in the stationary phase is increased from 0 to 4 g where they reach near saturation (Figure 10A). Equation (13) explains the trend observed in Figure 10A. The effect of the

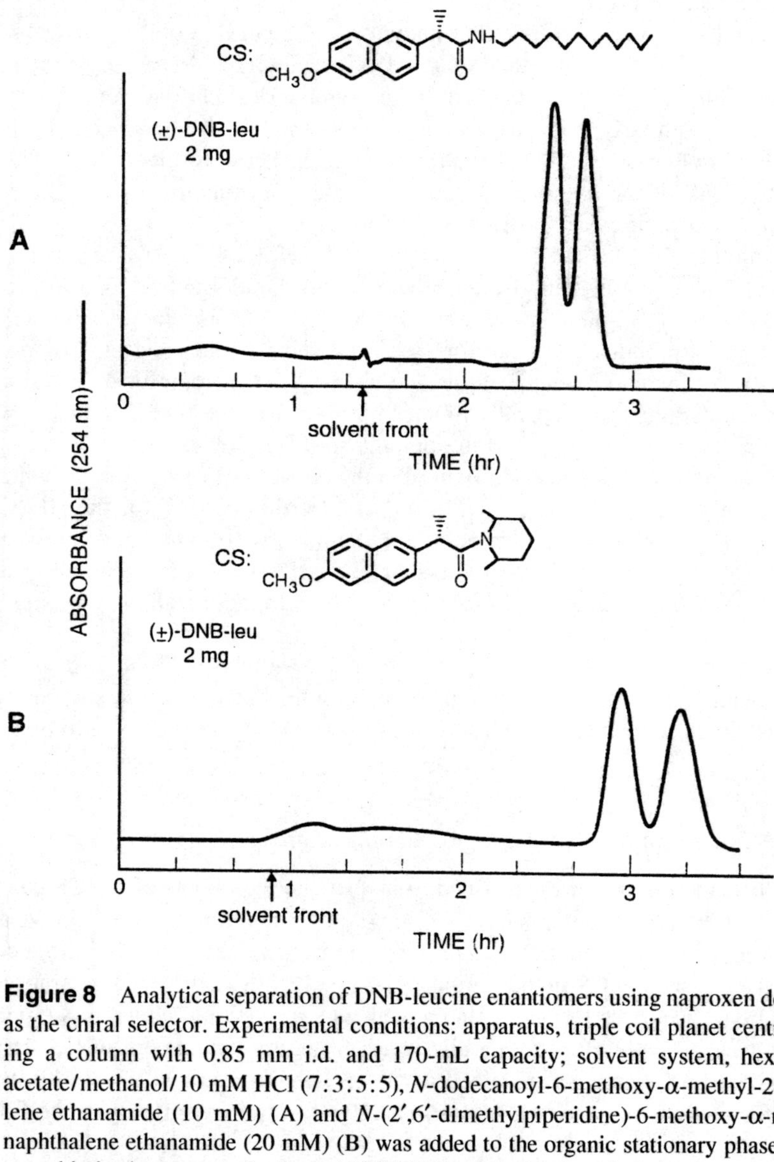

Figure 8 Analytical separation of DNB-leucine enantiomers using naproxen derivatives as the chiral selector. Experimental conditions: apparatus, triple coil planet centrifuge using a column with 0.85 mm i.d. and 170-mL capacity; solvent system, hexane/ethyl acetate/methanol/10 mM HCl (7:3:5:5), *N*-dodecanoyl-6-methoxy-α-methyl-2-naphthalene ethanamide (10 mM) (A) and *N*-(2′,6′-dimethylpiperidine)-6-methoxy-α-methyl-2-naphthalene ethanamide (20 mM) (B) was added to the organic stationary phase (80 mL) as a chiral selector; sample, 2 mg (±)-DNB-leucine dissolved in 1 mL lower phase; flow rate, 1 mL/min; revolution, 1000 rpm; stationary phase retention, 42% (A) and 57% (B) of the total column capacity.

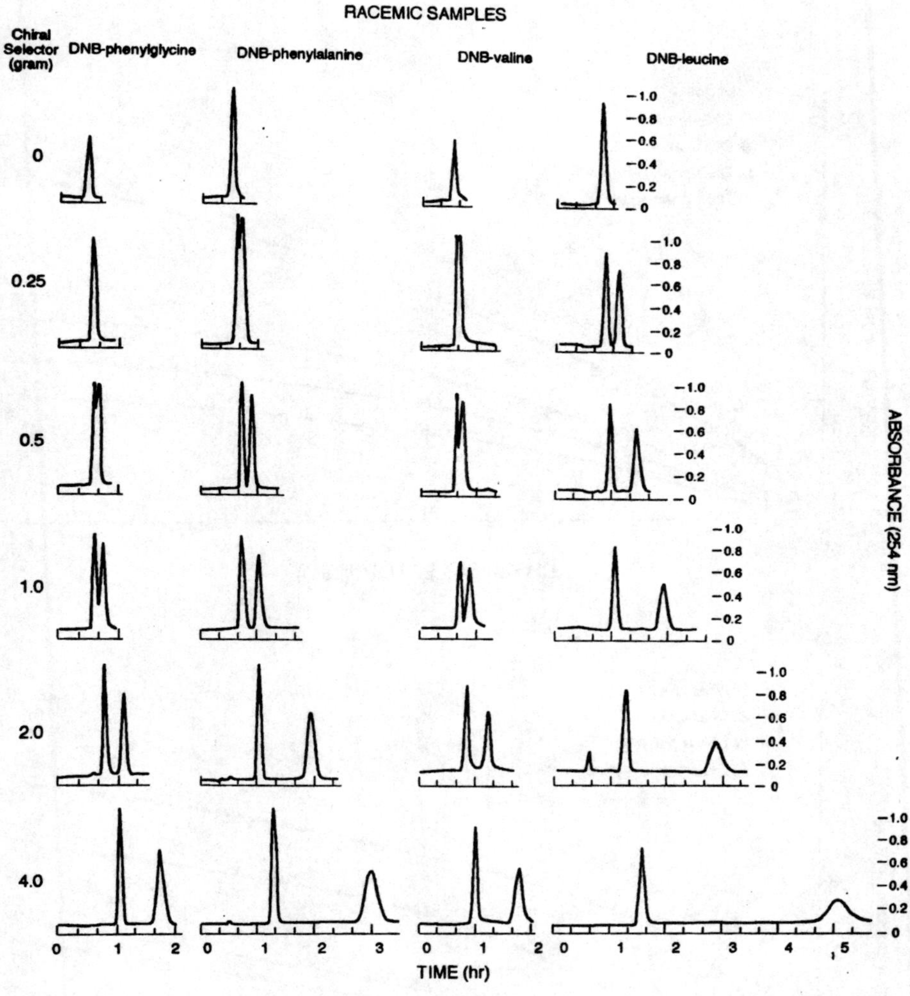

Figure 9 Effects of the amount or concentration of CS on the separation of DNB–amino acid racemates. In all resolved chromatograms, the first peak represents (−)-enantiomer and the second peak (+)-enantiomer. Experimental conditions: apparatus, commercial high-speed CCC centrifuge (Ito multilayer coil separator/extractor) with 10-cm revolution radius; separation column, multilayer coil consisting of 1.6 mm i.d. PTFE tubing with a total capacity of 330 mL; sample, racemic DNB–amino acid mixture consisting of DNB-phenylglycine, DNB-phenylalanine, DNB-valine, and DNB-leucine each 5–10 mg dissolved in 2 mL solvent (1 mL of each phase); solvent system, hexane/ethyl acetate/methanol/10 mM HCl (8:2:5:5 by volume); stationary phase, upper organic phase with CS ranging from 0 to 4 g in 200 mL as indicated; mobile phase, lower aqueous phase; flow rate, 3 mL/min; revolution, 800 rpm.

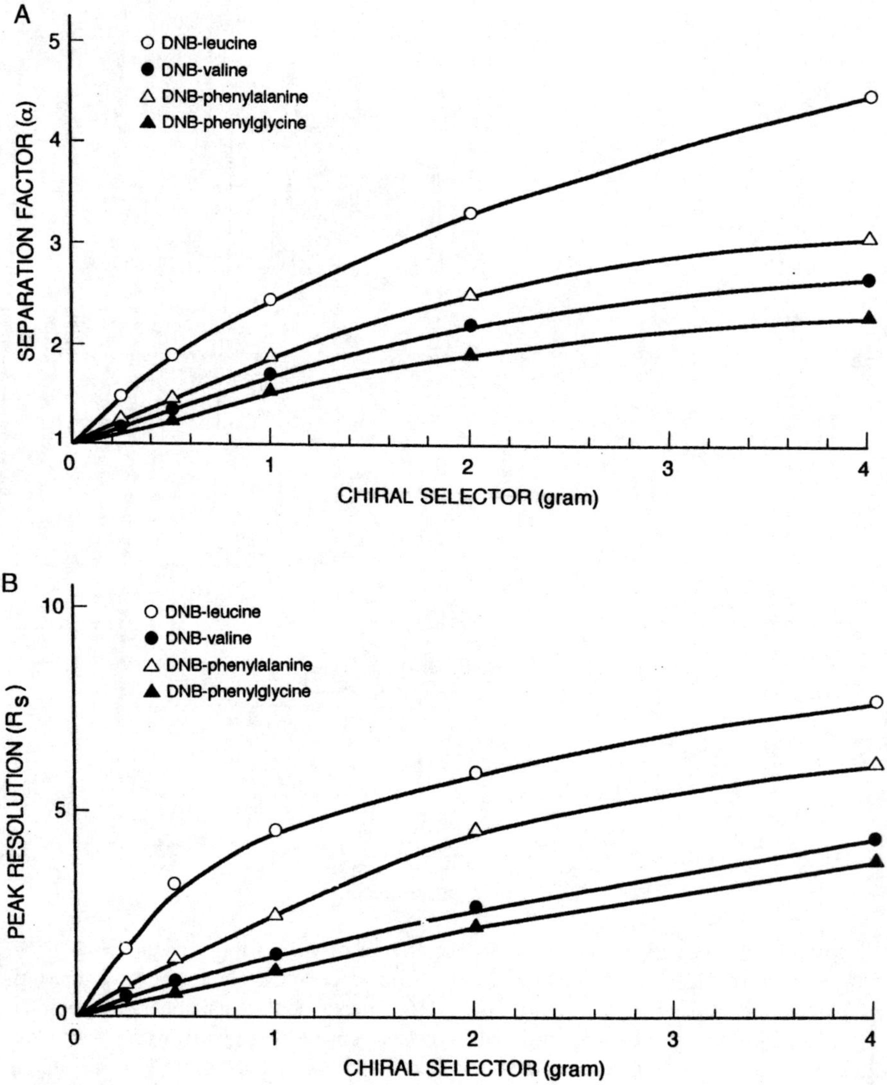

Figure 10 Effects of net amounts of CS on the selectivity factor, α (A) and the resolution factor, R_s (B) of DNB–amino acid racemates. For the experimental conditions, see Figure 7.

amounts of CS on the peak resolution are similarly shown in Figure 10B. In contrast with the separation factor (Figure 10A), the peak resolution (R_s) for all of the racemates increases nearly in proportion to the net amount of CS up to 4 g, indicating that the resolution can be improved by further increasing the CS concentration in the stationary phase. The above results clearly indicate an important technical strategy for the present method, i.e., the best peak resolution is attained by saturating the CS in the stationary phase in a given column, and the resolution is further improved by using a longer and/or wider bore coiled column that can hold greater amounts of CS in the stationary phase.

The effect on the CS distribution within the stationary phase was also studied. A given amount of CS was dissolved in various volumes of the stationary phase ranging from 50 to 200 mL so that the CS concentration was increased while the amount of CS remained constant. These CS-containing stationary phases were each placed at the beginning portion of the column and the separations were performed with the same solvent system used in the previous experiment. Figure 11 illustrates a set of chromatograms of four DNB–amino acid racemates according to the similar format used for Figure 9 [12]. The results clearly showed that the retention times of all four racemates are almost identical regardless of the volume of the stationary phase to which the CS was dissolved, indicating that the net amount of CS determines their α values. The peak resolution, on the other hand, was found to vary according to the CS distribution in the column as shown in Table 2: The 50-mL group shows substantially lower resolution that the 200-mL group. This may be explained as follows: It can be considered than two columns are serially connected. The first column is the chiral active column, the second column is a classical achiral column. The 50-mL group is subjected to chiral separation in a fourfold shorter column than the 200-mL group. The retention times and separation factors are identical because the solutes have seen the same mass of CS. The resolution factors differ because the effective column lengths are different.

Effects of Hydrophobicity of the Solvent System

In this series of experiments, the hydrophobicity of the two-phase solvent system composed of hexane/ethyl acetate/methanol/10 mM HCl was modified by changing the volume ratio between hexane and acetyl acetate, while the ratio of methanol and 10 mM HCl was kept constant. In all experiments, a given amount (1 g) of CS was dissolved in 200 mL of organic stationary phase. The results of the experiments are illustrated in Figure 12 [12]. As expected, the retention times of both DNB-valine and DNB-leucine racemates increased steadily with the reduced hydrophobicity of the solvent system when the hexane/ethyl acetate volume ratio was decreased as indicated in the left margin.

The separation factor and peak resolution for these separations are graphi-

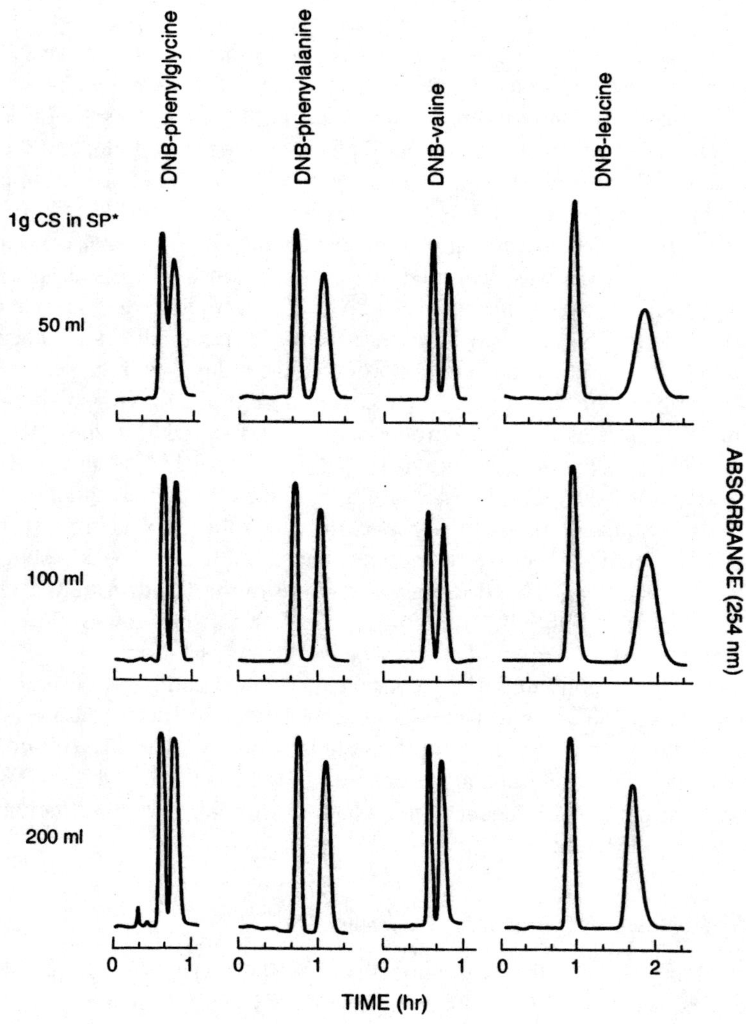

Figure 11 Effects of CS distribution in the stationary phase on the separation of DNB–amino acid racemates. Experimental conditions: apparatus, commercial high-speed CCC centrifuge with 10-cm revolution radius; column, multilayer coil of 1.6 mm i.d. Tefzel tubing with 320-mL capacity; sample, (±)-DNB-valine and (±)-DNB-leucine each 5–10 mg; solvent system, hexane/ethyl acetate/methanol/10 mM HCl (8:2:5:5 by volume); stationary phase, upper organic phase containing 1 g of CS in 50, 100, and 200 mL; mobile phase, lower aqueous phase; flow rate, 3 mL/min; revolution, 800 rpm.

Table 2 Effects of CS Distribution on Separation of DNB–Amino Acid Racemates

Racemic sample	CS-containing stationary phase[a] (mL)	Retention time (min) A_-	A_+^d	α^b	R_s^c
DNB[e]-phenylglycine	50	35	44	1.60	0.8
	100	36	46	1.55	1.2
	200	36	46	1.59	1.3
DNB-phenylalanine	50	44	64	1.80	1.8
	100	42	61	1.84	2.0
	200	43	64	1.88	2.2
DNB-valine	50	35	47	1.71	1.2
	100	35	46	1.68	1.4
	200	34	44	1.65	1.5
DNB-leucine	50	53	106	2.56	3.0
	100	54	110	2.60	3.5
	200	50	98	2.55	3.7

[a] CS, chiral selector, *N*-dodecanoyl-L-proline-3,5-dimethylanilide.
[b] α, separation factor computed from Eq. (13).
[c] R_s, peak resolution computed with α and the peak efficiency.
[d] A_-, A_+, (−) and (+)-DNB–amino acid, respectively.
[e] DNB, 3,5-dinitrobenzoyl.

cally illustrated in Figure 13 where these two parameters were plotted against the hexane/ethyl acetate volume ratio [12]. Quite different curves are observed between these two parameters. The α values of both racemates decrease sharply as the hydrophobicity of the solvent system decreases. The R_s values of these two racemates, on the other hand, show convex curves each with a maximum value at different locations: DNB-leucine shows the maximum value at the 9:1 volume ratio whereas that of more polar DNB-valine is around 7:3; in both cases the mean partition coefficient of the racemates is about 0.7. These results indicate that the best peak resolution is attained by optimizing the hydrophobicity of the solvent system so that the mean partition coefficient value for each racemate falls in a suitable range, e.g., between 0.6 and 0.8 in the above case. Although the use of a hydrophobic solvent system may yield a higher α value for the racemate, the peak resolution could be reduced due to an excessively short retention time.

2. Preparative Separation of Enantiomers

The preparative capability of the present system was investigated on the separation of DNB-leucine enantiomers by varying the CS concentration from 10 to 60 mM in the solvent system composed of hexane/ethyl acetate/methanol/10

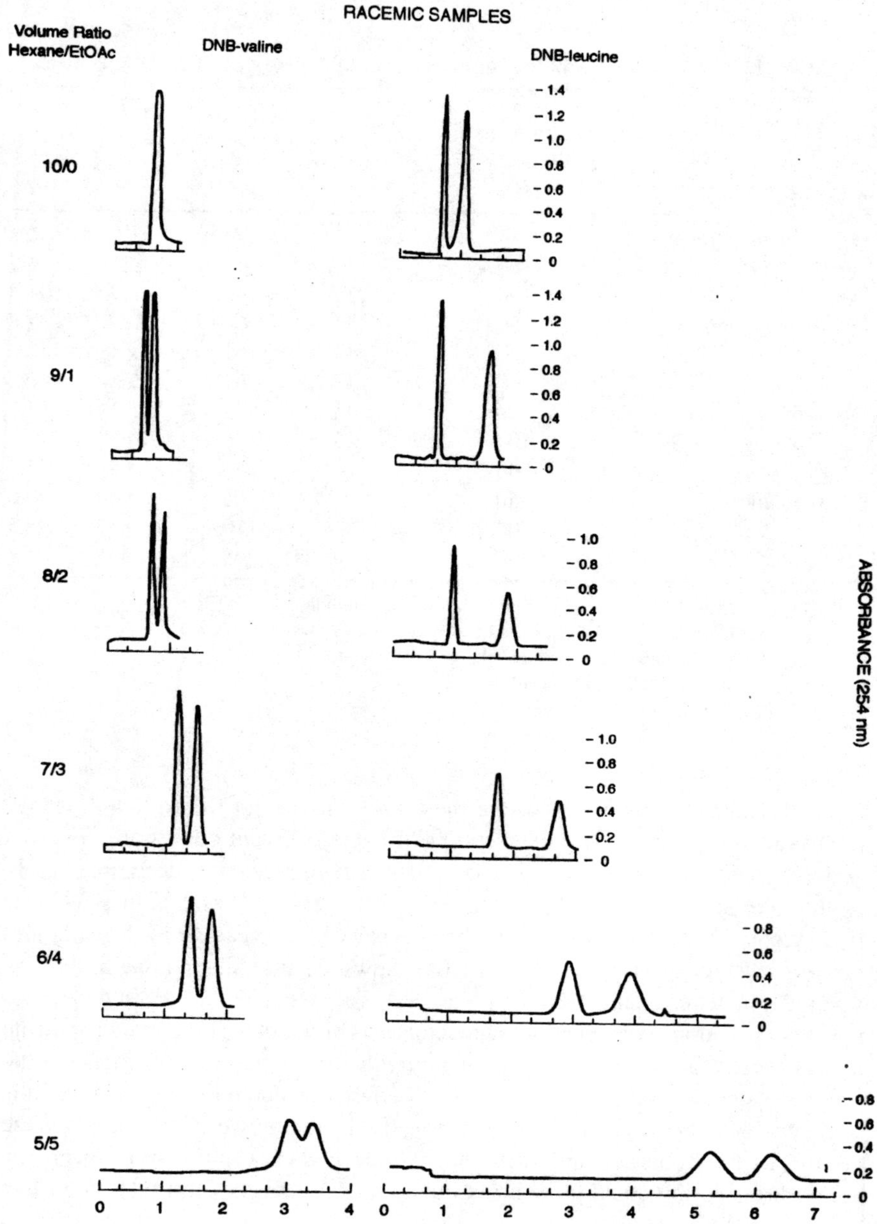

Figure 12 Effects of hydrophobicity of the solvent system on separation of DNB–amino acid racemates. Sample, (±)-DNB-valine and (±)-DNB-leucine each 5–10 mg; solvent system, hexane/ethyl acetate/methanol/10 mM HCl at volume ratios of 10:0:5:5, 9:1:5:5, 8:2:5:5, 7:3:5:5, 6:4:5:5, and 5:5:5:5 from the top to the bottom rows. In each solvent system 1 g of CS was introduced in 200-mL organic stationary phase. Other experimental conditions are identical to those described in Figure 7.

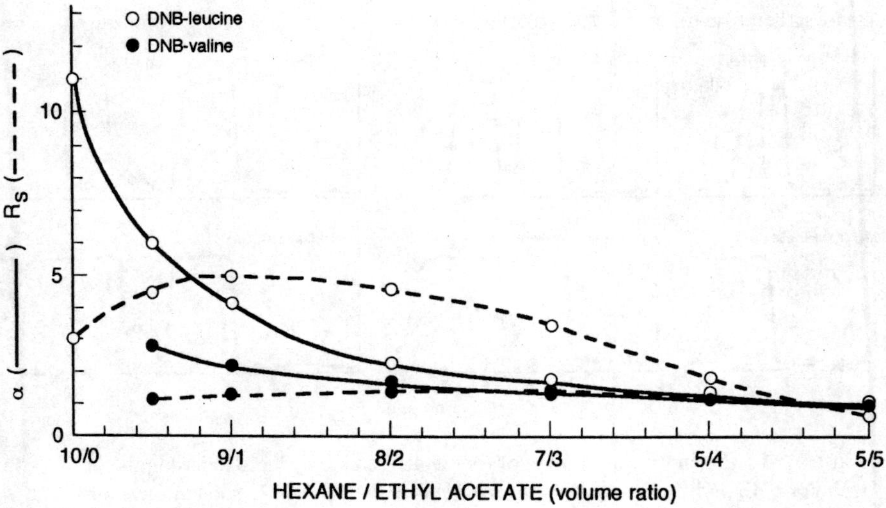

Figure 13 Effects of hydrophobicity of the solvent system on the α and R_s values of the racemates of DNB-valine and DNB-leucine. Experimental conditions are described in Figure 10.

mM HCl (6:4:5:5). The results shown in Figure 14 indicate that the sample loading capacity is mainly determined by the concentration (or the net amount) of chiral selector in the stationary phase [11], i.e., the higher the CS concentration (or amount) in the stationary phase, the greater the peak resolution and the sample loading capacity. The maximum sample size of 1 g was completely resolved in 9 hr with a 330-mL capacity column, the same column used for the analytical separation. The above results indicate that the standard HSCCC column can be used for both analytical and preparative separations simply by adjusting the amount of CS in the stationary phase. If the CS molecule is stable enough, the method has an additional advantage in that the column can be used repeatedly to separate a variety of enantiomers by dissolving appropriate chiral selectors in the stationary phase. The larger scale separation can be achieved by using a longer and/or greater inner diameter coiled column and also by applying the pH-zone–refining CCC technique described below.

B. pH-Zone–Refining CCC for Chiral Separation

pH-zone–refining CCC is a powerful preparative technique that is comparable to displacement chromatography and isotachophoresis. It yields a succession of

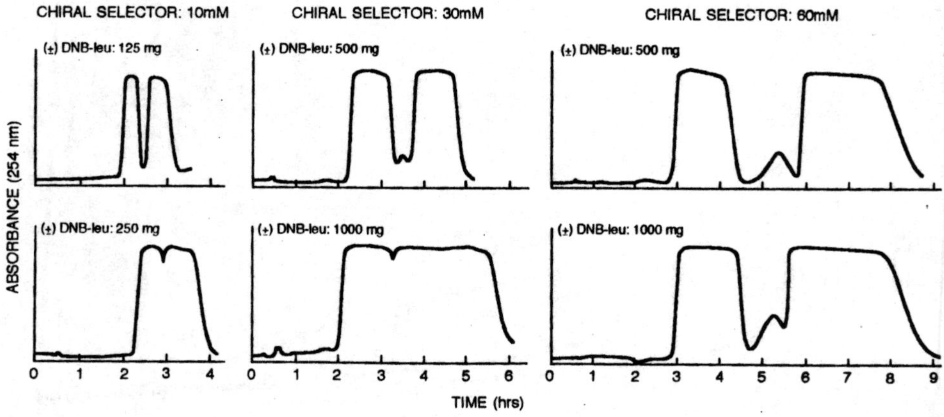

Figure 14 Preparative separation of (±)-DNB-leucine by the standard high-speed CCC technique. Experimental conditions: apparatus same as Fig. 7; solvent system, hexane/ethyl acetate/methanol/10 mM HCl (6:4:5:5) where the organic stationary phase containing CS at 10–60 mM as indicated; samples, (±)-DNB-leucine, 125–1000 mg dissolved in 10–45 mL of solvent consisting of equal volumes of each phase; flow rate and revolution were same as those described in Fig. 7; retention of the stationary phase, 65% of total column capacity.

highly concentrated rectangular solute peaks with minimum overlap where impurities are concentrated at the peak boundaries [13,19]. This technique was applied to the resolution of DNB–amino acid racemates using a binary two-phase solvent system composed of methyl *tert*-butyl ether and water where trifluoroacetic acid (retainer acid) and *N*-dodecanoyl-L-proline-3,5-dimethylanilide (chiral selector) were added to the organic stationary phase and ammonia (eluter base) to the aqueous mobile phase.

Figure 15A shows a typical chromatogram obtained by pH-zone–refining

---

Figure 15 Chiral separation of DNB–amino acid racemates by pH-zone–refining CCC. (A) Separation of (±)-DNB-leucine. (B) Separation of (±)-DNB-valine. Experimental conditions: apparatus as in Fig. 7; solvent system, methyl *tert*-butyl ether–water; stationary phase, upper organic phase containing 40 mM CS and 40 mM (A) or 80 mM (B) TFA; mobile phase, lower phase to which aqueous ammonia was added at 20 mM; sample, 2 g (±)-DNB-leucine (A) and (±)-DNB-valine dissolved in 50 mL solvent consisting equal volumes of each phase; flow rate, 3.3 mL/min; revolution, 800 rpm; analysis, chirality by analytical HSCCC and pH by a portable pH meter. Experimental conditions for the analysis as described in Fig. 7; stationary phase retention, 65% (A) and 72% (B) of total column capacity.

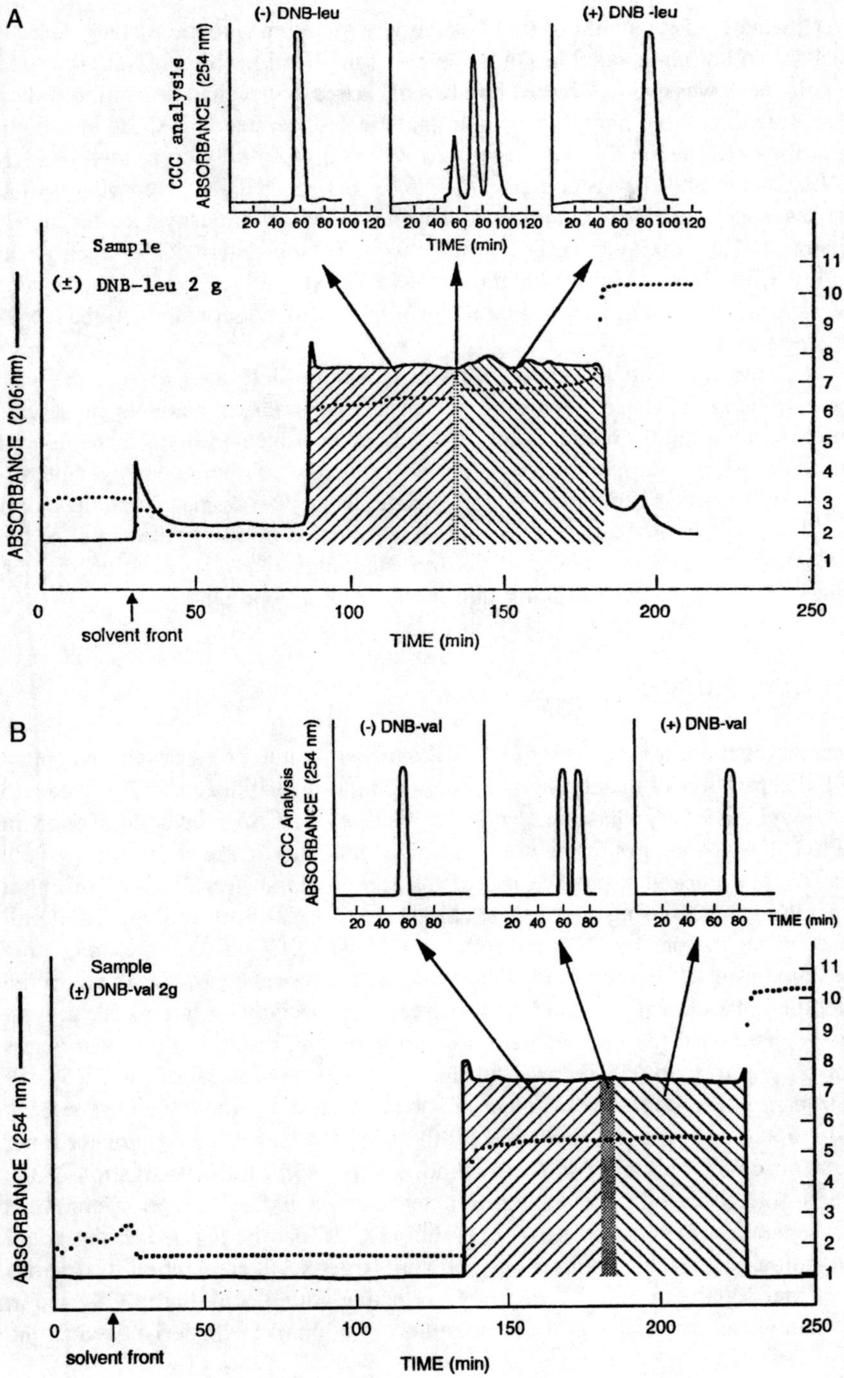
A
(-) DNB-leu
(+) DNB -leu
CCC analysis
ABSORBANCE (254 nm)
20 40 60 80 100 120
TIME (min)
Sample
(±) DNB-leu 2 g
ABSORBANCE (206 nm)
pH
1 2 3 4 5 6 7 8 9 10 11
0 50 100 150 200 250
solvent front
TIME (min)
B
(-) DNB-val
(+) DNB-val
CCC Analysis
ABSORBANCE (254 nm)
20 40 60 80
TIME (min)
Sample
(±) DNB-val 2g
ABSORBANCE (254 nm)
pH
1 2 3 4 5 6 7 8 9 10 11
0 50 100 150 200 250
solvent front
TIME (min)

CCC, where a 2-g amount of (±)-leucine was eluted in a single rectangular UV peak (solid line) in about 3 hr [11]. The pH of the fractions (dotted line) revealed that the peak was evenly divided into two pH zones with a sharp transition. When peak fractions were analyzed by the analytical scale standard CCC technique described earlier, the first zone (pH 6.5) was almost entirely composed of (−)-DNB-leucine and the second zone (pH 6.8) of (+)-DNB-leucine, whereas the narrow zone boundary contained both isomers and an impurity (see the upper diagram). This mixing zone is estimated to be no more than 5% of each peak. Figure 15B shows a similar separation of (±)-DNB-valine [19] obtained under the experimental conditions similar to those applied to the separation of the DNB-leucine racemate.

Compared with the standard CCC technique described earlier, the pH-zone–refining CCC technique allows separation of larger amounts in shorter times. In addition, the method uses relatively polar solvent systems that can hold the chiral selector for a much longer period within the column, reducing contamination in the purified fractions. In both techniques, however, leakage of the chiral selector into the eluate can be completely eliminated by filling the outlet of the column with a proper amount of CS-free stationary phase so as to absorb the chiral selector molecules leaking into the flowing mobile phase.

V. CONCLUSION

Despite the modest efficiency of CCC, the technique can be successfully applied to the separation of enantiomers by the addition of a suitable chiral selector to the liquid stationary phase in analogy to binding the CS to the solid support in HPLC. The advantages of the use of a liquid stationary phase are as follows: (1) The method permits repetitive use of the same column for a variety of chiral separations by choosing appropriate chiral selectors. (2) Both analytical and preparative separations can be performed in a standard CCC column by adjusting the amount of CS in the liquid stationary phase. Since the CS can be recovered after the separation, the method is cost-effective, especially for large-scale preparative separations. (3) The separation factor and peak resolution of enantiomers can be improved by an increase in the amount or concentration of CS in the stationary phase, and/or adjustment of solvent system hydrophobicity. The peak resolution can be further increased by the use of a longer and/or greater inner diameter coiled column. (4) The method is very useful for investigation of the enantioselectivity of a CS including determination of the formation constants and separation factors. (5) pH-zone–refining CCC can be applied to the chiral separation of ionizable compounds, allowing larger scale preparative separations in shorter separation times. The highly selective amino acid–based CS used in this study was very efficient for the enantioresolution of DNB-derivatized amino acid. It would not be so efficiently applied to other chiral compounds. CS requirements and synthesis are the main problem of chiral separations. The universal CS molecule may not exist.

GLOSSARY

Chiral:	(1) non-superimposable on its flat-mirror image. *Chiral axis*: an axis holding ligands in such a way that the resulting ensemble is not superimposable on its mirror image. *Chiral center*: the asymmetrical carbon atom as well as N^+ and P atoms with four different substituents. (2) related to chiral molecules, chiral separations, chiral chromatography, chiral selector, chiral phase.
Chirality:	The spatial property of a molecule of being non-superimposable on its flat-mirror image.
D-, L-:	Designation of the enantiomers of natural products, amino acids, sugars, hydroxy acids, related to glyceraldehyde.
Diastereoisomers:	Isomers differing in physical and chemical properties.
Enantiomers:	A pair of non-superimposable mirror image molecules.
Enantioselectivity:	Preferential involvement in the chromatographic process of one enantiomer over the other resulting in separation of the enantiomeric pair.
R-, *S*-:	Approved and unambiguous designation of the absolute configuration of a chiral center, also called the *Cahn-Ingold-Prelog* system.
Racemic mixture:	A mixture containing exactly the same number of the two molecules of an enantiomeric pair. It has no optical activity.
Racemate:	An equimolecular mixture of a pair of enantiomers.
Stereoisomers:	Molecules having the same constitution but differing in the arrangement of the atoms or groups of atoms in space, e.g., enantiomers, diastereoisomers, cis-trans isomers.
+, −:	Designation of the chiral molecules referring to their optical activity. Not unambiguously related to either the D-, L- or the *R*-, *S*-designations.

REFERENCES

1. M. Zief and L. J. Crane, *Chromatographic Chiral Separation*, Chromatographic Science Series, Marcel Dekker, New York, Vol. 40, 1988.

2. S. Allenmark, *Chromatographic Enantioseparation*, 2nd ed., Ellis Horward, New York, 1991.
3. W. H. Pirkle and J. M. Finn. *J. Am. Chem. Soc. 103*:3964 (1981).
4. Y. Ito, in *Countercurrent Chromatography: Theory and Practice* (N. B. Mandava and Y. Ito, eds.), Chromatographic Science Series, Vol. 44, Marcel Dekker, New York, 1988, pp. 79–442.
5. W. D. Conway, *Countercurrent Chromatography: Principle, Apparatus and Applications*, VCH, New York, 1990.
6. Y. Ito, in *Chromatography*, Journal of Chromatography Library, Part A, 5th ed. (E. Heftmann, ed.), Elsevier, Amsterdam, 1992, pp. A69–A107.
7. T. Tanimura, J. J. Pisano, Y. Ito, and R. L. Bowman, *Science 169*:54 (1970).
8. Y. Ito and R. L. Bowman, *J. Chromatogr. Sci. 8*:315 (1970).
9. A. P. Foucault (ed.), *Centrifugal Partition Chromatography*, Chromatographic Science Series, Vol. 68, Marcel Dekker, New York, 1995.
10. Y. Ito, *CRC Crit. Rev. Anal. Chem. 17*:65 (1986).
11. Y. Ma, Y. Ito, and A. Foucault, *J. Chromatogr. A 704*:75(1995).
12. Y. Ma and Y. Ito, *Anal. Chem. 67*:3069 (1995).
13. Y. Ito, K. Shinomiya, H. M. Fales, A. Weisz, and A. L. Scher, in *Modern Countercurrent Chromatography* (W. D. Conway and R. J. Petroski, eds.), ACS Symposium Series, Vol. 593, Washington, DC, 1995, pp. 154–183.
14. T. Takeuchi, R. Horikawa, and T. Tanimura, *J. Chromatogr. 284*:285 (1984).
15. S. Oya and J. K. Snyder, *J. Chromatogr, 370*:333 (1986).
16. B. Domon, K. Hostettmann, K. Kovacevic, and V. Prelog, *J. Chromatogr. 250*:149 (1982).
17. D. W. Armstrong, R. Menges, and I. W. Wainer, *J. Liq. Chromatogr. 13*:3571 (1990).
18. L. Oliveros, P. Franco Puertolas, C. Minguillon, E. Camacho-Frias, A. Foucault, and F. J. Le Goffic, *J. Liq. Chromatogr. 17*:2301 (1994).
19. Y. Ito and Y. Ma, *J. Chromatogr. A 753*:1 (1996).
20. Y. Ma and Y. Ito, unpublished data.

8

Preparative Scale Separations of Natural Products by Countercurrent Chromatography

Geewananda Gunawardana
Abbott Laboratories, Abbott Park, Illinois

James McAlpine
Phytera Inc., Worcester, Massachusetts

I. INTRODUCTION

The origins of countercurrent chromatography (CCC) go back to 1950 when the Craig countercurrent instrument was developed [1,2]. The early instruments were cumbersome and delicate; their operation required close attention and the experiments lasted for several days. The technique was widely used for separation and purification of natural products, particularly alkaloids, which are ideally suited for separation using chloroform, methanol, and dilute acid systems, one of the few solvent systems that could be used on the Craig instrument. Despite the shortcomings in instrumentation, CCC remained an important tool for the natural products chemist. This continued interest led to the development of droplet countercurrent (DCC) and rotational locular countercurrent (RLCC) instruments and then the high-speed countercurrent chromatography instruments (HSCCC) of today, which are much more efficient and user-friendly than their predecessors. Even though DCCC and RLCCC instruments are still available, often used for sample enrichment [3], they are being replaced by HSCCC instruments. CCC has been used extensively as an efficient chromatographic method in natural product isolations. However, utilization of the potential of CCC in other areas of chemistry has been limited. There are a few reasons for this slow acceptance of CCC:

Besides being a preparative method, CCC has been perceived as a complicated technique that requires an understanding of hydrodynamics, ability to interpret phase diagrams, and a good knowledge of partition coefficients in order to perform a simple separation. The difficulties involved in automating HSCCC instruments also contributed to their limited use. The advancements in the instrumentation and the greater understanding of the technique would make CCC a useful tool with a variety of applications.

A survey of the literature on CCC, published during 1997 and the first half of 1998, shows that about 25% of citations deal with the application of this technique for the preparative scale isolation of natural products. While this ascertains the effectiveness of the technique as a preparative method, closer examination of the authors of these publications reveals that this method is practiced by only a limited number of researchers. One of the reasons for the reluctance of the chemists, in general, to add CCC to their armamentarium of separation and purification tools is that unless one has gained first hand experience in the art of CCC, often from one of the few pioneering laboratories, it is not a trivial matter to obtain the necessary know-how from the literature to set up and operate a CCC. While there are volumes of information available on the theoretical aspects and applications of CCC, there is very little information available for a chemist who would like to utilize this technique to solve an isolation purification problem. This chapter is an attempt to fill that gap. No attempt is made to provide an exhaustive compilation of the literature on the use of CCC in isolation of natural products. Instead, sufficient entry points into the current literature are provided so that detailed information on specific areas may be easily accessed.

Natural products continue to play an important role in pharmaceutical, cosmetics, flavor, and dietary supplement industries. The isolation of compounds of interest from natural sources, such as plant extracts, microbial fermentation, or animal tissues, e.g., marine macroorganisms, presents a number of difficulties. Often the compounds of interest are present as minor components of extremely complex mixtures. Even though adsorption chromatography, high-performance liquid chromatography (HPLC), for example, has the sufficient resolution power to separate such mixtures at analytical scale, the scale-up of these methods into preparative scale may present formidable practical difficulties. The irreversible adsorption of minor components or denaturation on solid phase are other common problems encountered in solid phase chromatography. Furthermore, the costs involved in the scale-up of adsorption chromatography from analytical scale to preparative can be prohibitively expensive. As seen in a majority of recent publications on CCC, this technique has been used for rapid enrichment of compounds of interest for final purification by repeated CCC or other chromatographic methods such as HPLC. Countercurrent and adsorption chromatography are compared in the following section.

II. BASIC PRINCIPLES

Despite the implication of the name, CCC does not involve two liquid phases flowing in opposite directions. Several other names that describe this technique have been proposed [4], but CCC, the one coined by Yoichiro Ito [5], a pioneer in the development of this technique, remains the widely accepted name. In reality, CCC is a liquid chromatography technique with a liquid stationary phase and an immiscible liquid mobile phase. Just as in classical liquid chromatography (LC), the solute is partitioned between the stationary phase and the mobile phase repeatedly during the chromatography and eluted from the column in the same phase in which they were introduced.

There are similarities and differences between classical LC and CCC: both require the same chromatographic hardware such as pump, injector, detector, recorder, and column. The two methods differ fundamentally in the nature of the column. In LC, the solute separation or the exchange takes place at the surface of a solid support where, as in CCC, this occurs throughout the entire volume of the stationary phase. As a result, a CCC column would have a higher capacity (loadability) than an LC column of comparable volume making it suitable for preparative scale isolation.

Advantages of CCC columns over LC columns are as follows:

1. As there is no solid support in CCC, no irreversible adsorption takes place making the quantitative recovery of the solute possible. Furthermore, the chance for denaturation, the resulting tailing of peaks, and the loss of minor constituents during chromatography remain very low.
2. The nature of the column can be varied indefinitely by changing the solvent system and can be adapted to separate a wide variety of compound classes.
3. The column is generally used only once and as a result there is no need to be concerned about damaging the column by loading crude samples.
4. With the same solvent system, the column can be eluted in both normal and reverse modes to match the polarity of the compounds of interest. Also, the slow eluting compounds can be recovered by pumping out the stationary phase instead of eluting them with the mobile phase.
5. The common CCC columns can be adapted from analytical scale to 5- to 10-g scale without any hardware changes. Therefore, the method development and scale-up can be simplified. In preparative scale purification work for which most HSCCC instruments are used, the operation cost of CCC is much lower than that of preparative LC. While all LC columns have a limited lifetime, the HSCCC column which is usu-

ally used only once can be prepared each time with relatively inexpensive solvents with good column-to-column reproducibility.

Disadvantages of CCC columns over LC columns are as follows:

1. CCC may be referred to as a slow-resolution method compared to other LC methods. Even though the resolution power itself is comparable to that of other LC methods, the rate of generation of theoretical plates in the CCC column is limited and this results in longer separation times.
2. The process of selection of a suitable solvent system, which is comparable to the selection of a column and the development of a method in LC, remains less defined.
3. CCC instrumentation has not reached the same level of sophistication as other chromatographic instruments. In particular, the instrument design did not lend itself to automation, making CCC a labor-intensive technique. Some of these problems have been addressed in newer instruments.

III. APPARATUS

The DCCC and RLCCC instruments, in which the stationary phase is held in place by gravity alone while the mobile phase is percolated through, require lengthy elution times and the latter has limited resolution capability. Furthermore, only the solvent systems that form droplets when percolated can be used in the former. In order to overcome these shortcomings, a variety of CCC instruments have been developed in which centrifugal force is used to retain the stationary phase, allowing for faster elution and improved resolution. These instruments have been referred to as centrifugal countercurrent partition (CCPC), planetary coil countercurrent (PCC), centrifugal partition (CPC), high-performance centrifugal partition (HPCPC), horizontal flow-through coil planet centrifuge (HFTCPC), toroidal coil countercurrent (TCC), and multilayer coil separator extractor (MLCCC) chromatographs by various authors. They can be collectively referred to as high-speed countercurrent chromatographs (HSCCCs).

The HSCCC instruments can be divided into two types depending on the configuration of the column: (1) hydrodynamic and (2) hydrostatic instruments [6–8]. The hydrodynamic instruments devised by Yoichiro Ito, referred to as "Ito" or planetary coil countercurrent (PCC) instruments, consist of one or more coils of tubing that are in synchronous planetary motion in either a horizontal or a vertical plane. A single coil is balanced with a counterweight. The main advantages of these instruments are the absence of rotary seals and the simplicity of the column, which consists of a single length of tubing making them relatively maintenance-free. In these instruments, the two solvent phases experience complex centrifugal forces resulting in successive mixing and demixing zones while remaining in contact with each other throughout the entire column.

The hydrostatic instrument consists of cartridges with a series of channels connected by ducts engraved in plates of inert polymer resembling a droplet system. They are referred to as centrifugal partition chromatographs and the only commercially available instrument is produced by Sanki Engineering of Japan. In the Sanki instrument, the cartridge is spun around its axis and the two solvents experience a uniform centrifugal force. During the operation, the ducts contain only the mobile phase, and as a result, the two phases remain in contact with each other in the channels only. The hydrostatic instruments have much better solvent retention properties, and as a result a wider selection of solvents can be used.

IV. INSTRUMENT SETUP

As the HSCCC unit represents a chromatographic column, it can be incorporated into an existing LC system with a few modifications. Most HSCCC instrument manufacturers offer only the HSCCC unit, even though some may assemble a complete system using components from other vendors. To set up a basic system for preparative work, a suitable pump, an injector (a minimum of 5 mL sample loop is required for preparative work), and a fraction collector are needed. A UV detector and a recorder may be added, but preparative work can be performed with other monitoring systems such as thin-layer chromatography (TLC) or biological assays. Even though the current HSCCC instruments are not designed for interfacing with a computer for fully automated operation, a semiautomated system can be assembled with readily available components. A programmable multichannel gradient HPLC pump with a flow capacity of up to 10 mL/min and a fraction collector with a programmable power outlet are needed for such an assembly. When appropriately connected and programmed, the stationary phase fill, elution with a gradient system if necessary, fraction collection, turning off of the rotor at the end of the run, cleaning of the coil, and introduction of stationary phase for the next experiment can be performed unattended. With such a setup, only the sample injection has to be made manually. Even though it is feasible, the dedication of an autosampler cannot be justified as an average HSCCC run can take a few hours to complete.

V. SOLVENT SELECTION

The most important aspect of HSCCC is the selection of a solvent system for the separation at hand as this is similar to the combined selection of a column and the development of a method in LC. The separation of solutes in CCC depends on the differences of their partition coefficients between two immiscible solvent systems used as the stationary phase and the mobile phase. Ideally, the solutes of interest, not the total mixture, should partition in a 1:1 to 1:5 ratio in favor of the stationary phase. Generally PCC requires a smaller ratio than centrifugal partition chromatography for efficient separation.

Knowledge of existing solvent systems plays an important role in selecting a suitable solvent system: If the class of the compounds to be purified is known, based on chemotaxonomy, etc., a solvent system that has been used for the purification of that class of compounds can be selected from the literature reports as a starting point, and subsequently optimized by changing the individual components.

A procedure we have successfully employed for the selection of a solvent system for the separation of an unknown mixture follows. Three solvents have to be selected: a solvent in which the test sample is readily soluble, a solvent that is immiscible with the first one, and a solvent that is miscible with the two other solvents. The methylene chloride/methanol/water system is a common example. Then the solvents are combined according to Table 1. The solvents are mixed well and the systems that settle within 10–15 sec are identified. Obviously, some combinations will not result in a biphasic system and they should be discarded along with those that take more than 15 sec to settle. Equal volumes of the upper and lower phases from each system are then transferred to vials containing the test sample (5% w/v), mixed well, and allowed to settle. Then the two phases are examined by a suitable TLC system and the partition is estimated based on intensities of the TLC spots. If the compounds have any biological

Table 1 Preparation of a Solvent System (mL)

System no.	Solvent A	Solvent B	Solvent C
1	5	0	5
2	5	1	4
3	5	2	3
4	5	3	2
5	5	4	1
6	4	5	0
7	3	5	1
8	2	5	2
9	1	5	3
10	0	5	4
11	4	1	5
12	3	2	5
13	2	3	5
14	1	4	5
15	0	5	5

activity, equal amounts of the two phases are dried down and tested with the bioassay. A commercially available liquid handler can be programmed to perform all liquid mixing and transfer steps. Once the combinations that do not result in biphasic systems are identified for a given solvent system, they can be eliminated from the grid for routine use. The systems that give a smaller amount of one phase need not be discarded as an individual phase can be prepared by consulting a phase diagram. A ternary system can be further optimized by the addition of modifiers, i.e., solvents that can further alter the polarities of the two phases without affecting the miscibility of the two phases. A wealth of information is available on preparing biphasic solvent systems [8]. A number of other approaches for the selection of solvent systems such as the Oka approach, the expanded Margraff approach, the HBMW approach, and the multisolvent approach have been discussed in detail by Foucault [9].

The partition ratio can be monitored during optimization by several methods. In a bioassay-guided isolation, the corresponding bioassay can be used. However, the presence of several bioactive congeners could complicate the situation. A simple TLC assay works well for most compounds. HPLC or more sophisticated assay methods can be used, but generally TLC itself is adequate for general preparative work.

A list of solvent systems commonly used in CCC along with the preferred mode is given in Table 2. The solvent systems are divided into three groups

Table 2 Common Solvent Systems Used in CCC

Solvents	Run mode
Lipophilic	
n-Heptane/dichloromethane/water	N
n-Heptane/ethyl acetate/water	N
n-Heptane/ethyl acetate/methanol/water	N
n-Heptane/dichloromethane/acetonitrile	N
n-Hexane/methanol/water	N
n-Hexane/pentanol/water	N
Toluene/dichloromethane/water	N
Hexane/toluene/dichloromethane/methanol/water	N
Intermediate	
Dichloromethane/methanol/water	N
Ethyl acetate/butanol/acetonitrile/water	N
Ethyl acetate/*n*-propanol/water	N
Hydrophilic	
n-Butanol/water	N/R
n-Butanol/methanol/water	R
n-Butanol/acetic acid/water	R
n-Butanol/*n*-propanol/water	R
n-Pentanol/methanol/water	R

according to the polarity of the organic phase or the compounds that can be separated using them [10]. Certain solvents that are no longer used under normal laboratory settings for health and environmental reasons have been used extensively in CCC in the past. Of these solvents, benzene can be replaced with toluene while carbon tetrachloride and chloroform can be replaced with dichloromethane with minor adjustments provided proper precautions are taken in handling and disposing such solvents.

If ionizable compounds are separated, the pH of the solvent system should be controlled for maintaining good peak shape. Acetic acid, trifluoroacetic acid, triethylamine, and phosphate buffers are commonly used for this purpose.

VI. SAMPLE PREPARATION

The enrichment of the compounds of interest by fractionation prior to chromatography improves the loading capacity but is not mandatory for efficient separation of compounds. Even though CCC is ideally suited for separating mixtures containing compounds of vastly different polarity, the optimization of a solvent system for the compounds of interest in such mixtures becomes difficult. For example, the removal of large quantities of oils in fermentation extracts by partition with hexane or the removal of water-soluble tannins, etc., in plant extracts, allows for selection of optimal solvent systems for the purification of compounds of medium polarity. The sample may be dissolved in the mobile phase or in a mixture of the two phases, and should be filtered to remove any particulate matter. Ideally, the total volume of the sample should be less than 2–3% of the total column volume in order to obtain reasonable resolution, but larger sample volumes can be injected depending on the complexity of the mixture and the degree of purification expected. At the other extreme, if CCC is used in the extraction mode, i.e., for sample concentration or the like, several column volumes of sample may be "injected" into the column [11].

An often raised question is the loading capacity of CCC. Unfortunately, there is no satisfactory way to predict the capacity of a given system as it depends on a number of factors. Some of the more important factors include the volume of the stationary phase, the complexity of the mixture, and the differences in the partition properties of the individual components. Generally, most CCC instruments are capable of handling samples that differ in size by an order of magnitude. For example, we have purified fractions weighing several milligrams to 5 g on a Pharma-Tech HSCCC instrument with a column capacity of 0.9 L in isolating 9-dihydro-13-acetylbaccatin III (**1**) [12]. A safe practice in preparative scale work is to start with a smaller sample load and increase it in following runs depending on the results.

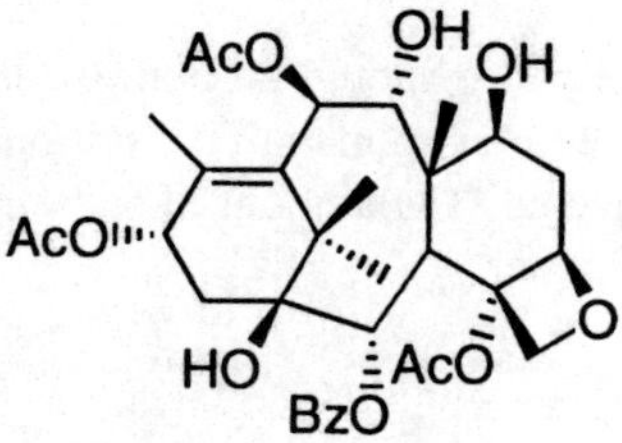

1

VII. OPERATION

A. Selection of the Stationary Phase and the Run Mode

Once the solvent system and the sample size are decided, the CCC run mode has to be selected. For Sanki centrifugal partition chromatography instruments, which are spun in one direction only, the selection of the run mode is quite straightforward: the ascending mode must be selected when the upper phase is used as the mobile phase and the descending mode when the lower phase is used as the mobile phase. On the other hand, for PCC instruments, the inlet and the direction of rotation have to be selected from eight possible combinations. Apparently, only two of these combinations would work satisfactorily. The manufacturer would identify the two ends of the coil as ''head'' and ''tail'' depending on the handedness of the coil and the direction of column rotation. It is less confusing if the instrument is spun in the forward mode only as done with Pharama-tech instruments. If the coil is spun in the forward mode and the upper phase is used as the mobile phase, it should be pumped in tail-to-head mode and this is referred to as the normal (N) mode. On the other hand, if the lower phase is used as the mobile phase it should be pumped in head-to-tail mode, reverse (R) mode, for optimum stationary phase retention. For solvent systems containing lipophilic and intermediate polarity organic phases, the normal mode (N) is more common, whereas for systems containing polar organic phases the reverse mode (R) is usual (see Table 2).

B. Introducing the Solvents

The stationary phase should be pumped into the coil in ascending mode for the Sanki instrument and in normal mode for PCC instruments to eliminate the formation of vapor locks. If a new instrument is used for the first time, it is important to note the coil volume. Once the coil is filled, it should be spun at about 500–600 rpm. A steady unchanged flow indicates a filled coil devoid of gaps. If a solvent system is used for the first time, the mobile phase should be pumped in the correct mode at the desired flow rate while the column is spun until the stationary phase loss is stabilized, before the sample is injected. The amount of stationary phase displaced should be noted for future reference.

C. Counterbalance

As the stationary phase is replaced, the counterweight has to be adjusted in instruments with a single coil. The guesswork involved in this process can be avoided as follows: Obtain the weight of the empty coil spool and the counterweight without the removable weights, and record this information along with the coil volume for this instrument. Calculate the ratio of the stationary to mobile phase in the equilibrated column from the column volume and the displaced volume during the run. Weigh about 20 mL of the two phases mixed in this ratio. If the solvent has not been used before, assume this as 4:1. Then, calculate the weight of the two solvents in the coil and add this to the weight of the spool, and this should equal the weight of the counterweight.

D. Operating Parameters

The three user-adjustable parameters are temperature, flow rate, and rotation speed, which in turn determine two passive parameters, stationary phase retention and the pressure. Theoretically, the increase in temperature should decrease the viscosity of solvents and increase both stationary phase retention and separation. Increased temperature also should allow the use of solvent systems that cannot be used at room temperature due to slow settling rates. However, the effect of temperature on preparative scale separations can be expected to be minimal, and no actual data could be found for natural products separations. Furthermore, the use of volatile solvents limits the range of operating temperatures for CCC.

The increase in flow rate would result in increase in peak broadening, and a 1- to 2-mL/min flow rate is considered satisfactory for most separations. As the number of theoretical plates and the resolution are shown to increase with rotational speed, it is preferable to maintain a high speed of rotation as long as the pressure is maintained within the limits. The Sanki rotary seal and the tubing used in PCC instruments have a pressure rating of about 60 bar.

E. Injection of Sample

If a solvent system is used for the first time, the sample should not be injected until the column has reached equilibrium, i.e., a steady flow of mobile phase is achieved with minimal leaching of stationary phase. If the behavior of a solvent system is well understood, the sample can be injected after the mobile phase is pumped for a few minutes and the fraction collection started. The injection of the sample will cause some loss of the stationary phase and an increase in pressure. The rotational speed or the flow rate should be adjusted at this time.

F. Detection

The stationary phase of most solvent systems tends to leach out during the run, limiting the use of a UV detector. With the Sanki instruments, which tend to have better solvent retention, a UV detector with a preparative cell can be used satisfactorily. The use of the evaporative light scattering detector (ELSD) has been demonstrated successfully by Thiébaut [13], and Schaufelberger [14]. As ELSD is a destructive detection method, a stream splitter must be installed for preparative separations. The feasibility of coupling CCC with both mass [15] and nuclear magnetic resonance (NMR) [16] spectrometers have been demonstrated as well.

G. Elution and End of Experiment

The flow of two to three column volumes of mobile phase is adequate to elute most of the compounds. Further elution becomes inefficient as this would result in severe peak broadening and lengthy run times. The compounds retained in the stationary phase after that may be recovered by stopping the rotation and pumping methanol, or water if a buffered system has been used, in the opposite direction. Fraction collection can be continued until more than one column volume of solvent has been pumped. It is possible to make multiple injections into the same column depending on the nature of the sample. If a complex extract is being fractionated whereby a significant amount of the mixture would be retained in the stationary phase, multiple injections may not reproduce the same resolution. However, if all components are eluted after separation, as in the case of chiral separations, multiple injections can be performed without loss of resolution.

Another variation of elution, referred to as the dual-mode CCC, has been demonstrated by Thiébaut to give improved separation of compounds retained in the stationary phase compared to normal mode [17]. In this technique, instead of backflush, the phase roles are reversed during the run. The theoretical basis for the improved separation has been discussed and the efficiency of the method is demonstrated by separating mixtures of polyoxypropylene glycol, fatty acid, and antibiotic combinations. Another technique referred to as dual CCC, which is a modification of the instrumentation, is discussed later.

The CCC columns should be stored with an inert solvent such as methanol and the practice of pushing the stationary phase out at the end of a run with air or nitrogen should be discouraged. The use of a programmable HPLC pump allows for the collection of fractions of the stationary phase under controlled conditions while cleaning the column with a solvent suitable for storage. The introduction of the next stationary phase into a column filled with a solvent is less problem-prone than filling an empty column.

VIII. TROUBLESHOOTING

Besides problems common to all LC methods, the other most frequently encountered problem in CCC is the loss of the stationary phase, which can occur under various circumstances.

1. *Siphoning of the solvents*. If the coil configuration was changed, check to make sure that a forbidden mode has not been selected (see Section VII). If the configuration is correct, most probably a vapor block or an air gap has formed in the column. If the sample has not been injected, continue to pump the stationary phase until the air gap is eliminated. Mixing of certain solvents can be exothermic, e.g., methanol–water, and such solvents should be thoroughly mixed and allowed to equilibrate before they are introduced into the column. If inorganic salts are used as buffers, the solvent should be carefully checked for possible precipitation of salts after mixing.
2. *Stationary phase does not retain before sample injection*. If the density difference of the two phases is low, the continuous loss of a small amount of stationary phase during the run should not be a problem. However, if the loss is significant, the spin rate should be increased and the flow rate reduced until the stationary phase is retained sufficiently. Change of mode should also be considered. Make sure the column is spinning; in quieter Sanki instruments this may not be so obvious.
3. *Stationary phase does not retain after sample injection*. This can be due to particulate matter in the sample, overloading, highly viscous samples, or poor solubility of the sample in one of the phases. The spin rate should be increased for a few minutes after reducing or completely stopping the flow. If the problem persists on reverting to normal conditions, the experiment has to be aborted and the sample recovered (see Section VI).
4. *Increased pressure*. The pressure of the system depends on the solvent flow and spin rates. A significant pressure buildup can result with slow settling and viscous solvent systems and also if the sample forms an emulsion. A suitable combination of flow and spin rates should be selected in order to maintain the pressure. If the high pressure persists, the connecting tubing should be examined for twists, particularly with multicoil instruments.

IX. SELECTED EXAMPLES

Even though CCC is not a widely used chromatographic technique, a vast amount of information is available on various aspects of this subject. In this chapter, no attempt is made to review the complete literature on applications of CCC with natural products. Instead, a few selected examples from recent literature are discussed to illustrate the capabilities of this technique and to serve as entry points for further exploration of various aspects of CCC.

Application of CCC in isolation of antibiotics, marine natural products, and natural products in general have been reviewed [18]. The same text also has a chapter on HSCCC separation of Chinese medicinal herbs. More recent reviews on the use of CCC in isolation of antibiotics [19], lipids [20], proteins [21,22], and peptides [23] as well as a review of CCC in natural products in general [24] have appeared. A review on the CCC instrumentation, solvent selection, and some recent applications [25] and an account of the industrial applications of CCC [11] have been published.

A. Terpenes

The significance of taxane diterpenes as cancer chemotherapeutic agents has prompted a number of investigators to attempt their purification by CCC from extracts of various yew species. Two very closely related taxanes, taxol (**2**) and cephalomannine (**3**), have been separated using hexane/ethyl acetate/methanol/ethanol/water (10:14:10:2:13) system. The amount of ethanol present in the system was shown to be crucial for the separation of the two compounds, and taxol was isolated in 90% yield in better than 98% purity [26]. The recycling of CCC eluents has been shown to give an improved resolution coefficient of the same two compounds from 0.7 to 1.27 [27]. In another investigation, a solvent system consisting of *n*-hexane/ethyl acetate/ethanol/water has been used to purify taxol and related analogs in two steps [28]. In the first step, using the ratio 1:1:1:1, the compounds were separated into two groups. Then the individual compounds were purified using the solvents in the ratios 3:3:2:3 and 4:4:3:4.

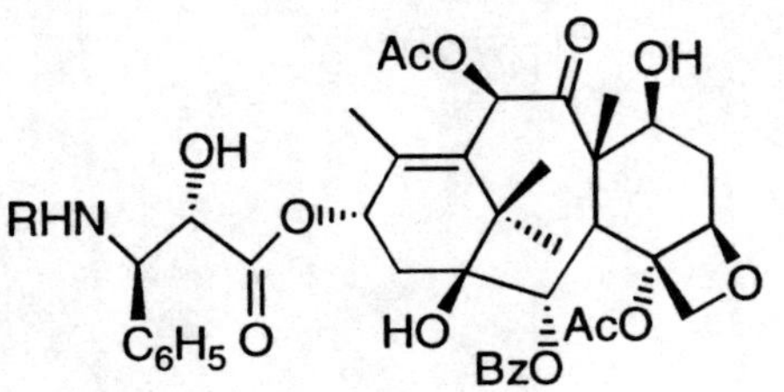

2 R = COC_6H_5
3 R = $COCH(CH_3){:}CH(CH_3)$

The insect antifeedent azadirachtin A (**4**) has been isolated in a two-step process from neem seed extracts using hexane/*t*-butyl methyl ether/methanol/water in 1:3:1:2 and 4:5:4:5 ratios [29].

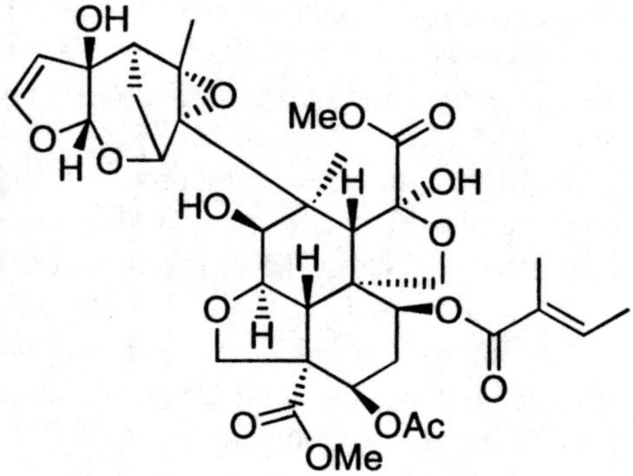

4

B. Aromatic Compounds

The preparative separation and purification of quercetin (**5**), kaemferol (**6**), and isorhamnetin (**7**) from the extracts of *Ginkgo biloba* leaves using chloroform/methanol/water (4:3:2) has been described [30]. A fraction containing the phenolic constituents of the needles of Norway spruce (*Picea abies*) has been separated using a chloroform/methanol/isopropanol/water (5:6:1:4) system. Three acylated compounds, isorhamnetin 3-*O*-(6″-Ac)glucoside (**8**), syringetin 3-*O*-(6″-Ac)glucoside (**9**), and kaempferol 3-*O*-(6″-Ac)glucoside (**10**), were isolated in the pure form, whereas several other compounds produced distinct but overlapping peaks that were subsequently purified by HPLC [31].

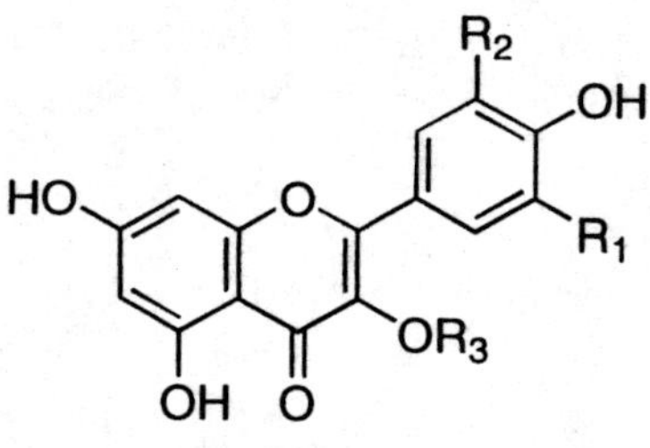

5 $R_1 = R_3 = H$, $R_2 = OH$
6 $R_1 = R_2 = H$, $R_3 = OH$
7 $R_1 = R_3 = H$, $R_2 = OMe$
8 $R_1 = H$, $R_2 = OMe$, R_3 = O-6'-acetylglucose
9 $R_1 = R_2 = OMe$, R_3 = O-6'-acetylglucose
10 $R_1 = R_2 = H$, R_3 = O-6'-acetylglucose

The catechin (**11**)–containing fractions of the hot water extract of green tea leaves were separated by CCC using water/chloroform (1 : 1 v/v) and water/ethyl acetate (1 : 1 v/v) solvent systems. The resultant fractions were further purified by column chromatography and HPLC to yield various catechin derivatives [32].

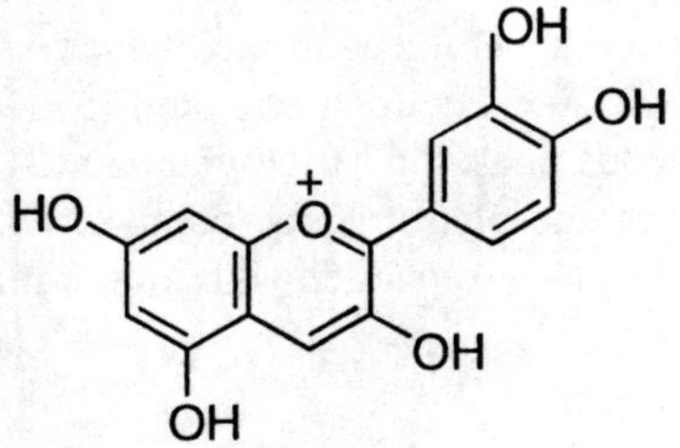

11

Four anthocyanins, cyanidin-3-glucoside (**12**), cyanidin-3-rutinoside (**13**), delphinidin-3-glucoside (**14**), and delphinidin-3-rutinoside (**15**), from grapes (*Vitis vinifera*) and black currant (*Ribes nigrum*) have been isolated by gradient elution centrifugal partition chromatography using ethyl acetate/butanol/water systems on Sanki LLB-M and LLI-7 instruments. In one example, the stationary phase composition was ethyl acetate/butanol/water (5:5:90), while the mobile phase composition was changed linearly from ethyl acetate/butanol/water (77:15:8) to ethyl acetate/butanol/water (40:46:14). Each phase was acidified with 0.2% trifluoroacetic acid [33]. The use of multiple CCC instruments connected in series to separate a number of catechin glycosides has been reported [34]. Two kaempferol sophorosides have been isolated from the methanolic extract of saffron and subsequently purified as their acetates by HPLC [35].

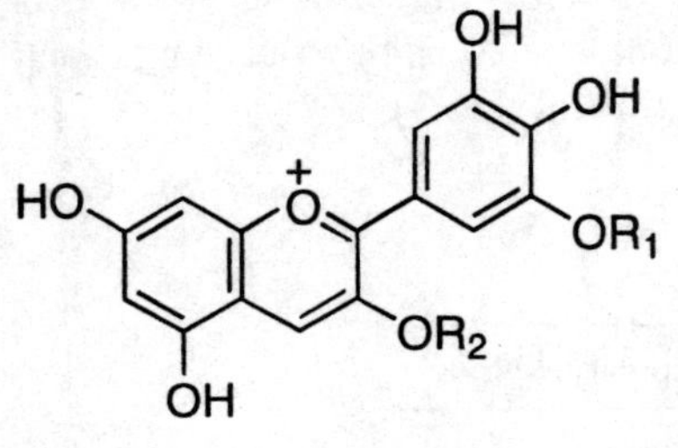

12 R_1 = H, R_2 = glucose
13 R_1 = H, R_2 = rutinose
14 R_1 = OH, R_2 = glucose
15 R_1 = OH, R_2 = rutinose

C. Alkaloids

The pH zone–refining method is a new technique used for the separation of compounds with ionizable functions. This recently developed method has wide applications in natural products chemistry, particularly in separation of alkaloids as shown by the following examples: A 3-g sample of a crude alkaloid extract of *Crinum moorei* was separated on a multilayer coil separator with a column volume of 300 mL using methyl *t*-butyl ether and water. Hydrochloric acid (5–10 mM) was used as the retainer acid in the aqueous phase and triethylamine (5–10 mM) was used as the eluent base in the organic phase. Three compounds, crinine (**16**), powelline (**17**), and crinamidine (**18**), were separated with minimum overlap [36].

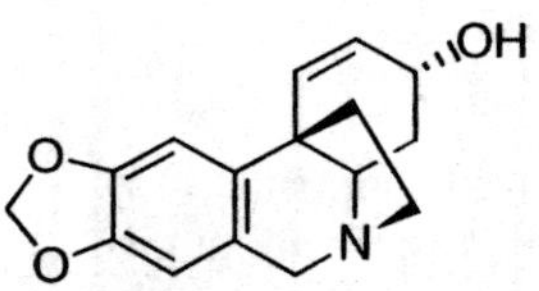

16

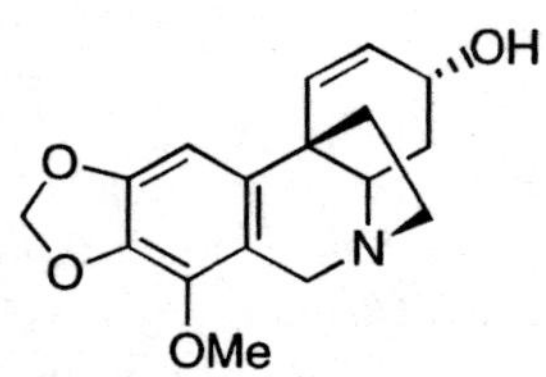

17

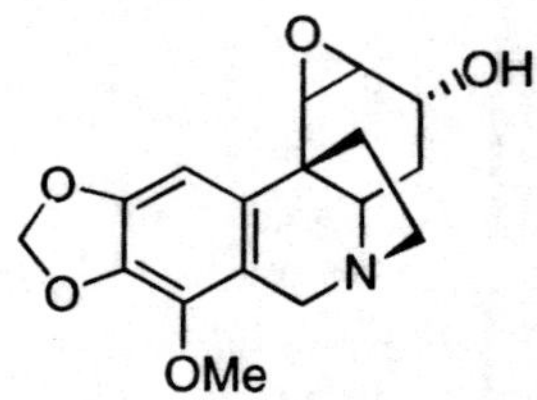

18

On a similar approach, the alkaloidal fraction of *Senecio fuberi*, which contains the three alkaloids squalidine (**19**), platyphylline (**20a**), and neoplatyphylline (**20b**), has been completely separated by HSCCC using chloroform/0.07 M sodium phosphate/0.04 M citrate buffer (pH 6.2–6.45) (1:1) [37]. The lower layer was used as the mobile phase in head to tail mode at 750 rpm. It is noteworthy that the last two compounds separated are cis-trans isomers that have not been separated by conventional chromatographic methods. Furthermore, it has been shown that the retention times for these two compounds can be reduced by changing the pH of the mobile phase from 6.2 to 6.45 (Table 3).

Table 3 Effect of pH on Separation

	Platyphylline		Neoplatyphylline	
Buffer pH	R_t(min)	R_v(mL)	R_t(min)	R_v(mL)
6.21	92	184	70	104
6.38	70	140	56	112
6.45	52	104	42	84

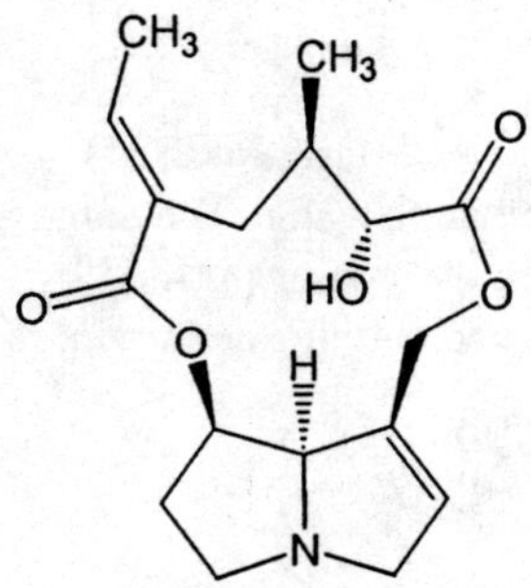

19

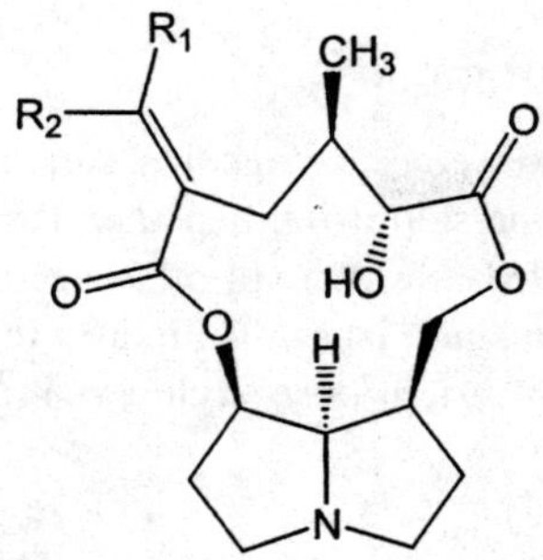

20a $R_1 = H$, $R_2 = CH_3$
20b $R_1 = CH_3$, $R_2 = H$

The alkaloids of *Sophora flavescens*, sophocarpine (**21**) derivatives, have been separated using chloroform/methanol/phosphate buffer (0.23 mM, pH 5.4, 27:20:13) [38].

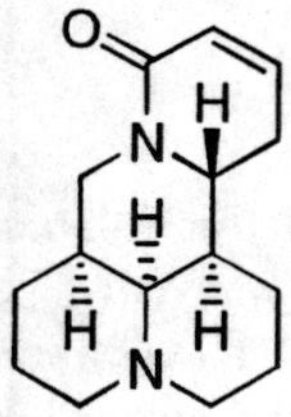

21

A series of isoquinoline alkaloids including the novel compound ancistrobertsonine (**22**) has been isolated from the extracts of *Ancistrocladus robertsoniorum* [39].

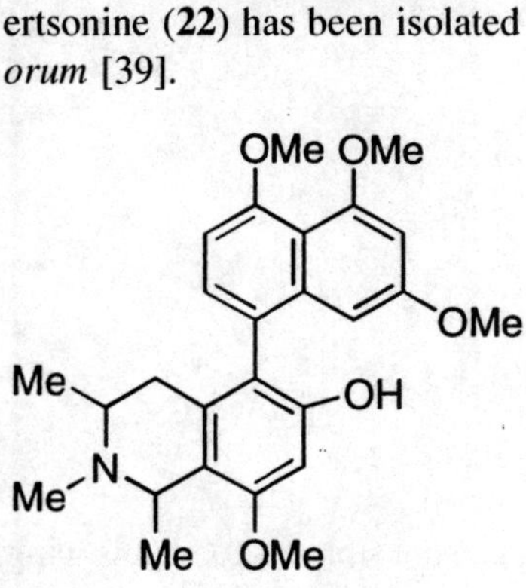

22

D. Acetogenins

The annonaceous acetogenins, squamocin (**23**) and rolliniastatin-2 (**24**), were isolated in one step using heptane/ethyl acetate/methanol/water. The same system also yielded four pairs of other acetogenins which were later purified by HPLC: squamocin/molvizarin, rolliniastatin-2/asimicin, neoannonin/atemoyin, and iso-desacetyluvaricin/desacetyluvaricin [40].

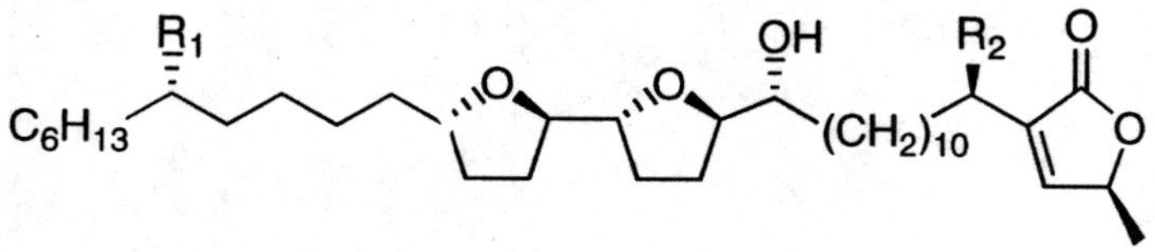

23 R_1 = OH, R_2 = H
24 R_1 = H, R_2 = OH

E. Microbial Metabolites

Cephalosporin C (**25**) and desacetyl cephalosporin C (**26**) from fermentation broths were separated using an aqueous solvent system consisting of polyethylene glycol (PGE-600, 15% w/w) and ammonium sulfate (17.5% w/w) on a Pharma-tech model CCC-800 instrument rotating at 600 rpm. The solvent composition is not given, but the use of the upper layer of the mixture as the stationary phase has given better stationary phase retention, higher resolution, and higher number of theoretical plates compared to when the lower layer was used as the stationary phase (stationary phase retention 33.5% vs. 26.5%, resolution 0.60 vs. 0.43, and number of theoretical plates 270 vs. 160) [23].

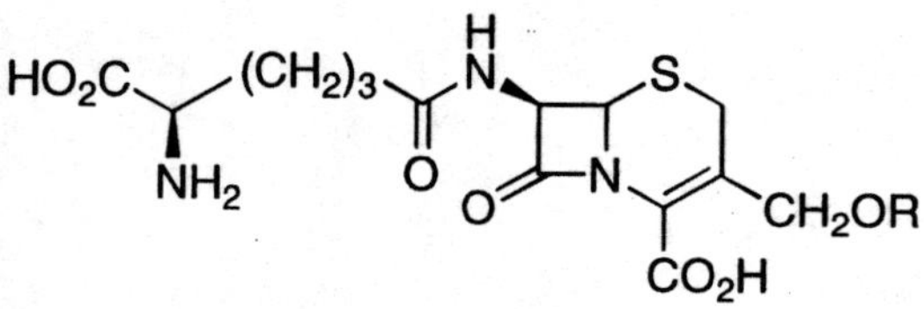

25 R = $COCH_3$
26 R = H

The *S. aureus* toxin, enterotoxin A, which is responsible for food poisoning, has been detected by CCC in contaminated mushrooms [41]. The isomers of 3-oxo-Δ5-steroid have been purified from crude *E. coli* lysate using a system consisting of PEG-3350 and potassium phosphate buffer (pH 7), each at 12.5%

(w/w) in water [42]. Three novel cytochalasins, phomacin A (**27**), B (**28**), and C (**29**), have been purified by a combination of CCC and HPLC [43].

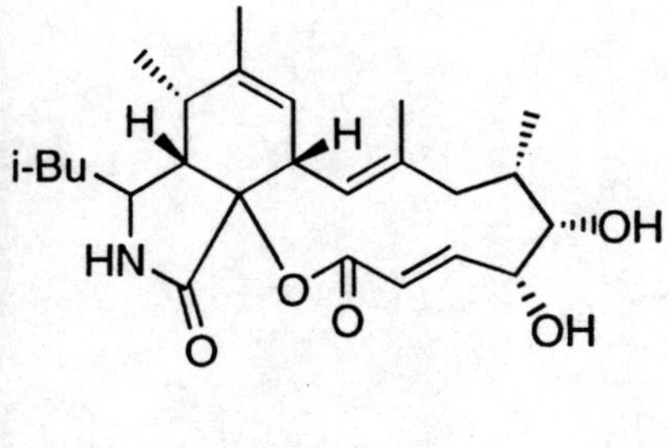

27

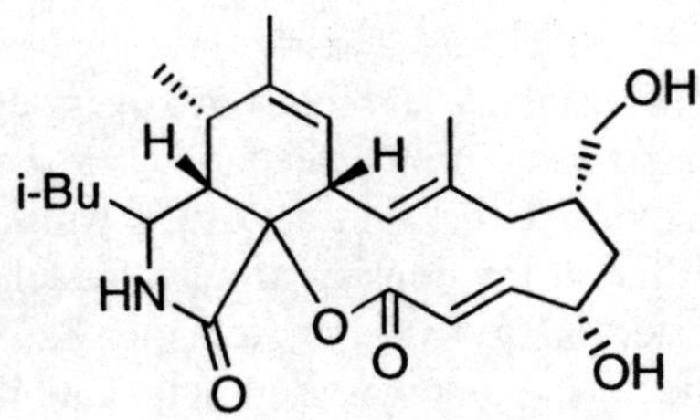

28

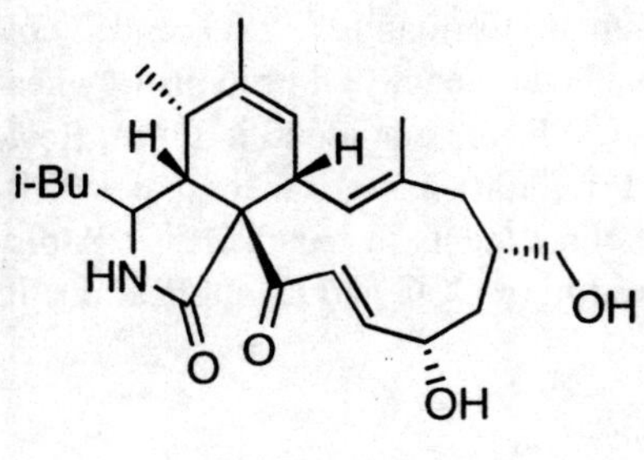

29

A novel erythromycin analog, 6-desmethyl-6-ethylerythromycin A (**30**), produced by a genetically engineered strain of *Saccharopolyspora erythrea*, has been isolated by a series of CCC separations followed by final HPLC purification. The solvent system containing dichloromethane/methanol/phosphate buffer (1:

1:1) at pH 6 and 8 has been used successfully to enrich and fractionate erythromycin-containing fermentation broths [44].

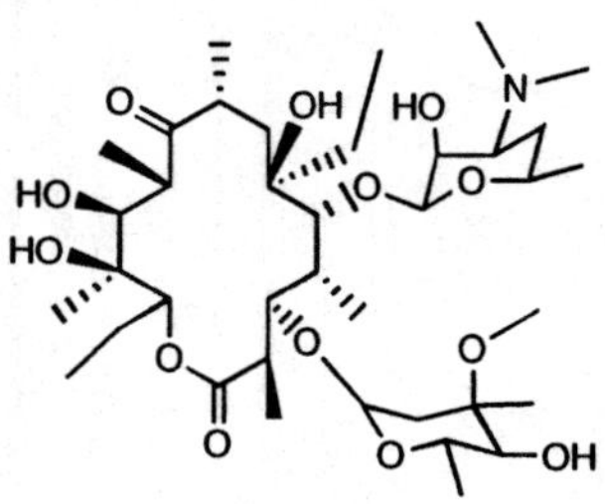

30

F. Flavor and Aroma Chemistry

Countercurrent chromatography has been extensively used in the investigation of aroma- and flavor-producing components in cosmetics and foods. An overview of CCC in flavor analysis and the isolation of some aroma precursors in rose flowers has been published [45,46]. The dichloromethane extract of the whisky new distillate has been separated into 17 different fractions of aroma character using a pentane/dichloromethane/ethanol/water (30:5:18:12) system on a Sanki CPC-B92 instrument rotating at 500 rpm. The upper layer was used as the mobile phase. The individual components of the fractions were identified by GC/MS and it was demonstrated that the method is suitable for separating less polar small molecules according to their functionalities [47]. Seven novel aroma precursors, carotenoid glycosides, from safron have been isolated from the methanol extract [48]. A solvent system consisting of *N,N*-dimethylformamide, PEG, and Ficoll (polysucrose) has been used for analyzing the heterogeneity of proteins in wheat flower [49]. Two novel terpene glycosidic esters have been identified from Reisling wine using a combination of CCC and HPLC [50]. Two further studies demonstrate the potential of CCC in isolation of constituents in wine [51,52]. A discussion about the concepts and general operation of CCC and its applications in flavor chemistry has been published [53].

X. DUAL COUNTERCURRENT CHROMATOGRAPHY

Lee and co-workers reported a further modification of the PCCC apparatus to produce a true countercurrent and its application to the separation of biologically active lignans of *Schisandra rubriflora* [54]. Later, they termed the technique dual countercurrent chromatography, or DuCCC, and reviewed the literature in which two other applications are described [55]. The instrument consists of a multilayer PCCC fitted with three-way adapters at the two ends and the middle of the coil, allowing for the introduction of the sample to a position midway in the coil and the pumping of the two phases in opposite directions through the coil while fractions are collected at both terminals. The polar compounds were eluted with the lower phase of the hexane/ethyl acetate/methanol/water (10:5:5:1) system while the less polar compounds eluted with the upper phase. As there is no stationary phase in this technique, a crude extract containing compounds of a wide range of polarities have been separated successfully. The technique appears to have significant potential as a preparative tool and perhaps may find wide applications once refined and automated instrumentation is commercially available.

REFERENCES

1. L. C. Craig, W. Hausmann, P. Ahrens, and E. J. Harfenist, *Anal. Chem. 23*:1326 (1951).
2. L. C. Craig and D. Craig, in *Techniques in Organic Chemistry, Volume 3, Separation and Purification*, Part 1 (A. Weissberger, ed.), Interscience, New York, 1956, p. 247.
3. A. Marston and K. Hostettmann, *J. Chromatogr. A 658*:315 (1994).
4. A. Berthod, *Instr. Sci. Technol. 23*:75 (1995).
5. Y. Ito, in *Advances in Chromatography*, Vol. 24 (J. C. Giddings, E. Grushka, and J. Cazes, eds.), Marcel Dekker, New York, 1984, p. 181.
6. Y. Ito and W. D. Conway (eds.), *Countercurrent Chromatography: Apparatus, Theory and Applications*, VCH, Weinheim, 1990.
7. N. B. Mandava and Y. Ito, *Countercurrent Chromatography*, Chromatographic Science Series, Vol. 44, Marcel Dekker, New York, 1989.
8. J. M. Sorensen and W. Arlt, *Liquid–Liquid Equilibrium Data Collection*, Frankfurt/Main, Germany, 1980.
9. A. P. Foucault ed., *Centrifugal Partition Chromatography*, Chromatographic Science Series, Vol. 68, Marcel Dekker, New York, 1995.
10. A. Berthold, *J. Chromatogr. 550*:677 (1991).
11. I. A. Sutherland, L. Brown, S. Forbes, G. Games, D. Hawes, K. Hostettmann, E. H. McKerrell, A. Marston, D. Wheatley, and P. Wood, *J. Liq. Chromatogr. Relat. Technol. 21*(3):279 (1998).
12. G. P. Gunawardana, U. Premachandran, N. S. Burres, D. N. Whittern, R. Henry, S. Spanton, and J. B. McAlpine, *J. Nat. Prod. 55*:1686 (1992).
13. S. Drogue, M. C. Rolet, D. Thiébaut, and R. Rosset, *J. Chromatogr. 538*(1):91 (1991).
14. D. E. Schaufelberger, T. G. McCloud, and J. A. Beutler, *J. Chromatogr. 538*(1):87 (1991).
15. K. Zhengrong, K. L. Rinehart, R. M. Milberg, and W. D. Conway, *J. Liq. Chromatogr. Relat. Technol. 21*(1/2):65 (1998).

16. M. Spraul, U. Braumann, J. Renault, P. Thepenier, and J. Nuzillard, *J. Chromatogr. A 766*(1/2):255 (1997).
17. M. Agnely and D. Thiébaut, *J. Chromatogr. A 790*(1/2):17 (1997).
18. Y. Ito and W. Conway (eds.), *High-Speed Countercurrent Chromatography*, Chem. Anal., Vol. 132, John Wiley and Sons, New York, 1996.
19. H. Oka, K. Harada, and Y. Ito, *J. Chromatogr. A 812*(1/2):35 (1998).
20. F. Le Goffic, *Lipid Technol. 9*(6):148 (1997).
21. K. Shinomiya, Y. Kabasawa, and Y. Ito, *J. Liq. Chromatogr. Relat. Technol. 21*(11): 1727 (1998).
22. Y. Shibusawa and Y. Ito. *Prep. Biochem. Biotechnol. 28*(2):99 (1998).
23. M. Ying and Y. Ito, *Anal. Chim. Acta 352*(1–3):411 (1997).
24. J. McAlpine, *Meth. Biotechnol. 4 (Natural Products Isolation)*, 247 (1998).
25. A. P. Foucault and L. Chevolot, *J. Chromatogr. A 808*(1/2):3 (1998).
26. F-Y. Chiou, P. Kan, I-M. Chu, and C-J. Lee, *J. Liq. Chromatogr. & Technol. 20*: 57 (1997).
27. Q.-Z. Du, C.-Q. Ke, and Y. Ito, *J. Liq. Chromatogr. Relat. Technol. 21*(1/2):157 (1998).
28. X. Cao, Y. Tian, T. Zhang, and Y. Ito, *Prep. Biochem. Biotechnol. 28*(1):79 (1998).
29. H. E. Hummel, D. F. Hein, Y. Ma, Y. Ito, and E. F. Chou, *Toegepaste Biol. Wet. (Univ. Gent) 62*(2a):213 (1997).
30. F.-Q. Yang, T.-Y. Yang, B.-X. Mo, L.-J. Yang, Y.-Q. Gao, and Y. Ito, *J. Liq. Chromatogr. Relat. Technol. 21*(1/2):209 (1998).
31. R. Slimestad, A. Marston, and K. Hostettmann, *J. Chromatogr. A 719*:438 (1996).
32. F. Shahidi and R. Amarowicz, *Polyphenols 94*:185 (1995).
33. J. Renault, P. Thepenier, M. Zeches-Hanrot, L. L. Men-Olivier, A. Durand, A. Foucault, and R. Margraff, *J. Chromatogr. A 763*:345 (1997).
34. Q.-Z. Du, C.-Q. Ke, and Y. Ito, *J. Liq. Chromatogr. Relat. Technol. 21*(1/2):203 (1998).
35. M. Straubinger, M. Jeuzussek, R. Weibel, and P. Winterhalter, *Nat. Prod. Lett. 10*(3): 213 (1997).
36. Y. Ma, Y. Ito, E. Sokolosky, and H. M. Fales, *J. Chromatogr. A 685*:259 (1994).
37. D. G. Cai, M. J. Gu, G. P. Zhu, J. D. Zhang, T. Jin, T. Y. Zhang, and Y. Ito, *ASC Symp. Series 593*:87 (1995).
38. Y. Liming, R. Fu, T. Zhang, X. Li, J. Deng, and X. Zhang, *Beijing Ligong Daxue Xuebao* (Chinese) *17*(2):224 (1997).
39. G. Bringmann, F. Teltschik, M. Schaffer, R. Haller, S. Bar, S. A. Robertson, and M. A. Ishakia, *Phytochemistry 47*(1):31 (1997).
40. P. Duret, A.-I. Waechter, R. Margraff, A. Foucault, R. Hocquemiller, and A. Cave, *J. Liq. Chrom. Rel. Technol. 20*(4):627 (1997).
41. A. Rasooly and Y. Ito, *J. Liq. Chromatogr. Relat. Technol. 21*(1/2):93 (1998).
42. L. Qi, Y. Ma, and Y. Ito, *J. Liq. Chromatogr. Relat. Technol. 21*(1/2):83 (1998).
43. K. A. Alvi, B. Nair, H. Pu, R. Ursino, C. Gallo, and U. Mocek, *J. Org. Chem. 62*(7): 2148 (1997).
44. D. L. Stassi, S. J. Kakavas, K. A. Reynolds, G. Gunawardana, S. Swanson, D. Zeidner, M. Jackson, H. Liu, A. Buko, and L. Katz, *Proc. Natl. Acad. Sci. USA 95*: 7305 (1998).

45. P. Winterhalter, H. Knapp, M. Straubinger, S. Fornari, and N. Watanabe, *ACS Symp. Series 705*:181 (1998).
46. H. Knapp, M. Straubinger, S. Fornari, N. Oka, N. Watanabe, and P. Winterhalter, *J. Agric. Food Chem. 46*(5):1966 (1998).
47. T. Taniguchi, N. Miyajima, and H. Komura, *Dev. Food Sci. 37B*:1767 (1995).
48. M. Straubinger, B. Bau, S. Eckstein, M. Fink, and P. Winterhalter, *J. Agric. Food Chem. 46*(8):3238 (1998).
49. J. L. Den Hollander, B. I. Stribos, M. J. van Buel, K. C. A. M. Luyben, and L. A. M. Wielen, *J. Chromatogr. B 711*(1/2):223 (1998).
50. B. Bonnlaender, B. Baderschneider, M. Messere, and P. Winterhalter, *J. Agric. Food Chem. 46*(4):1474 (1998).
51. N. Watanbe, M. Messere, and P. Winterhalter, *Nat. Prod. Lett. 10*(1):39 (1997).
52. P. Winterhalter, M. Messerer, and B. Bonnlander, *Vitis 36*(1):55 (1997).
53. L. van der Weilen, M. van Buel, and J. den Hollander, *Neth. Chem. Mag*. (The Hauge) *9*:322 (1997).
54. Y. W. Lee, Q. C. Fang, Y. Ito, and C. E. Cook, *J. Nat. Prod. 52*:706 (1989).
55. Y. W. Lee, in *High-Speed Countercurrent Chromatography* (Y. Ito and W. D. Conway, eds.), Chemical Analyses Series, Vol. 132, John Wiley and Sons, New York, 1996, p. 93.

Index